Studies in Fuzziness and Soft Computing 298

For further volumes:
http://www.springer.com/series/2941

Rudolf Seising, Enric Trillas, Claudio Moraga,
and Settimo Termini (Eds.)

On Fuzziness

A Homage to Lotfi A. Zadeh – Volume 1

Springer

Editors
Dr. Rudolf Seising
European Centre for Soft Computing
Edificio de Investigación, 3ª Planta.
Mieres
Asturias
Spain

Prof. Claudio Moraga
European Centre for Soft Computing
Edificio de Investigación, 3ª Planta.
Mieres
Asturias
Spain

Prof. Enric Trillas
European Centre for Soft Computing
Edificio de Investigación, 3ª Planta.
Mieres
Asturias
Spain

Prof. Settimo Termini
Department of Mathematics and Informatics
Università di Palermo
Palermo
Italy

ISSN 1434-9922 e-ISSN 1860-0808
ISBN 978-3-642-35640-7 e-ISBN 978-3-642-35641-4
DOI 10.1007/978-3-642-35641-4
Springer Heidelberg New York Dordrecht London

Library of Congress Control Number: 2012954156

Printed on acid-free paper

Springer is part of Springer Science+Business Media (www.springer.com)

Lotfi Zadeh and three unknown persons in the 1950s as an instructor at Columbia University in New York.

A Foreword

When I started looking at the material in this awesome volume, the first thought that came to mind was "social network." While this term has been greatly overused nowadays by the media, this volume is clearly a social network with Lotfi Zadeh at the center. The term is even more appropriate in the case of Zadeh, who in addition to being a thinker of historical note, is an extremely social human being. In addition to providing inspiring technical ideas that have allowed many people in this network to carve out impressive careers of their own, Lotfi has often provided advice on matters both professional and personal to members of this network. Lotfi was never too busy to listen to the problems of others. I often observed that Lotfi had more patience listening to other's social problems than technical matters. These pieces help to provide views of Zadeh as if looking into a big house through different windows.

This volume, in addition to providing insights to the individual contributors' experiences with Lotfi either socially or technically, even more interestingly it provides the opportunity to experience in many cases, another dimension of the contributors. While I have known most of the contributors to this volume for many years, this is one of few, if not only occasion, I have had to read their writings on a non-technical and more personal subject. In many cases, I found this to be a rewarding and an eye opening experience as I am sure other readers of this book will find.

The inclusion of pictures tremendously enhances the pleasure of this volume. Not only are there pictures of Lotfi but enjoyable pictures of other members of the community. The pictures in this volume almost span the life of the idea of fuzziness. They include black and white pictures vintage pictures from the pre-digital days that are almost invaluable. These pictures inspire warm memories. For me, it was quite notable to observe the consistency of Lotfi's physical appearance over the long history that these pictures cover.

Lotfi spent many years on the outside trying to convince people of the value of his idea of fuzzy sets before the successful applications in Japan showed its usefulness. It is worth noting that these pioneering applications in Japan occurred at a time when Japan was a rapidly raising star in the world's technological and economic order; a fact that amplified and accelerated the worlds appreciation for fuzzy sets. In many ways Lotfi is still an outsider, in this case in his own fuzzy set community. Most of the applications of fuzzy sets are based on the Mamdani-Sugeno model. This paradigm is a kind of disjunctive approach, as we get more information we add possibilities. Zadeh's perspective, as conveyed with his paradigm of restriction-based semantics, is a kind of conjunctive approach, as we get more information we reduce possibilities.

Even now as he marches into his nineties and is unable to attend conferences and interact with fellow attendees as he so enjoys, Lotfi continues to build a social network. This time using the latest technology, the Internet, he has built a social network around his inspired idea of the Berkeley Initiative in Soft Computing. Every day I receive messages from people around the world via this network of interrelated scholars. These messages usually involve interesting ideas rather then simply being announcements of conferences as is the case with some other groups. The most interesting and challenging are those that come from Lotfi, particularly those related to his attempt to deal with the issue of causality.

The editors Rudolf Seising, Enric Trillas, Claudio Moraga and Settimo Termini are to be congratulated for coming up with such a wonderful idea to help celebrate a life as rich and human as Zadeh's in this manner.

Ronald R. Yager
New York City, July, 2012

Fig. 0.1. Ron R. Yager lecturing at the 14th International Conference on Information Processing and Management of Ucertainty in Knowledge-based Systems (IPMU) 2012 in Catania (Sicily), Italy, on July 11, 2012.

Foreword by the Editors

It is without any kind of doubt that the work of Professor Lotfi A. Zadeh is of a great relevance in both the scientific and technological sides. His is a work that not only meant an important departure, but also a clear cut, from some old views on which thinking and research were anchored. He also spread his work all around the world through many lectures in many countries, and offered and continues to offer exciting new views to open-eyed people wishing to do research without being blocked by some old formal ways of looking at some theoretic or practical problems, impeding to pose them in a path allowing for its treatment.

It could be said that Zadeh opened a new paradigm in, at least, Science and Technology that was initially marked by the then surprising possibility of controlling physical systems whose behavior is empirically described by a set of linguistic rules with imprecise terms, but not by an 'exact' system of differential equations, when it exists, being usually computationally difficult to solve to obtain good enough values of its outputs. A new wave of researchers arose that, today, is followed by hundreds of researchers and engineers located in almost all parts of the world. Of this wave those authors contributing to this book are but a sample.

At the very beginning of the 'fuzzy adventure', just the presentation of the idea of 'Fuzziness' not only provoked violent oppositions, as it is often the case for innovative ideas, but also triggered, and, in a sense 'forced', a rethinking of a few crucial and debated problems aroused at the beginning of last Century in the field of the foundations of Mathematics and Logic. Also some papers tried to axiomatize the notion of 'fuzzy set' in order to take it as the starting point of a subsequent building of Mathematics. Besides remembering these first reactions to the then new emerging notion, we can today certainly affirm that the notion of Fuzziness stands as one of the *really* new concepts that have recently enriched the world of Science in the same good company of the ones of Computation, Information and Complexity, but also of Bohr's Complementarity. Science grows not only through technical and formal advances on one side and useful applications on the other side, but also by introducing and assimilating new concepts in its corpus. These, in turn, produce new developments and applications. Fuzziness has done all these things and will remain as one of the few new concepts aroused in the XX Century.

It is not usual that the founder of a new line of research can see in his life both the theoretical growing of it, as well as the success of some technological applications actually important from both the economical and the business points of view, as those coming from Fuzzy Logic and Soft Computing. This is the case with Professor Zadeh, an electrical engineer passionated by posing problems in mathematical terms that not only introduced the theoretical basis of Fuzzy Logic, but who also contributed a lot in its technological side with the insights coming from some of his many and single authored papers. Zadeh is not only well credited as the introducer of Fuzzy Logic and Soft Computing, but also and around twelve years ago, of the new field of 'Computing with Words' from which a new frontier for Computer Sciences is clearly visible and that can allow to afford yet unanswered questions in Philosophy, Linguistics, Science, Sociology, Technology and, last but not least, Industry. To qualify Zadeh as the 'father of fuzzy logic' is a good short description of its personality.

It is well known how broad is the spectrum covered by the work of Lotfi Zadeh. In a way, this seems to be reflected in the variety of contributions building this book. Some authors that contribute to this book chose to speak of personal meetings with Lotfi; others, about how particular papers of Zadeh opened for them a new research horizon. There are contributions documenting results obtained following ideas of Zadeh, thus implicitly acknowledging the inspiration he gave for those achievements. Finally, there are contributions of several 'third generation fuzzysists/softies' who were first lead into the world of Fuzziness by a disciple of Lotfi Zadeh, who, following his example, took care of opening for them a new road in science.

This book just aims at homaging both Professor Lotfi A. Zadeh's personality and work, once he surpassed his ninety years and is happily creative. His gentle attitude towards all people he met, as well as his wide tolerance with those that tried to contradict, and sometimes to blame, his contributions, is a characteristic of him that helped to approach many people to his ideas. Zadeh never refused the contact with and the offer of advise to young or yet inexpert people; never refused to gentle discuss in either a public or a private space on his thinking, and this in both spoken or written form. Zadeh is always in the opposite side of those kinds of great researchers who like to be distant and elevated from other people. For short, Zadeh is a nice human being who, aside of taking an exquisite care of his creature, likes to be in a close intellectual contact with people of all conditions.

The four editors of this book in homage to Professor Zadeh, all of them working from time ago in different areas of Fuzzy Logic or Soft Computing, would like to thank the multitude of authors contributing to its two volumes. This amount of people is a clear signal of the world-wide recognition reached by Zadeh's ideas.

Rudolf Seising, Enric Trillas, Claudio Moraga, Settimo Termini,
Mieres (Asturias, Spain), and Palermo (Italy),
October, the 30th, 2012

Genesis of the Book

I. Pre-history

When we started planning this book, born from discussions by the editors at the *European Centre for Soft Computing* (ECSC), we wrote the following letter to more than 500 scientists in the field of Soft Computing whose e-mail addresses we knew :

Dear colleagues,

In 2012 it will be 50 years that Professor Lotfi A. Zadeh used the word "fuzzy" for the first time in a scientific paper:

"..., we need a radically different kind of mathematics, the mathematics of fuzzy or cloudy quantities which are not describable in terms of probability distributions."[1]

It is also not to be forgotten that in about three and a half years, the theory of Fuzzy Sets and Systems (FSS) will be 50 years old, and that in this year 2011 its founder Lotfi A. Zadeh celebrated his 90th anniversary! It is our opinion that this 50 years long development of a now well-known theory that is used in technology, economics and other fields should have a mirror in the scientific literature. To this end we would like to edit a book entitled "On Fuzziness".

At this remarkable point of time we think that it is important to have a printed collection of documents showing the history, the present stage and the future expectations from the own views of the protagonists.

We will publish this documentation in a book and we invite you as well as other protagonists in the field of FSS, to contribute to this "homage" to the life-long work of Lotfi A. Zadeh. Furthermore, we also would like to invite and encourage scientists and researchers who have not been enthusiastic with FSS but who accompanied with their criticisms the genesis and the development of that field to participate in this book project, since, without their contribution, both the history and the prospect for its future would remain incomplete.

[1] Zadeh, Lotfi A.: From Circuit Theory to System Theory, *Proceedings of the IRE*, May 1962, pp. 856-865: 857.

Hence, we ask you to contribute with a short paper "on fuzziness" (about five (5) pages) from your personal point of view. We would like to ask you to mention in this non-technical contribution to the book how you did arrive to the field of FSS and to present your views and expectations "on fuzziness".

We also kindly ask you to include, if available, one or two photographs from the times that you will mention in your contribution.

We do not want to publish papers glorifying Lotfi A. Zadeh, because no one likes this kind of papers, nor he would like to see such a book.

We hope that you will contribute to this book and that you will help us to create a very good document on the history, the presence and the future views on our area of science and technology.

Please, send us your contribution as a Word-file before January 15, 2012!

When the first reactions appeared, we did not expect that we would have to create a two-volumes-book, but after some weeks it became clear that we would have to work with the manuscript of a collection of many pages. At the end of this procedure we had to distribute all the contributions on two volumes and it was almost impossible to find reasonable partitions of the different paper types. As a most sensible and fair solution we chose the alphabetical order relating to the first authors of each contribution. To have two volumes of almost the same size the first includes the papers "A - Ma" and the second includes the papers "Me – Z" and a *Postscriptum* of four special papers (see below).

II. Historical Troubles

Already one of the first examples that Lotfi Zadeh used in his seminal article "Fuzzy Sets" was the "class of all real numbers which are much greater than 1" – others were as we all know the "class of all beautiful women" and the "class of all tall men". He wrote that these classes "were not classes or sets in the usual mathematical sense of these terms" and "that it was a fact that such imprecisely defined 'classes' played an important role in human thinking, especially in the fields of pattern recognition, communication of information and abstraction."[2]

Today we know that they also play an important role in finishing book manuscripts. Most authors wrote that they would send their manuscripts "before the end of [x]" where $x \in \{$ January, February, March, ..., December $\}$ and also the year could have been 2011 or 2012. Some authors asked for waiting some time by using fuzzy concepts as the following examples show: "Give me a couple of days please." or "I need few more days." or "Certainly 10 days should be enough." or "Please wait for me. This weekend I will finish."

[2] Zadeh, Lotfi A.: Fuzzy Sets and Systems. In: Fox, Jerome (Ed.): *System Theory*, Microwave Research Institute Symposia, Series XV. Broooklyn, New York: Polytechnic Press, 1965. pp. 29–37: 29.

We got e-mails including the sentence "I will try to finish mine before he finishes his :-)." – And until we worked with this book manuscript for over one year, we are sure that the meaning of the following sentence is pretty fuzzy: "I will do my best."

Concerning the requested contribution of "about five (5) pages" we got – indeed – papers of 5 pages but as the reader of the book will notice very quickly, there are also a couple of shorter papers and there are many longer papers. We cede it to our interested readers to find the right membership function of the class of papers of "about five (5) pages" in these two volumes.

III. More Historical Troubles

There are always exceptions! For some of the submitted contributions to this book we would not find anybody who would say that it has "about five (5) pages". Thus, the membership values of these papers as an element to the set of "about five (5) pages'-papers" is almost zero. Even one of the editors used to think that fuzzified on page-numbers! We considered that these papers deserved not to be reduced, since they represent a comprehensive review of the past/present and a dream of the future. How was to handle these contributions? – We decided to have a "Postscriptum" at the end of this book (volume II) and we put these four contributions into this part.

IV. Figures and Photographs

There are two kinds of figures or pictures in these two volumes: usually authors of scientific papers use pictures, paintings, statistics, etc. to illustrate their findings and results in figures. Consequently, there are many of those figures in this book but we also asked the authors to look for old photographs that show themselves with Lotfi Zadeh and/or with other protagonists of the fuzzy community. Many of the authors went into cellars, attics, garages or any other crawl space where they assumed that they have such pictures – lost from view. They opened boxes, folders, binders, photo albums and yearbooks – may be for the first time since many years or decades – and therefore we received a huge amount of unknown pictures.

We are very glad that we can publish such photographs in these two volumes because some of them are important contemporary documents or at least nice memorabilia. Most of the photograph are privately owened by the authors and we publish them with their courtesy. Other photographs we have taken from the archive of one of the editors.[3]

[3] Figs. 0.1, 83.1, 86.1, 89.2 and 102.1 as well as the photographs that show Lotfi Zadeh page 7 of volume I (Thanks to Lotfi Zadeh for this gift!) and the one on page 7 of volume II of this book.

V. Additional Thanks

The editors are most thankful to the authors for their willingness to write their papers, to Prof. Dr. Janusz Kacprzyk for accepting the book in his series *Studies in Fuzziness and Soft Computing*, to Prof. Ron R. Yager for writing the Foreword, and last but not least to the Springer Verlag (Heidelberg) and in particular to Dr. Thomas Ditzinger, Leontina Di Cecco, and Holger Schäpe for helping this edition find its way to the publisher's list.

We thank the reviewers of the papers very much, particularly for their help we thank Luis Argüelles, Christian Borgelt, Lluis Godo, and Alejandro Sobrino; special thanks for proofreading a big number of contributions go to Brian R. Gaines!

VI. End

Finally, after having survived to all that without a single nervous attack, the last pending paper arrived, the last pictures were selected, and the editors could exclaim 'Good heavens! The book is ended!'. But then one of them, in low voice, added 'Not yet. The last section deserves a few lines with wishes for Lotfi'. Thus,

In the name of all those who contributed to this book, the editors would like to finally add: 'Long life to Professor Zadeh!'

Fig. 0.2. The editors of this book at the *Second Saturday's Scientific Conversations* (SSC) in Palazzo Steri, Palermo, Sicily, May 14, 2011. May be in this moment they were agreed to prepare the book in hand!

RS+ET+CM+ST,
Mieres (Asturias, Spain), and Palermo (Italy),
October, the 30th, 2012

Contents of Volume I

Contents of Volume II

List of Contributors – Volume I

Argüelles Méndez, Luís, Affiliated Researcher at the European Centre for Soft Computing in Mieres, Asturias, Spain.

Atanassov, Krassimir, Department of Bioinformatics and Mathematical Modelling, Institute of Biophysics and Biomedical Engineering, Bulgarian Academy of Sciences, Sofia, Bulgaria.

Balas, Marius M., Faculty of Engineering, Department of Automation and Applied Informatics, "Aurel Vlaicu" University of Arad, Arad, Romania.

Balas, Valentina E., Faculty of Engineering, Department of Automation and Applied Informatics, "Aurel Vlaicu" University of Arad, Arad, Romania.

Barro, Senén, Centro de Investigación en Tecnoloxías da Información (CITIUS), Universidade de Santiago de Compostela, Galicia, Spain.

Batyrshin, Ildar, Research Program of Applied Mathematics and Computations, Mexican Petroleum Institute, México, D. F.

Bezdek, James C., EE and CS, University of Melbourne, Australia.

Bilgiç, Taner, Department of Industrial Engineering, Bogazi çi University, Istanbul, Turkey.

Bloch, Isabelle, Telecom ParisTech – CNRS LTCI – Paris, France.

Bocklisch, Steffen F., Faculty of Electrical Engineering and Information Technology, Systems Theory, Chemnitz University of Technology, Germany.

Bocklisch, Franziska, Faculty of Behavioural and Social Sciences, Cognitive and Engineering Psychology, Chemnitz University of Technology, Germany.

Bouchon-Meunier, Bernadette, Laboratoire d'Informatique de Paris 6 (LIP6), Université Pierre et Marie Curie-CNRS, Paris, France.

Bugarín Diz, Alberto José, Centro de Investigación en Tecnoloxías da Información (CITIUS), Universidade de Santiago de Compostela, Galicia, Spain.

Bustince, Humberto, Departamento de Automática y Computación, Pamplona, Universidad Publica de Navarra, Spain.

Carlsson, Christer, Institute for Advanced Management Systems Research, Abo Akademi University, Turku, Finland.

Castillo, Oscar, Research Chair of Graduate Studies, Tijuana Institute of Technology, Tijuana, Mexico.

Cat, Jordi, Department of History and Philosophy of Science, Indiana University, USA.

Chang, Elizabeth, Curtin University of Technology, IT School of IS and Curtin Business School Institute for Digital Ecosystems & Business Intelligence, Perth, Australia.

Ćirić, Miroslav , Department of Mathematics and Computer Science, Faculty of Science and Mathematics, University of Nis, Serbia.

Coletti, Giulianella, Dipartimento di Matematica e Informatica, Università di Perugia, Italy.

De Maio, Carmen, Dept. Computer Science, University of Salerno, Fisciano (SA) Italy.

Deshpande, Ashok, Adjunct Professor, College of Engineering Pune (COEP); India.

De Tré, Guy, Ghent University, Department of Telecommunications and Information Processing, Ghent Belgium.

Dillon, Tharam S., Department of Computer Science and Computer Engineering, LaTrobe University, Melbourne Australia.

Di Martino, Ferdinando, Dipartimento di Costruzioni e Metodi Matematici in Architettura, Università degli Studi di Napoli "Federico II", Napoli, Italy.

Dong, Fangyan, Tokyo Institute of Technology, Nagatsuta, Midori-ku, Yokohama, Japan.

Eklund, Patrik, Department of Computing Science, Umeå University, Sweden.

Esteva, Francesc, Institut d'Investigació en Intel·lència Artificial (IIIA), Consejo Superior de Investigaciones Científicas (CSIC), Barcelona, Spain.

Fedrizzi, Mario, Department of Computer and Management Sciences, University of Trento, Italy.

Fedrizzi, Michele, Department of Computer and Management Sciences University of Trento, Trento, Italy.

Felix, Rudolf, F/L/S Fuzzy Logik Systeme GmbH, Dortmund, Germany.

Fenza, Giuseppe, Department of Computer Science, University of Salerno, Fisciano (SA), Italy.

Freksa, Christian, Cognitive Systems Research Group, Department of Mathematics and Informatics, University of Bremen, Germany.

Freytes, Hector, The ALOPHIS Group, Department of Philosophy, University of Cagliari, Italy.

García-Honrado, Itziar, University of Oviedo. Oviedo (Asturias), Spain. Affiliated researcher at the European Centre for Soft Computing, Mieres, Spain.

García-Lapresta, José Luis, Department of Applied Economics, University of Valladolid, Spain.

Galán, M. Ángeles, Department of Applied Mathematics, University of Málaga, Spain.

Gil, María Ángeles Dpto. Estadistica e Investigacion Operativa y Didactica de la Matematica, Universidad de Oviedo, Asturias, Spain.

Gil Aluja, Jaime, President of the Real Academia de Ciencias Económicas y Financieras de España, Spain.

Giuntini, Roberto, The ALOPHIS Group, Department of Philosophy, University of Cagliari, Italy.

Godo, Lluís, Institut d'Investigació en Intel·lència Artificial (IIIA), Consejo Superior de Investigaciones Científicas (CSIC), Barcelona, Spain.

Gomide, Fernando, Department of Computer Engineering and Automation, School of Electrical and Computer Engineering, University of Campinas, Campinas, São Paulo, Brazil.

Gottwald, Siegfried, Emeritus Professor, Abteilung Logik und Wissenschaftstheorie, Institut für Philosophie, Universität Leipzig, Germany.

Guadarrama, Sergio, Electrical Engineering and Computer Science Department (EECS), University of California, Berkeley, CA, USA. Affiliated researcher at the European Centre for Soft Computing (ECSC), Mieres, Spain and at the International Computer Science Institute (ICSI).

Hanss, Michael, Institut für Technische und Numerische Mechanik, Universität Stuttgart, Germany.

Heister, Hanns-Werner, Rosengarten (Hamburg), Germany

Held, Pascal, Fakultät für Informatik, Otto-von-Guericke-Universität Magdeburg, Germany.

Helgason, Cathy M., Departments of Neurology and Psychiatry, University of Illinois College of Medicine, USA.

Helgesson, Robert, Department of Computing Science, Umeå University, Sweden.

Hirota, Kaoru, Department of Computational Intelligence and Systems Science, Interdisciplinary Graduate School of Science and Engineering, Tokyo Institute of Technology, Yokohama, Japan.

Hsueh, Nien-Lin, Department of Information Engineering and Computer Science, Feng Chia University, Taiwan.

Hussain, Omar K., School of Information Systems, Curtin University, Perth, Australia.

Ignjatović, Jelena, Department of Mathematics and Computer Science, Faculty of Science and Mathematics, University of Nis, Serbia.

Jamison, K. David, Department of Mathematical and Statistical Sciences, University of Colorado Denver, Colorado, USA.

Jobe, Thomas H., University of Illinois College of Medicine Departments of Neurology and Psychiatry, USA.

Kacprzyk, Janusz, Systems Research Institute, Polish Academy of Sciences, and Warsaw School of Information Technology (WIT), Warsaw, Poland.

Kasabov, Nikola, Knowledge Engineering and Discovery Research Institute, Auckland University of Technology, New Zealand.

Keller, James, Electrical and Computer Engineering Department, University of Missouri, Columbia, MO, USA.

Kerre, Etienne E., Department of Applied Mathematics and Computer Science, Ghent University, Belgium.

Kiseliova, Tatiana, Department of Exact and Natural Sciences, Iv. Javakhishvili Tbilisi State University, Georgia.

Klir, George J., Distinguished Professor Emeritus of Systems Science, Binghamton University (State University of New York at Binghamton), Binghamton, New York , USA.

Koczy, Laszlo T., Szechenyi Istvan University (Gyor) and Budapest University of Technology and Economics, Hungary.

Kortelainen, Jari, Department of Electrical Engineering and Information Technology, Mikkeli University of Applied Sciences, Finland.

Kovalerchuk, Boris, Department of Computer Science, Central Washington University, Ellensburg, WA, USA.

Kreinovich, Vladik, Department of Computer Science, University of Texas at El Paso, USA.

Kruse, Rudolf, Fakultät für Informatik, Otto-von-Guericke-Universität Magdeburg, Germany.

Kwiatkowska, Mila, Department of Computing Science, Thompson Rivers University, Kamloops, BC V2C 0C8 Canada.

Lawry, Jonathan, Department of Engineering Mathematics, University of Bristol, UK.

Ledda, Antonio, The ALOPHIS Group, Department of Philosophy, University of Cagliari, Italy.

Lee, Chang-Shing, Department of Computer Science and Information Engineering, National University of Tainan, Taiwan.

Lee, Jonathan, Department of Computer Science and Information Engineering, National Central University, Taiwan.

Lee, E. Stanley, Department of Industrial & Manufacturing Systems Engineering, Kansas State University, Manhattan, Kansas, USA.

Lin, Chin-Teng, Brain Research Center, EE&CS Departments, National Chiao-Tung University, Taiwan.

Lodwick, Weldon A., Department of Mathematical and Statistical Sciences, University of Colorado Denver, Colorado, USA.

Loia, Vincenzo, Department of Computer Science, University of Salerno, Fisciano (SA) Italy.

Luhandjula, Monga K., Department of Decision Sciences, University of South Africa, Unisa, Pretoria, South Africa.

Maccarone, Maria Concetta, Istituto di Astrofisica Spaziale e Fisica Cosmica di Palermo, Istituto Nazionale di Astrofisica, Palermo, Italy.

Magdalena, Luis, European Centre for Soft Computing, Mieres, Asturias, Spain.

Marques Pereira, Ricardo A., Department of Computer and Management Sciences, University of Trento, Italy.

Martin, Trevor, Artificial Intelligence Group, Department of Engineering Mathematics, University of Bristol, UK.

Maturo, Antonio, Faculty of Social Sciences of the University of Chieti — Pescara, Italy.

Mayor, Gaspar, Department of Mathematics and Computer Science University of the Balearic Islands, Palma (Mallorca), Spain.

Moewes, Christian, Fakultät für Informatik, Otto-von-Guericke-Universität Magdeburg, Germany.

Paoli, Francesco, The ALOPHIS Group, Department of Philosophy, University of Cagliari, Italy.

Perfilieva, Irina, Centre of Excellence IT4 Innovations, Division of the University of Ostrava, Institute for Research and Applications of Fuzzy Modeling, Ostrava, Czech Republic.

Salerno, Saverio, Dipartimento Ingegneria Elettronica e Ingegneria Informatica, University of Salerno, Fisciano (SA) Italy.

Sergioli, Giuseppe, The ALOPHIS Group, Department of Philosophy, University of Cagliari, Italy.

Sessa, Salvatore, Dipartimento di Costruzioni e Metodi Matematici in Architettura, Università degli Studi di Napoli "Federico II", Napoli, Italy.

Teytaud, Olivier, Department of Computer Science and Information Engineering, National University of Tainan, Taiwan and TAO team, Inria Saclay IDF, LRI UMR (CNRS - Universite Paris-Sud), France.

Vantaggi, Barbara, Dipartimento di Scienze di Base e Applicate per L'Ingegneria, Sezione di Matematica, La Sapienza Università di Roma, Italy.

Ventre, Aldo G.S., Dipartimento di Architettura "Luigi Vanvitelli", Seconda Università degli Studi di Napoli, Italy.

Wang, Mei-Hui, Department of Computer Science and Information Engineering, National University of Tainan, Taiwan.

Yager, Ronald R., Machine Intelligence Institute, Iona College, New Rochelle, NY, USA.

Zadrożny, Sławomir, Systems Research Institute, Polish Academy of Sciences, Warsaw, Poland.

List of Contributors – Volume II

Aisbett, Janet, Faculty of Science and IT, The University of Newcastle, Callaghan, NSW, Australia.

Bordogna, Gloria, IDPA, Italian National Council of Research, Milano, Italy.

Dubois, Didier, Institut de Recherche en Informatique de Toulouse (IRIT), CNRS and Université Paul Sabatier, Toulouse, France.

Gaines, Brian R., University of Calgary and University of Victoria, Canada.

Gibbon, Greg, School of Design Communication and IT, University of Newcastle, Australia.

Kandel, Abraham, Department of Computer Science and Engineering, The University of South Florida, Tampa, Florida, USA.

Last, Mark, Department of Information Systems Engineering, Ben-Gurion University of the Negev, Israel.

Melin, Patricia, Chair of Graduate Studies, Computer Science, Tijuana Institute of Technology, Tijuana, Mexico.

Mendel, Jerry M., Signal and Image Processing Institute, Ming Hsieh Department of Electrical Engineering, Viterbi School of Engineering, University of Southern California, Los Angeles, CA, USA.

Moraga, Claudio, Emeritus Researcher, Unit: Fundamentals of Soft Computing, European Centre for Soft Computing, Mieres, Asturias, Spain.

Mordeson, John N., Department of Mathematics, Creighton University, Omaha, Nebraska, USA.

Nakama, Takehiko, European Center for Soft Computing in Mieres, Asturias, Spain.

Navarro, María G., Instituto de Filosofía, Centro de Ciencias Humanas y Sociales, CSIC, Spain.

Niskanen, Vesa A., University of Helsinki, Department of Economics & Management, Helsinki, Finland.

Novák, Vilém, Centre of Excellence IT4 Innovations, Division of the University of Ostrava, Institute for Research and Applications of Fuzzy Modeling, Ostrava, Czech Republic.

Nurmi, Hannu, Department of Political Science and Contemporary History, University of Turku, Finland.

Olivas, José Angel, Department of Information Technologies and Systems, University of Castilla-La Mancha, Spain.

Ovchinnikov, Sergei, Mathematics Department, San Francisco State University, San Francisco, CA, USA.

Pal, Sankar K., Center for Soft Computing Research, Indian Statistical Institute, Calcutta, India.

Palm, Rainer, AASS Research Centre, Department of Computer Science, Örebro University, Örebro, Sweden.

Pasi, Gabriella, Department of Informatics, Systems and Communication, Universita' degli Studi di Milano Bicocca, Milano, Italy.

Pedrycz, Witold, Department of Electrical and Computer Engineering, University of Alberta, Edmonton, Canada.

Peeva, Ketty, Faculty of Applied Mathematics and Informatics, Technical University of Sofia, Bulgaria.

Petry, Frederick E., Geospatial Sciences and Technology Branch, Naval Research Laboratory.

Prade, Henri, Institut de Recherche en Informatique de Toulouse (IRIT) , CNRS and Université Paul Sabatier, Toulouse, France.

Pykacz, Jarosław, Institute of Mathematics, University of Gdansk, Poland.

Radojevic, Dragan, Mihajlo Pupin Institute, University of Belgrade, Belgrade, Serbia.

Rakus-Anderson, Elisabeth, Department of Mathematics and Science, Blekinge Institute of Technology, Karlskrona, Sweden.

Reformat, Marek, Department of Electrical and Computer Engineering at University of Alberta, Canada.

Reusch, Bernd, Professor emeritus, Fakultät für Informatik, Technische Universität Dortmund, Germany.

Rickard, John T., Research Director, Distributed Infinity, Inc., Larkspur, CO USA.

Rosch, Eleanor, Department of Psychology, UCB Program in Cognitive Science, Group in Religious Studies and Group in Buddhist Studies, University of California, Berkeley, USA.

Ruspini, Enrique, Unit: Collaborative Soft Intelligent Systems, European Centre for Soft Computing, Mieres, Spain.

Sala, Antonio, Instituto Univ. de Automática e Informática Industrial (AI2), Dept. Ingeniería de Sistemas y Automatica, Universitat Politecnica de Valencia, Spain.

Sánchez, Elie, Marseille, France.

Sanchez, Daniel, European Centre for Soft Computing, Mieres, Spain and Department of Computer Science and A.I., University of Granada, Spain.

Scozzafava, Romano, Dipartimento di Scienze di Base e Applicate per L'Ingegneria, Sezione di Matematica, La Sapienza Università di Roma, Italy..

Seising, Rudolf, Adjoint Researcher, Unit: Fundamentals of Soft Computing, European Centre for Soft Computing in Mieres, Spain.

Sobrino, Alejandro, Faculty of Philosophy, University of Santiago de Compostela, Spain. Affiliated Researcher, European Centre for Soft Computing, Mieres, Asturias, Spain.

Sowa, John F., VivoMind Research, Croton on Hudson, New York, USA.

Syropoulos, Apostolos, Xanthi, Greece.

Tabacchi, Marco Elio, Scientific Director at Istituto Nazionale di Ricerche Demopolis, Italy. Associate Researcher at Università degli Studi di Palermo, Italy.

Tamir, Dan E., Department of Computer Science, Texas State University, San Marcos, Texas, USA.

Tanscheit, Ricardo, Department of Electrical Engineering Pontifical Catholic University of Rio de Janeiro, Rio de Janeiro, RJ, Brazil.

Termini, Settimo, Professor of Theoretical Computer Science at the University of Palermo. Affiliated Researcher at the European Centre for Soft Computing, Mieres, Asturias, Spain.

Terricabras, Josep-Maria, Professor of Philosophy at the University of Girona.

Trillas, Enric, Emeritus Researcher, Unit: Fundamentals of Soft Computing, European Centre for Soft Computing, Mieres, Asturias, Spain.

Torra, Vicenç, Institut d'Investigació en Intel·ligència Artificial (IIIA), Consejo Superior de Investigaciones Científicas (CSIC), Bellaterra, Catalonia, Spain.

Türkşen, I. Burhan, Department of Industrial Engineering, TOBB Economics and Technology University, Ankara Turkey, Department of Mechanical and Industrial Engineering, University of Toronto, Ontario, Canada

Urtubey, Luis Adrian, Facultad de Filosofía, Ciudad Universitaria Córdoba, Argentina.

van Deemter, Kees, Computing Science Department, King's College, University of Aberdeen, Scotland.

Várkonyi-Kóczy, Annamária R., Institute of Mechatronics and Vehicle Engineering, Óbuda University, Budapest, Hungary.

Vojtáš, Peter, Department of Software Engineering, School of Computer Science, Charles University Prague, Czech Republic.

Walker, Carol L., New Mexico State University, USA.

Walker, Elbert A., Department of Mathematical Sciences, New Mexico State University, USA.

Watada, Junzo, Graduate School of Information, Production and Systems, Waseda University, Hibikino, Wakamatsu, Kitakyushu, Japan.

Werbos, Paul J., National Science Foundation and Center for Large-Scale Integration and Optimization of Networks (CLION), FedEx Institute of Technology, Memphis TN, USA.

Wierman, Mark J., Creighton University, Omaha, NE, USA.

Yager, Ronald R., Machine Intelligence Institute, Iona College, New Rochelle, NY, USA.

Zimmermann, Hans-Jürgen, Emeritus Professor, RWTH Aachen and INFORM GmbH, Germany.

Part I

On Fuzziness: A – Ma

1

From Japanese Art to Fuzzy Logic (A Personal Voyage on Soft Computing)

Luis Argüelles

"When a traveller leaves he never returns: the experiences obtained from his voyage changes him forever into a different man"

(Chinese traditional proverb)

1.1 The Initial Thoughts

Every journey, even the longer ones, begins with a simple step and they usually result into personal changes, new experiences and developments through time. Looking back to the past, my journey on fuzzy logic really started in 1977 even without realizing it, when I was interested in the works of North American architect Frank Lloyd Wright. In that time, I soon learned about Wright's interest on traditional Japanese architecture and art, so I became interested, too. In a public library I found a jewel of a book titled "Katsura Daitokuji" [3] where the imperial Villa of Katsura and the Daitokuji monastery, both at Kyoto, were shown with abundance of graphical material.

An intrinsic feature of traditional architecture in Japan is the existence of moving walls, both internal and external. External ones help to open the building towards the exterior, usually a Zen garden, so space flows in a bidirectional way and no definite limits do exist for the building as a three dimensional solid. On the other hand, internal moving walls help to create divisions inside the internal space, and as an example, a living room can be divided by an internal wall, resulting into two smaller living rooms or a smaller living room and a small room. After the division is made, a living room is "less" living room than before, but, since the walls are movable, in some way living rooms are at the same time large and small, or a combination of large, medium and small. These multivalued spaces fascinated me and created a personal interest on the Japanese way of thinking and culture that has accompanied me for the rest of my life.

About ten years later I had the opportunity to read an interesting article in Byte magazine about Fuzzy-Logic. It was 1987 or 1988, I don't remember exactly, but at that time Byte was the *de-facto* standard technical magazine for people interested on in-depth articles related to software, algorithms and computing in the field of

R. Seising et al. (Eds.): *On Fuzziness: Volume 1*, STUDFUZZ 298, pp. 3–9.
DOI: 10.1007/978-3-642-35641-4_1

small-scale computers. Advanced enough to give you a competitive edge in Spain, a country that was wakening in computer science or, better said, whose private companies were starting to demand sophisticated technical skills in their information technologies departments.

I still remember that the article started with a variation of the Sorites Paradox by means of introducing the concept of age in human beings, from young to old ones, and the impossibility to find a precisely defined point of change between both classifications, suggesting values in age that were "old" and "young" at the same time. This multiple and simultaneous membership of an element x to two different sets A and B with different membership degrees captivated me. I must recognize that I didn't understand well the complete article, but the initial part dedicated to the exposition of fuzzy sets, or at least the "age paradox" triggered a light in my mind. In fact, I usually use this example in these times when I'm asked about what fuzzy logic is from people not scientifically related to this field of mathematics.

1.2 Professional Development: Mining

In my professional career I have had the privilege and luck to have hold the position of head of the R+D department of a private coal mining company from 1987 to 2010, and I say "privilege and luck" because aside results, the company always encouraged me to study, learn and research on new things. It's not usual that a private company pays an employee for studying, but at this point I must share with the reader an important milestone on my voyage: The president of the coal company, Efrén Cires, had previously worked as a technical executive in IBM in Madrid. At the end of the sixties last century he attended a seminar in California about fuzzy logic taught by Lofti Zadeh, so he always motivated me to read about fuzzy logic.

Even so, coal mines and mining in general are always generating organizational and technical challenges, and in the mid-nineties we experienced the problem of a growing complexity system of coal transport inside and outside the mine involving trains, coal wagons and single and multiple rail tracks. The R+D department was soon directed towards developing models based on discrete event simulation techniques [1] . Since these models are not directly related to fuzzy-logic, I shall dedicate some lines in order to expose what is known as the "Joe the barber problem" [6] that will help to put into perspective the basics of discrete event driven simulation.

Joe owns a barbershop that has only one barber chair, that is, only one customer can be serviced at a given time. From experience and data acquisition, Joe knows that customers arrive at his shop every 15 ± 6.5 minutes. This distribution is constant throughout the day. If no customers are waiting, Joe will tend to take his time cutting hair. As customers arrive and fill up the shop, Joe will speed up his hair cutting because he knows that if a new customer observes a long queue he will leave the shop. The time it takes Joe to cut hair is given by the distribution shown on table 1.1:

Table 1.1. Time distribution for servicing a hair cutting

People in the queue	Time to give hair cut (minutes)
0	18
1	16
2 or 3	14
4 to 5	13
More than 15	12

Joe would like to know on average: a) the perçentage of time he is busy, b) how many clients can enter his shop (barbershop's capacity), c) maximum content of people in the queue, and d) average contents of the queue. The following short program, written in a simulation computer language called GPSS/H, offers a reply to these questions:

```
SIMULATE TIME       FUNCTION Q(WAIT),D7
0,18/1,16/2,14/3,14/4,13/5,13/6,12
GENERATE            15,6.5      ;people arrive
QUEUE               WAIT        ;wait in seats
SEIZE               JOEB        ;engage Joe for hair cut
DEPART              WAIT        ;leave the seat
ADVANCE             FN(TIME)    ;cut hair
RELEASE             JOEB        ;free the barber
TERMINATE                       ;leave the shop
GENERATE            480*5       ;simulate for five days
TERMINATE           1           ;end of simulation
START               1
END
```

the solutions are a: 99,5%, b: 158, c: 4 and d: 1,75. Now, if we substitute barber chairs for train locomotives, waiting chairs for coal storage units and so on, we soon end up with mining problems that can only be solved using simulation at computers, although recent developments suggest that intelligent models based on fuzzy-logic techniques can be also applied [7].

For researching the topic, we hired Prof. John Sturgul, then lecturing at the School of Mines in the Idaho University and now at the University of Adelaide, a world-class expert on mining simulation. After working together for about two years, the School of Mines of the University of Oviedo in Spain got interested in our research, so I introduced John to Prof. Maria Teresa Alonso, at that time sub director of the School. I remember one day in an informal meeting in Teresa's office that we started to speak about fuzzy-logic. John was interested and Teresa suggested to arrange an introductory course on fuzzy-logic for the students. Three months later, on November 1998 the first course ever taught on fuzzy sets applied to engineering in the North of Spain was ready. The theoretical part was easy to organize just taking introductory material from the known "red book" by Klir and Yuan [4], but for the practice classes things were not so clear: Not all the students had programming skills in computer languages such as Pascal or C, so I decided to use the LISP language as the computing platform for the course, specifically the freely available implementation *NewLisp*, written by Lutz Müller [5]. With this idea on mind I wrote FuzzyLisp, a layer of code composed by a set of functions that allows to implement small to medium scale fuzzy-logic based projects with a minimum of computer language knowledge. As an example, we can define in LISP three fuzzy sets, fm1, fm2, fm3 for representing the concept of young, mature and old persons as follows:

```
(setq fm1 '(young 0 0 40)),
(setq fm2 '(mature 0 40 80)),
(setq fm3 '(old 40 80 80))
```

then, building a linguistic variable named "age" from these three triangular membership functions is immediate:

```
(setq age '(fm1 fm2 fm3))
```

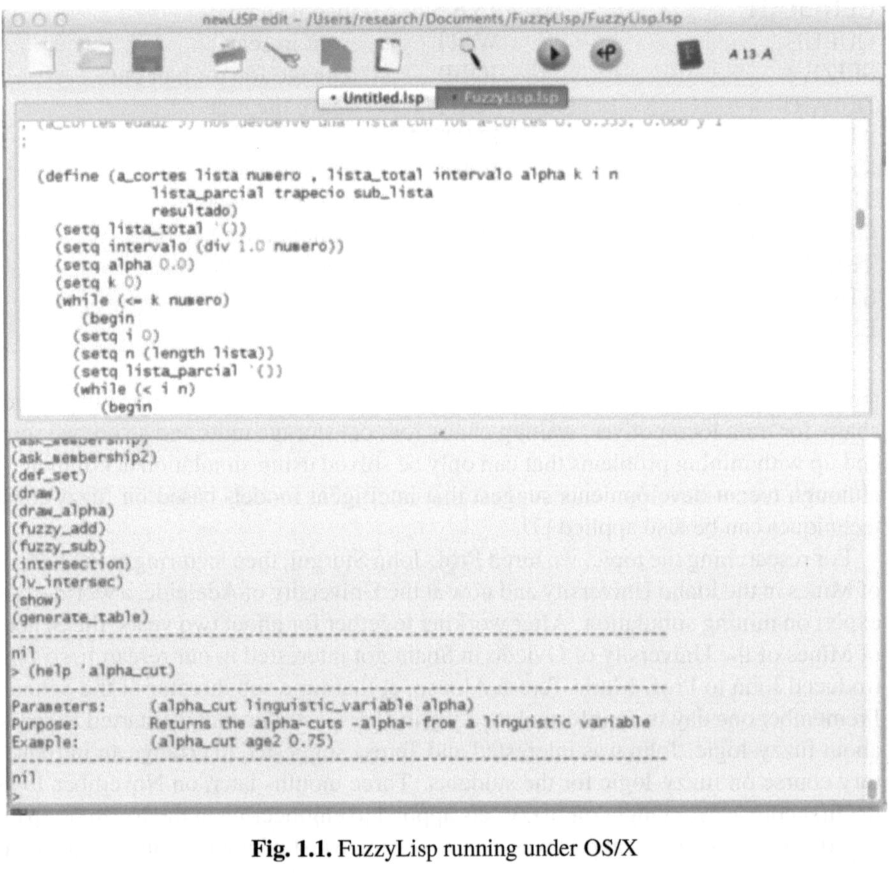

Fig. 1.1. FuzzyLisp running under OS/X

From these LISP data structures, FuzzyLisp was powerful enough to handle alpha-cuts calculations, fuzzy number arithmetic, fuzzy expert rules, and fuzzification, inference and defuzzification procedures. A screen capture of fuzzy-Lisp can be seen in figure 1.1.

In those years the community of Fuzzy-Logic in Spain was relatively small, but Internet was starting to help to bring together people with similar enthusiasms, so soon I contacted Dr. Teresa de Pedro and Dr. Ricardo García Rosa, then at the Institute of Automatics of the Spanish National Scientific Research Council (CSIC). They were researching on autonomous driving cars [2] and a presentation of their prototype and results was planed for TV and other media on June, 2000. I was invited to that presentation. Lofti Zadeh was also invited. Interestingly, Lofti and me discussed not about fuzzy-logic, but the rivalry between Airbus and Boing airplanes while going by car from the Institute to Madrid's downtown, but this is a story for another paper.

1.3 A Shift of Paradigm

Those were exciting moments in my professional career. Some months before I had developed a fuzzy-logic based system for assessing the difficulty of observation of double stars while using medium and small scale telescopes, and was at the same time developing a model of artificial pancreas to be used on diabetic rats successfully tested in the Faculty of Medicine at the University of Oviedo. For the time being, the relationship with people at the Institute of Automatics was getting stronger and about autumn, 2002, the Institute and the company I was working for signed a four years agreement for researching on an intelligent model applied to a winning system for coal mines, merging fuzzy-logic based technologies with Virtual Reality environments.

We finished the project in 2007, achieving all the goals established in the first white reports where all the required specifications were described. Meanwhile those years, an important movement in fuzzy-logic in Spain was taking form. In an informal conversation, Zadeh suggested Prof. Enric Trillas to promote the creation of sort of an European Centre of research on intelligent systems, ultimately giving birth to the European Centre for Soft Computing, ECSC, in the city of Mieres. At the end of the mining project at the company, Teresa de Pedro and Ricardo García introduced me to Prof. Trillas while visiting the Centre. I think it was a case that can be described in figurative language as a chemical bond between two persons, at least on my part.

Three years later, in May 2010, I left the mining company I had worked for along twenty-three years of my life. In some way it was a *liberation*. Simultaneously, Prof. Trillas invited me to become an affiliated researcher at the Centre. I accepted immediately. First, it represented a quantum leap in scientific ambience and academic advance from the experiences enjoyed at the private company. Second, if you know Prof. Trillas, even if you know him only a bit, you already know it is almost impossible to decline one of his offerings.

My personal voyage on soft computing is nowadays focused on artificial conscience at the Centre. I don't know how many years I will be able to travel following this direction, being well aware that maybe it's not a travel for measuring it on years but decades or even centuries. The only thing I'm absolutely sure is that now, the voyage is more intense and alive than never before.

Fig. 1.2. From left to right: Luis Argüelles, Teresa de Pedro and Lofti Zadeh. Madrid, 2000

Acknowledgement. In this paper, I've mentioned some names of people that have caused important points of inflexion in my scientific career. There are more, of course, but their importance is not comparable. Also, it is usually surprising to realize that relatively few people is what really bends and shapes our travel in life. My acknowledgement goes to all of them.

References

1. Argüelles, L.: Designing new interfaces for GPSS/H. In: Mine Simulation, Balkema (1997)
2. Autopia: Programa Autopía: Vehículos automatizados, `http://www.car.upm-csic.es/autopia/index.html` (acceded March 2012)

3. Borrás, M.L.: Katsura Daitokuji. Tudor (Ediciones Polígrafa), Barcelona (1970)
4. Klir, G.J., Yuan, B.: Fuzzy Sets and Fuzzy Logic: Theory and Applications. Prentice Hall, Upper Saddle River (1995)
5. Müller, L.: NewLisp, `http://www.newlisp.org/` (acceded March 2012)
6. Sturgul, J.: Mine Design: Examples Using Simulation. Society for Mining Metallurgy, Liteleton (1999)
7. Zhang, R., Phillis, Y.A., Kouikoglou, V.S.: Fuzzy Control of Queuing Systems. Springer, London (2010)

2

On Zadeh's Intuitionistic Fuzzy Disjunctions and Conjunctions

Krassimir Atanassov

Abstract. Two new operations, called "Second Zadeh's intuitionistic fuzzy disjunction and conjunction" are introduced on the basis of the Second Zadeh's intuitionistic fuzzy implication. Some of their properties are studied.

2.1 Introduction

During the last ten years a lot of operations were defined over Intuitionistic Fuzzy Sets (IFSs; see [3]) and logics. Here, we will discuss three new operations, generated by Zadeh's implication, introduced in fuzzy set theory (see, e.g., [10]). Its First and Second IFS-analogues were introduced in [5, 6, 8]. In [7], on the basis of the First Zadeh's IF-implication, we constructed the First Zadeh's conjunction and disjunction. Now, on the basis of the Second Zadeh's IF-implication, we will construct new Zadeh's conjunction and disjunction.

In [10], 10 different fuzzy implications are discussed. Having in mind the classical logic equality

$$x \vee y = \neg x \to y, \tag{1}$$

where x and y are logical variables, $\vee$ - disjunction, $\to$ - implication and $\neg$ - negation, we see that for any implication we can construct a disjunction and after this, using De Morgan's laws, a conjunction (or vice versa).

2.2 Definition and Algebraic Properties of the Second Zadeh's Intuitionistic Fuzzy Disjunction and Conjunction

The intuitionistic fuzzy propositional calculus has been introduced more than 20 years ago (see, e.g., [1, 3]). In it, if x is a variable, then its truth-value is represented by the ordered couple

$$V(x) = \langle a, b \rangle,$$

so that $a, b, a+b \in [0,1]$, where a and b are the degrees of validity and of non-validity of x and there the following definitions are given.

R. Seising et al. (Eds.): *On Fuzziness: Volume 1*, STUDFUZZ 298, pp. 11–15.
DOI: 10.1007/978-3-642-35641-4_2

Below, we shall assume that for the two variables x and y the equalities: $V(x) = \langle a,b\rangle, V(y) = \langle c,d\rangle$ $(a,b,c,d,a+b,c+d \in [0,1])$ hold.

For two variables x and y operations "conjunction" (&), "disjunction" ($\vee$), "implication" ($\rightarrow$), and "(standard) negation" ($\neg$) are defined by:

$$V(x\&y) = \langle \min(a,c), \max(b,d)\rangle,$$

$$V(x \vee y) = \langle \max(a,c), \min(b,d)\rangle,$$

$$V(x \rightarrow y) = \langle \max(b,c), \min(a,d)\rangle,$$

$$V(\neg x) = \langle b,a\rangle.$$

In [4], the following two operations, which are analogues to operations "conjunction" and "disjunction", are defined

$$V(x+y) = \langle a,b\rangle + \langle c,d\rangle = \langle a+c-ac, bd\rangle,$$

$$V(x.y) = \langle a,b\rangle.\langle c,d\rangle = \langle ac, b+d-bd\rangle.$$

The two standard modal operators (see [9]) have the following intuitionistic fuzzy estimations (see [2]).

$$V(\Box p) = \Box V(p) = \langle \mu(p), 1-\mu(p)\rangle,$$

$$V(\Diamond p) = \Diamond V(p) = \langle 1-\nu(p), \nu(p)\rangle.$$

In [7], using (1), above form of disjunction and intuitionistic fuzzy form of First Zadeh's IF-implicarion, introduced by the author in [5, 6] with the form

$$V(x \rightarrow_{Z,1} y) = \langle \max(b, \min(a,c)), \min(a,d)\rangle,$$

we introduced First Zadeh's intuitionistic fuzzy disjunction and conjunction with the following forms of their estimations

$$V(x \vee_{Z,1} y) = \langle a,b\rangle \vee_{Z,1} \langle c,d\rangle = \langle \max(a, \min(b,c)), \min(b,d)\rangle.$$

$$V(x \wedge_{Z,1} y) = \langle a,b\rangle \wedge_{Z,1} \langle c,d\rangle = \langle \min(a,c), \max(b, \min(a,d))\rangle.$$

In [8], the following new implication was introduced

$$V(x \rightarrow_{Z,2} y) = \langle \max(b, \min(a,c)), \min(a, \max(b,d))\rangle.$$

It is shown that the two Zadeh's IF-inplications generate the classical negation ($\neg$).

Now, by analogy with the first case, we define a new disjunction

$$V(x \vee_{Z,2} y) = \langle a,b\rangle \vee_{Z,2} \langle c,d\rangle = \langle \max(a, \min(b,c)), \min(b, \max(a,d))\rangle.$$

We will call the new disjunction *"Second Zadeh's intuitionistic fuzzy disjunction"*.

We see also, that

$$V(x \rightarrow'_{Z,2} y) = \neg\langle a,b\rangle \vee_{Z,2} \langle c,d\rangle$$

$$= \langle b,a\rangle \vee_{Z,2} \langle c,d\rangle = \langle \max(b,\min(a,c)),\min(a,\max(b,d))\rangle = V(x \rightarrow_{Z,2} y),$$

i.e., the implication generates a disjunction that generates the initial implication.

Let us suppose below that De Morgan's laws are valid, i.e.,

$$x\&y = \neg(\neg x \vee \neg y). \tag{2}$$

We must immediately note that in IFS theory there are a lot of examples in which (2) is not valid, but this will be an object of discussions in future research. In the present case, as we mentioned above, the negations, generated by the implications coincide with the classical negation and by this reason De Morgan's laws are valid.

Therefore, using (2) and definition of $\vee_{Z,2}$, we can construct

$$V(x \wedge_{Z,2} y) = \langle a,b\rangle \wedge_{Z,2} \langle c,d\rangle = \langle \min(a,\max(b,c)),\max(b,\min(a,d))\rangle.$$

We will call the new conjunction *"Second Zadeh's intuitionistic fuzzy conjunction"*.

For both new operations, having in mind that $\wedge_{Z,2}$ is obtained from $\vee_{Z,2}$ by (2), we will check fistly that

$$V(\neg(\neg x \wedge_{Z,2} \neg y)) = \neg(\neg\langle a,b\rangle \wedge_{Z,2} \neg\langle c,d\rangle))$$
$$= \neg(\langle b,a\rangle \wedge_{Z,2} \langle d,c\rangle)) = \neg\langle \min(b,\max(a,d)),\max(a,\min(b,c))\rangle$$
$$= \langle \max(a,\min(b,c)),\min(b,\max(a,d))\rangle = V(x \vee_{Z,2} y).$$

Therefore, both operations are correctly defined.

We can immediately check the validity of the equalities

$$V(x \wedge_{Z,2} x) = V(x),$$
$$V(x \vee_{Z,2} x) = V(x),$$

i.e. the Idempotent Laws hold. However, the equalities

$$V(x \wedge_{Z,2} y) = V(y \wedge_{Z,2} x),$$
$$V(x \vee_{Z,2} y) = V(y \vee_{Z,2} x),$$
$$V((x \wedge_{Z,2} y) \wedge_{Z,2} z) = x \wedge_{Z,2} (y \wedge_{Z,2} z)),$$
$$V((x \vee_{Z,2} y) \vee_{Z,2} z) = x \vee_{Z,2} (y \vee_{Z,2} z)),$$
$$V((x \wedge_{Z,2} y) \vee_{Z,2} z) = (x \vee_{Z,2} z) \wedge_{Z,2} (y \vee_{Z,2} z),$$
$$V((x \vee_{Z,2} y) \wedge_{Z,2} z) = (x \wedge_{Z,2} z) \vee_{Z,2} (y \wedge_{Z,2} z)$$

are not valid. For example, if $V(x) = \langle 0.0,0.5\rangle, V(y) = \langle 0.0,1.0\rangle$, then

$$V(x \wedge_{Z,2} y) = \langle 0.0,0.5\rangle \neq \langle 0.0,1.0\rangle = V(y \wedge_{Z,2} x).$$

Therefore, both operations are not commutative and associative ones, and none is distributive with respect to the other.

In [1], the following relation is introduced for every $a,b,c,d \in [0,1]$ so that $a+b, c+d \in [0,1]$:

$$\langle a,b\rangle \leq \langle c,d\rangle \text{ if and only if } a \leq c \text{ and } d \geq b,$$

$$\langle a,b\rangle \geq \langle c,d\rangle \text{ if and only if } \langle c,d\rangle \leq \langle a,b\rangle.$$

The following inequalities are valid:

$$V(x.y) \leq V(x\&y) \leq V(x \wedge_{Z,1} y) \leq V(x \wedge_{Z,2} y),$$

$$V(x \vee_{Z,1} y) \leq V(x \vee_{Z,2} y) \leq V(x \vee y) \leq V(x+y).$$

Theorem. The following inequalities are valid

$$\text{(a) } V(\Box(x \vee_{Z,2} y)) \leq V(\Box x \vee_{Z,2} \Box y),$$

$$\text{(b) } V(\Diamond(x \wedge_{Z,2} y)) \geq V(\Diamond x \wedge_{Z,2} \Diamond y),$$

$$\text{(c) } V(\Box(x \wedge_{Z,2} y)) \leq V(\Box x \wedge_{Z,2} \Box y),$$

$$\text{(d) } V(\Diamond(x \vee_{Z,2} y)) \geq V(\Diamond x \vee_{Z,2} \Diamond y).$$

Proof. (a) First, we check the validity of the inequality

$$1 - \max(a, \min(b,c)) \geq \min(1-a, \max(a, 1-c)) \quad (3)$$

for arbitrary $a,b,c \in [0,1]$, such that $a+b \leq 1$. Let

$$X \equiv 1 - \max(a, \min(b,c)) - \min(1-a, \max(a, 1-c)).$$

If $b \leq c$, then

$$X = 1 - \max(a,b) - \min(1-a, \max(a, 1-c)).$$

If $a \leq 1-c$, then

$$X = 1 - \max(a,b) - \min(1-a, 1-c) = 1 - \max(a,b) - 1 + \max(a,c) \geq 0.$$

If $a > 1-c$, then

$$X = 1 - \max(a,b) - \min(1-a, a) = 1 - \max(a,b) - 1 + \max(1-a, a) \geq 0.$$

If $b > c$, then

$$X = 1 - \max(a,c) - \min(1-a, \max(a, 1-c)).$$

If $a \leq 1-c$, then

$$X = 1 - \max(a,c) - \min(1-a, 1-c) = 1 - \max(a,c) - 1 + \max(a,c) = 0.$$

If $a > 1-c$, then $c > 1-a \geq b > c$, which is impossible. Therefore, in all cases $X \geq 0$, i.e., (3) is valid.

Now, let x and y be two variables. Then,

$$V(\Box(x \vee_{Z,2} y)) = \Box(\langle a,b\rangle \vee_{Z,2} \langle c,d\rangle)$$

$$= \Box\langle \max(a,\min(b,c)),\min(b,\max(a,d))\rangle$$

$$= \langle \max(a,\min(b,c)),1-\max(a,\min(b,c))\rangle$$

(from (3))

$$\leq \langle \max(a,\min(1-a,c)),\min(1-a,\max(a,1-c))\rangle$$

$$= \langle a,1-a\rangle \vee_{Z,2} \langle c,1-c\rangle$$

$$= \Box\langle a,b\rangle \vee_{Z,2} \Box\langle c,d\rangle = V(\Box x \vee_{Z,2} \Box y).$$

(b)-(d) are proved by analogy.

2.3 Conclusion

In a next research, we will study the relations between the First and Second Zadeh's implications, conjunctions and disjunctions from one side, and the other intuitionistic fuzzy operations and operators from the other.

Acknowledgement. The author is grateful for the support provided by the projects DID-02-29 "Modelling processes with fixed development rules" and BIn-2/09 "Design and development of intuitionistic fuzzy logic tools in information technologies" funded by the National Science Fund, Bulgarian Ministry of Education, Youth.

References

1. Atanassov, K.: Two Variants of Intuitonistic Fuzzy Propositional Calculus, Preprint IM-MFAIS-5-88, Sofia (1988)
2. Atanassov, K.: Two Variants of Intuitionistic Fuzzy Modal Logic, Preprint IM-MFAIS-3-89, Sofia (1989)
3. Atanassov, K.: Intuitionistic Fuzzy Sets. Springer, Heidelberg (1999)
4. Atanassov, K.: Remarks on the Conjunctions, Disjunctions and Implications of the Intuitionistic Fuzzy Logic. International Journal of Uncertainty, Fuzziness and Knowledge-Based Systems 9(1), 55–65 (2001)
5. Atanassov, K.: Intuitionistic Fuzzy Implications and Modus Ponens. Notes on Intuitionistic Fuzzy Sets 11(1), 1–5 (2005), `http://ifigenia.org/wiki/issue:nifs/11/1/01-05`
6. Atanassov, K.: On Some Intuitionistic Fuzzy Implications. Comptes Rendus de l'Academie Bulgare des Sciences, Tome 59(1), 19–24 (2006)
7. Atanassov, K.: On Zadeh's Intuitionistic Fuzzy Disjunction and Conjunction. Notes on Intuitionistic Fuzzy Sets 17(1), 1–4 (2011)
8. Atanassov, K.: Second Zadeh's Intuitionistic Fuzzy Implication. Notes on Intuitionistic Fuzzy Sets 17(3), 11–14 (2011)
9. Feis, R.: Modal Logics. Gauthier-Villars, Paris (1965)
10. Klir, G.J., Yuan, B.: Fuzzy Sets and Fuzzy Logic. Theory and Applications. Prentice Hall, New Jersey (1995)

3

Fuzzy Systems towards Field Applications

Valentina E. Balas and Marius M. Balas

3.1 Introduction

We first heard about *fuzzy sets* and *fuzzy logic* when beginning our doctoral researches in the late '90s. These concepts were gaining a special popularity in Romania at that time, due to the pioneering activity of Grigore Moisil (1906-1973), Constantin Virgil Negoiţă (b. 1936), Dan Ralescu (b. 1950) and other mathematicians and engineers. More we read about the fuzzy paradigm we realized its preeminence in terms of knowledge representation for computers, its flexibility and versatility for a huge amount of scientific and technical domains.

That is why meeting Lotfi A. Zadeh in flesh and blood was one of our most memorable and inspiring experience. Thanks to his unlimited willingness to explain and to discuss all the aspects and implications of the fuzzy concepts, he reinforced our passion for this research subject and now we use to consider ourselves as members of the numerous international "fuzzy community" that is formed by the scientists and researchers sharing the same passion.

Fig. 3.1. Lotfi Zadeh visiting the "Aurel Vlaicu" University of Arad, in 2005, with Rector Lizica Mihut and Valentina Balas

R. Seising et al. (Eds.): *On Fuzziness: Volume 1*, STUDFUZZ 298, pp. 17–22.
DOI: 10.1007/978-3-642-35641-4_3

As any fundamental concept, the fuzzy sets offer lots of different facets, each one able to generate at its turn new concepts and new applicative fields. However, the main force of the fuzzy sets is their capability to represent into computers elements of world knowledge, linguistically expressed [1, 2].

Fig. 3.2. Lotfi Zadeh with Valentina Balas, in 2003, when "Aurel Vlaicu" University of Arad awarded him the Honoris Causa Doctorate

We briefly explored fuzzy expert systems able to take valid decisions out of uncertain information, fuzzy modeling which can follow qualitative descriptions and the fuzzy fusion of information [3], and each time with the feeling that we handle an exceptional tool that allows us to do everything we want. And here we get to the border limits of the fuzzy country: who needs solutions from the fuzzy systems has to know them from the very beginning. The fuzzy sets allow us to represent any problem, but cannot find solutions by themselves, even when structured as logical inference engines. An expert still must write the fuzzy rule base that supports the inference.

However this obstacle couldn't stop the expansion that created the nowadays fuzzy empire! Following a consequent policy of alliances, the fuzzy sets created a wide spread proposal of fuzzy hybrid intelligent systems, exploiting the slightest possible resource of the Artificial Intelligence! The neural networks, the genetic algorithms or the swarm systems have the ability to find by themselves solutions to problems and their symbiosis with the fuzzy sets just boosts their capacities and facilitates the designers' work. At this point, intelligent fuzzy or fuzzy hybrid systems may be used for any possible decision application that is provided with bus centered digital computer architecture, able to work in a programmable behavioral manner (*work station, personal computer* PC, *microprocessor* μP, *microcontroller* μC, *digital signal processor* DSP, etc.) This approach, that could be considered rather "desktop", has obvious disadvantages in the case of the field operating systems. The field and industrial intelligent applications need other kind of features: miniaturization, low energy consumption, low costs, reliability, high speed, etc. which means genuine electronic circuits.

Fig. 3.3. Lotfi Zadeh with Marius and Valentina Balas at the International Workshop on Soft Computing Application SOFA 2005, Arad, Romania

3.2 On the Hardware Implementations of the Fuzzy Systems

When speaking about electronic implementations for fuzzy systems we can point two historical milestones:

a) The invention of the Takagi-Sugeno fuzzy controllers, which replace the proper fuzzy sets, at the fuzzification of the output variables, by singletons. This way the structures simplify, encouraging the implementations, still preserving their fuzzy nature;

b) The development of different fuzzy integrated circuits, inspired by the Negoiţă-Ralescu α-cut theorem, that allow the convenient numerical modeling of the fuzzy sets membership functions, mostly in the Japan of '80s.

The key element that ensured from the very beginning the programmable operation of the digital computers, creating the *software* paradigm, is the fact that the architecture was build around a central spine bone: the data/address/control bus. However this architecture has a fundamental limitation: only one instruction may be executed at a time. The entire Central Processing Unit of the digital computer must basically wait for a instruction to complete before proceeding to the next one. Many techniques, such as the caches memories, the pipelines and different parallel architectures are trying to fight this inconvenient, but since the structure is inherently serial, the limitations of the digital computers' speed operation are immutable. The decision that makes possible the removal of these limitations is to implement fuzzy systems outside the conventional IT environment, in other words to conceive and build custom electronic circuits able to emulate or to reproduce the conventional software controlled behavior.

In some previous papers we proposed a special class of fuzzy systems, inspired by the Michio Sugeno's fuzzy controller, which is specially oriented towards the hardware implementations.

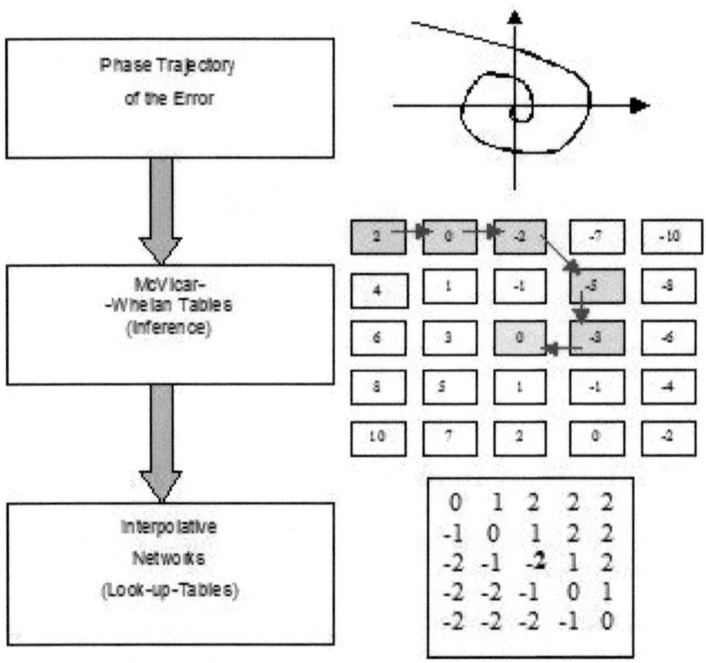

Fig. 3.4. The fuzzy-interpolative theoretical tools

The fuzzy-interpolative systems [4] are taking advantage of both linguistic and interpolative nature of the fuzzy systems, combining the advantages of their both sides:

a) The fuzzy sets and fuzzy logic theory are applied for the *conception* and the *development* stages of the control and decision algorithms;
b) Linear interpolation networks ensure the *implementation* stage [5].

3.3 On the FPGA Implementations of the Fuzzy Systems

Developing this approach we are questing for a new strategic expanding of the fuzzy paradigm over the field of the electronic circuits whose operation is programmed in the structural sense, not in the behavioral one, the so called *Field-Programmable Gate Arrays* FPGA. Such an approach is expected to create a class of *intelligent fuzzy circuits* that has the potential to significantly improve the performances of the future intelligent applications.

Compared to usual bus oriented architecture devices, a FPGA circuit can perform the same algorithm hundred times faster, with very low current consumption and working in extremely rough environment conditions. The FPGAs are replacing the virtual computing of the logic functions by the synchronized succession of the instructions and all their steps, with *a real wired circuit*, able to operate continuously, with no complicated synchronization constraints. The control software is now simply configuring a wired logic circuit [6, 7].

Many wonderful applications are already issued on behalf of the FPGA implementations of the intelligent hybrid systems, if we were to invoke only the speech recognition [8].

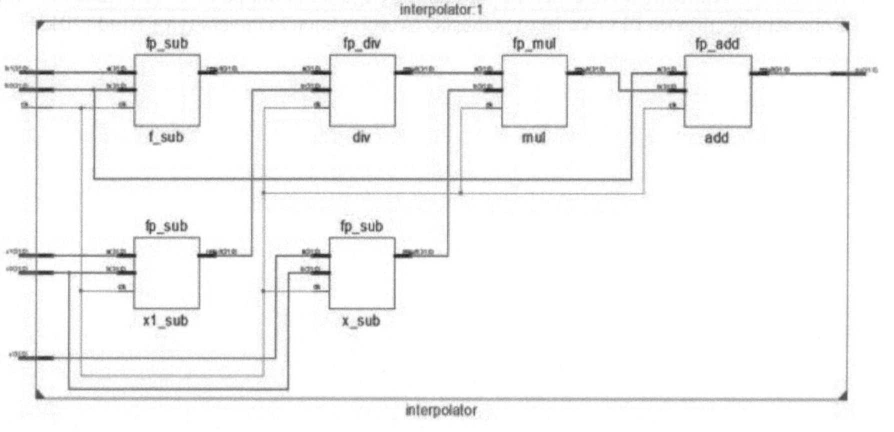

Fig. 3.5. A FPGA floating point interpolator written in VHDL

Our present goal, inspired by personal discussions with Lotfi Zadeh and Michio Sugeno, is to build a FPGA development platform for fuzzy-interpolative applications. An intermediary step is illustrated in Figure 3.5: a floating point interpolation block [9].

References

1. Zadeh, L.A.: Interpolative Reasoning as a Common Basis for Inference in Fuzzy Logic, Neural Network Theory and the Calculus of Fuzzy If/Then Rules. Opening Talk Proceedings of 2nd International Conference on Fuzzy Logic and Neural Networks, Iizuka, Japan, pp. XIII–XIV (1992)
2. Zadeh, L.A.: From Search Engines to Question-Answering Systems – The Problems of World Knowledge, Relevance, Deduction and Precisiation. In: Invited Conference, SOFA 2005, Arad, Romania (August 30, 2005)
3. Balas, V.E., Jain, L.C.: World Knowledge for Sensors and Estimators by Models and Internal Models. Journal of Intelligent and Fuzzy Systems 21(1-2), 79–88 (2010)
4. Balas, M.M.: The Fuzzy-Interpolative Methodology. In: Balas, V.E., Fodor, J., Várkonyi-Kóczy, A.R. (eds.) Soft Computing Based Modeling in Intelligent Systems. SCI, vol. 196, pp. 145–167. Springer, Heidelberg (2009)
5. Dale, S., Dragomir, T.-L.: Interpolative-Type Control Solutions. In: Balas, V.E., Fodor, J., Várkonyi-Kóczy, A.R. (eds.) Soft Computing Based Modeling in Intelligent Systems. SCI, vol. 196, pp. 169–203. Springer, Heidelberg (2009)
6. Toacşe, G., Nicula, D.: Electronică Digitală. Circuite Integrate Digitale, Limbajul de Descriere Hardware VHDL. Editura Teora, Bucuresti (1996)
7. Nicula, D., Toacşe, V.: Electronică digitală, Verilog HDL. Editura Tehnica (2005)
8. Rodriguez-Andina, J.J., Moure, M.J., Valdes, M.D.: Features, Design Tools, and Application Domains of FPGAs. IEEE Transactions on Industrial Electronics 54(4), 1810–1823 (2007)
9. Socaci, M.: Aplicaţie VHDL pentru calculul în virgulă mobilă pe FPGA (VHDL Floating Point Application on FPGA). Master thesis, "Aurel Vlaicu" University of Arad (2012)

4

Another Half Century of Progress in Fuzzy Logic?

Senén Barro

We often celebrate over the years those past events that continue to influence us today. Fuzzy set and fuzzy logic as well as the scientific-technological disciplines that are derived from them have made enormously valuable contributions to the world as a whole and to the fields of science and engineering in particular. It is, therefore, appropriate to celebrate the 50th anniversary of the day when fuzzy was first used not only as a word in a scientific paper, but also as the central concept of that particular paper. This concept went on to be significantly developed in both the theoretical and applied fields and subsequently has opened up new horizons in thinking, research and development. I think the best way to celebrate this important event is to reflect on what has been done with fuzzy logic[1] over the last half century and then to let this reflection serve to guide us into another half century of progress in which there is still much to be contributed by this field of study, research and development. My modest contribution to the project will follow along this line.

4.1 The Shadow of Zadeh Is Lengthened

Few scientists have initiated new fields of knowledge. Even fewer have been able to enjoy during their lifetime the success and overall recognition of their findings and contributions. Lotfi Zadeh is one of these scientists. He has been an active witness to the great development of fuzzy logic and its applications, though not everything has been a bed of roses. We all know the numerous attacks that have been leveled at this field. So as not to convert this article into a list of such critiques, allow me to cite the most widely known one, "The Paradoxical Success of Fuzzy Logic," by C. Elkan, which was aptly refuted by Prof. Zadeh and twenty other important scholars in the field [3]. The debate was taken up again years later by Trillas and Alsina [2] but this time with less notoriety.

The critiques have, at times, been subtle but at other times, quite vicious. Occasionally, their reflections have served to stir up the field and then after the storm there has been not only a calm but also a much clearer day than ever before. In other

[1] Fuzzy logic generally refers to a multi-faceted group of conceptual developments, some formalized, others not, as well as mathematical and logical advancements and approximations for solutions to uncountable problems in the most diverse fields. All of these use fuzzy logic as a synecdoche of common use and will be used as such in this paper.

R. Seising et al. (Eds.): *On Fuzziness: Volume 1*, STUDFUZZ 298, pp. 23–29.
DOI: 10.1007/978-3-642-35641-4_4

cases, criticisms have been made without solid arguments or scientific base. They were only supported by prejudices and fundamentalist positioning in favor of other theories, disciplines or orientations that tried to confront fuzzy logic as if it were some sort of species that could not coexist in the same ecosystem. We easily recall the unjust and barbed comment leveled at professor Zadeh by a colleague set against fuzzy logic who said, "Lotfi, I hope I live long enough to see you invited to the White House, where the President will present you with a medal 'for fooling the Japanese into thinking that fuzzy logic is a good idea'."[2]

Fig. 4.1. Professor Zadeh photographed with some of his collaborators during the time the author spent as a visiting researcher at Soda Hall, UC Berkeley, in 1997. We were young, full of energy and optimism. [Photograph taken by Fay, the wife of Professor Zadeh].

The figure of Professor Zadeh has affected the development, consolidation, use and even occasional abuse of fuzzy logic in two ways. First, his scientific and personal reputation gradually strengthened the commitment to fuzzy logic at a time when supporters were scarce in this new scientific-technological field. Second, his enormous charisma may also have worked to 'subconsciously' inhibit criticism and even 'sedate' the perennial questioning that is the hallmark of anyone involved in research. It may even have inhibited the rebelliousness characteristic of young researchers, a rebelliousness that can lead to the most disruptive advances. In sum, veneration for the founder and father of fuzzy logic has affected all of us: creating

[2] This anecdote was recounted in the article [1], which, by the way, was itself highly critical with one of the most well known applications of fuzzy logic, that of fuzzy control.

enthusiasm for this field of knowledge and technological development, along with the occasional blind (or at least scotomatous) advances.

Fig. 4.2. Lotfi Zadeh and Ebrahim Mamdani in Mieres, Spain, during a meeting of the Scientific Committee of the European Center for Soft Computing in 2007. Were they examining a fuzzy camera? I think not.

4.2 Fuzzy Trend and Modus Vivendi Fuzzy

We all know that trends in the scientific world play a significant influence on what is studied, what research is financed, especially with public money, and what is published. Fuzzy logic has also had its trends, which over the last decades have produced a notable increase in the number of researchers, of conferences and of scientific journals focused on this field. This isn't the appropriate place for a long list of statistics but one simple example of this is showed in figure 4.3.

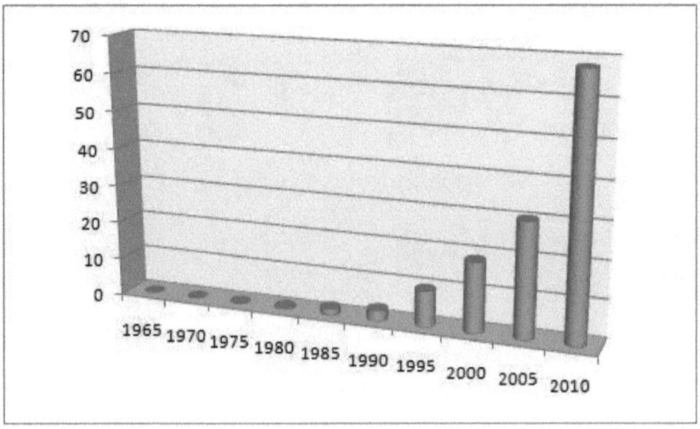

Fig. 4.3. Evolution of the number of publications that contain the word "fuzzy" in their title, keywords and abstracts from 1965 to the present – source: Scopus-

We can see that an exponential increase has been maintained from 1965 to the present. This increase is understandable during the first years. It is less logical, however, for the same exponential rate to continue throughout the four decades. The most paradoxical aspect, however, is not the exceptional increase in the number of related publications in the field but the thematic distribution, which I will now address.

4.3 The Filter of Innovation

While the number of publications on fuzzy logic has increased exponentially in the seminal journals, the same cannot be said of the publications and patents that reflect significant technological advances. Far fewer contributions of real innovation based on total or partial developments in fuzzy logic have run the full gap from R+D to a commercial product. There are many reasons for the poor relationship between the huge number of scientific articles that point to the practical use of fuzzy logic, including those that speak of specific applications in concrete areas, and the lack of patents, real functioning solutions and commercial innovations based on fuzzy logic One such cause is that among the articles published, a significant number of them are of little or no value. Some authors like D.D. Nauk propose a second cause for the disparity between publications and marketed products. In "GNU Fuzzy" – Proceedings

IEEE FUZZ-IEEE, pp. 1-6, 2007 – Nauk suggests that the inexistence of computer programs for the development of applications has greatly limited the commercial application or real use for fuzzy logic. This suggestion seems limited. It is difficult to think of a technological field where ample computer programs are not developed to provide the needed tools for analysis, simulation, design, synthesis..., unless, of course, there is not a demand for them. In my opinion, the great disparity between the number of papers related to fuzzy logic and its actual innovative practical use lies with its confinement. Fuzzy logic has been and continues to be principally used by specialized researchers within the field of fuzzy logic. It has not been incorporated into other diverse fields for potential application by their respective specialists. In other words, fuzzy logic is not being used by experts and specialists of other areas of R+D. They have not imported the theoretical, methodological or computational tools of fuzzy logic thereby limiting its usefulness for developing applications, products and services in the areas of their interest.

4.4 We Need to Clear Out the Great Forest of Fuzzy Logic

It is not intrinsically negative that an academic field produces a large number, or even a huge number, of publications. A priori, this should be good. In reality, the vast field of fuzzy logic has been built upon and given shelter to a significant number of scholars who have not opted for quality. These fuzzy logic researchers have made the quantity of publications their scientific and/or academic "modus vivendi." In my opinion, this has produced negative results. These investigators do not seem to have as their goal the advancement of knowledge and technological development. It appears that, at times, their only objective is the number of publications. We cannot attribute this phenomenon only to the field of fuzzy logic, but it seems to have happened there and continues to occur to an especially high degree.

At this point it is opportune to remember the famous saying of Isaac Newton: "If I have seen further it is by standing on the shoulders of giants."[3]. There are some, unfortunately, who stand on shoulders of giants only to avoid getting their feet wet from having to cross the puddles at hand. They are not looking ahead with advancement and progress in mind.

I feel that the field of fuzzy logic needs to experience a process of clarification, perhaps even an antiseptic cleansing. It is not an issue of questioning the fundamentals, so solidly built. We need to strip this theory laden building of all dispensable elements that tend to hide its valuable foundation. We must also protect it from the unnecessary weight that could cause an older building to collapse like a house of cards. Only then can we continue constructing a solid and secure building upon our well laid foundation. By allowing any part which is poorly constructed to remain, we threaten the stability of the entire building. If you will permit one more comparison, what I want to propose is that we clean the forest of nonproductive species, all

[3] This expression is originally attributed to Bernard of Chartres (XIIth Century) but was made famous by a letter that Newton wrote at the end of the XVIIth Century to Robert Hooke, another distinguished English scientist.

the underbrush that can encourage the spread of endless fires. This will give all the valuable trees the needed space for healthy growth and to produce more and better fruit.

4.5 Conclusion

The field of Fuzzy Logic is home to brilliant scholars and experts that have contributed significantly with relevant studies in each and every one of its sub disciplines. These investigators are especially entrusted with the task of doing to the field of fuzzy logic just what the Spanish Royal Academy of Language does with the Spanish language: cleanse, establish and give it splendor. This does not imply the need to create an academy of senior researchers assigned the task of scrutinizing every contribution to the field, judging what is valid or not. If, however, the practical and commercial applications of research are filtered through the relentless laws of the market, there should be a corresponding filter for the scientific community's publications. The filtering process should be the responsibility of scientific journal and conferences committees. In our case, the process of reviewing papers is not given the necessary time nor done with sufficient understanding for such a dynamic and specialized field as fuzzy logic. The editorial policies of the journals and conferences do

Fig. 4.4. Professor Zadeh and the author during his visit to Santiago de Compostela to participate in the "26th IEEE International Symposium on Multiple-Valued Logic (ISMVL 1996)", May 28-31, 1996, Santiago de Compostela, Spain. The building in the background, now remodeled, is in use today.

not help as much as they should to promote a standard of scientific excellence above all other considerations. They often seem to focus on factors of another nature. It would be a fine thing if a few prominent scholars were able to provide greater clarity in this field. If this could be done we would guarantee another half century of progress in fuzzy logic.

Acknowledgement. Author acknowledges the support of the Spanish Ministry for Economy and Innovation and the European Regional Development Fund (ERDF/FEDER) under grant TIN2011-29827-C02-02. I also appreciate the valuable help of Professor Alberto Bugarín when writing this paper.

References

1. Pease, R.A.: Third Thoughts on Fuzzy Logic. IEEE Micro. 15(4), 78–79 (1995)
2. Trillas, E., Alsina, C.: Comments to 'Paradoxes of Fuzzy Logic, Revisited'. International Journal of Approximate Reasoning 26(2), 157–159 (2001)
3. Several authors: A Fuzzy Logic Symposium on 'The Paradoxical Success of Fuzzy Logic'. IEEE Intelligent Systems 9(4), 2–49 (1994)

5

A Retrospective Glance from Russia at Wonderland of Fuzziness

Ildar Batyrshin

I have received my engineering diploma in Systems of Automated Control from the Moscow Institute of Physics and Technology (MIPT, sometimes referred to as "the Russian MIT") in 1975. This institute prepared specialists for high-tech governmental research institutes mainly located around the Moscow city. But I preferred to be free from duties of researcher working on prescribed tasks and returned to my city Kazan where I held assistant professor position in Department of Applied Mathematics of Kazan Institute of Chemical Technology (now it is Kazan National Research Technological University). I looked for some interesting areas of research and thanks to the department head Prof. Vladimir Skvortsov that drew my attention on the paper of Zadeh [39]. This paper was translated in Russian and published in 1974 in periodical collection of papers "Mathematics Today" in series "Mathematics and Cybernetics. News in Life, Science, and Engineering". The mathematics that was presented in this paper and the style of presentation of results were so unusual and different from the mathematics that I have studied in university so I read this paper as a truly absorbing book. I tried to find all of papers from the list of references, and then I followed the references of papers that I have found and so on, and so on. I plunged into the world of fuzziness that was enigmatic and wonderful. I started to work on several directions of research: entropy of fuzzy sets, fuzzy preference relations, fuzzy similarity relations and their applications in clustering and decision making. It may be interesting that my first results in fuzziness have been presented on October 1977 on Seminar on Applied Statistics led by Orlov A. I. in Central Economics and Mathematics Institute (CEMI) of Academy of Sciences of USSR and have been published in a book on statistics [5]. It should be noted that in statistics community of SU it was ambivalent attitude to fuzziness. Sometimes it was critics of the fuzziness from researchers that could not find the rationale for the concept of membership function but some researchers tried to substantiate the concept of fuzzy sets in terms of random sets [25]. In [5] I studied the possible axiomatizations of measure of entropy of fuzzy set [17] and relationships between them and metrics and it was established one-to-one correspondence between (symmetric) metrics defined by positive valuations [13] and measures of fuzziness satisfying an axiom of strict monotonicity for "sharpened versions". Further these results have been extended on Kleene algebras and it was shown that a metric De Morgan algebra M will be a Kleene algebra if and only if it can be defined a measure of entropy on M [8]. In

R. Seising et al. (Eds.): *On Fuzziness: Volume 1*, STUDFUZZ 298, pp. 31–38.
DOI: 10.1007/978-3-642-35641-4_5

[6] I have studied transitivity properties of strict preference relations, for example, it was shown that a fuzzy quasi-series satisfying transitivity in the form

$$\text{if } P(x,y) \geq 0 \text{ and } P(y,z) \geq 0 \text{ then } P(x,z) = max(P(x,y), P(y,z))$$

defines a fuzzy quasi ordering and hence defines a hierarchy of ordered partitions on the set of alternatives. These results have been inspired by Zadeh's paper [38] and [24]. The obtained results have been included later in my PhD dissertation [7]. In this dissertation I have also proposed and studied a general scheme of hierarchical clustering procedures invariant under numeration of objects and invariant under monotone transformations of similarity values. They were many attempts before in cluster analysis to develop clustering algorithms satisfying these properties of invariance but only by means of the concepts of fuzzy equivalence relation and transitive closure studied by Zadeh in [38] it was possible to develop a scheme of hierarchical clustering procedures satisfying both invariance properties [10].

It should be noted that in Soviet Union (almost till today in Russia) the official language of scientific publications was Russian. The abstracts of almost all scientific papers published abroad were published in Russian in special abstract bulletins. Many good scientific books published in English have been translated into Russian. Interesting papers with new ideas were translated and published in Russian in special volumes or in periodical collections of papers like "Mathematics Today" that distributed by subscription for pennies. Before signing by Soviet Union the Universal Copyright Convention in 1973 most of all foreign scientific journals have been copied (I can suppose that without permission) and could be found in almost all universities and research centers of USSR. For example, I found the copy of the journal *Information and Control* with Zadeh's paper "Fuzzy Sets" [36] in the library of Kazan Institute of Chemical Technology. But later it was almost impossible to find in the libraries of Soviet universities the foreign journals published after 1973 because the universities had not dollars for paying for them. Only 2-3 central libraries in Moscow and may be in Leningrad (now it is Saint Petersburg) had some of most important foreign journals in selected research areas. To find copies of the papers on fuzzy logic I used all possibilities to go to Moscow and to work in State Scientific Technical Library (SSTL).

SSTL was located in the center of Moscow near metro station Kuznetskij Most. In one block from this metro station it resided my good friend Valery Tarassov. I met him on one of Soviet conferences on fuzzy sets and decision making when he was the PhD student of Moscow State Technical University n.a. N. E. Bauman (MSTU). He was (and till now he is) very hospitable and his apartment was something like permanent center of fuzziness in Moscow during late 70's-90's. Many researchers in fuzziness coming to Moscow from different parts of SU found in his apartment friends and colleagues. The friendly conversations about fuzziness and other actual questions usually accompanied by bottles of vine, cognac, liqueur or vodka brought from all parts of SU often continued till the late night.

In one of such conversations it was arisen an idea to write a book on "almost all" topics of fuzziness. Several young researchers have been joined together for such ambitious project starting to write two-three chapters by each person [1]. At this time I was already a PhD student of Department of Applied Mathematics of Moscow Power Engineering Institute, Alexey Averkin and Alexander Blishun have been with Computer Center of Soviet Academy of Sciences as researcher and PhD student and Valery Tarassov was PhD student of MSTU. In one of my visits to SSTL I happened to meet a young man holding in his hands a bright yellow volume of "Fuzzy Sets and Systems" journal. It was Valery Silov coming to Moscow from Sevastopol city, Ukraina, for several days and spending some time in the libraries looking for papers on fuzzy logic. We moved together from library to Tarassov apartment and after several drinks he joined us with willingness to write some chapters of the book on fuzziness [1]. Note that further he wrote very interesting book on application of fuzzy cognitive maps to modeling dynamic multi-objective fuzzy systems [33].

The idea of writing book [1] was supported by Prof. Dmitry Pospelov, editor of book series "Problems of Artificial Intelligence" in Nauka ("Science") Publishing House where have been published later this book. Dmitry Pospelov was a head of Department of Problems of Artificial Intelligence of Computer Center of Soviet Academy of Sciences and at the same time a professor of MIPT. He served later as a President of Associations for Artificial Intelligence of USSR and Russia and Chairman of Council of Soviet Association for Fuzzy Systems (SAFS). He served also as a chair of many workshops and conferences on artificial intelligence and fuzziness in USSR and later in Russia. The book [1] contained the following chapters:

1. Methods of formalization of fuzziness.
2. Fuzzy relations and their application in analysis of complex systems.
3. Measures of fuzziness of fuzzy sets.
4. Fuzzy measures and integrals.
5. Fuzzy numbers, equations and approximation of linguistic values.
6. Fuzzy logic and approximate reasoning.
7. Generation and recognition of fuzzy languages.
8. Fuzzy algorithms.
9. Fuzzy models of optimization and decision making.
10. Methods of construction of membership functions.

The total list of references of all chapters contained about 600 works on fuzziness and related topics. This book was published more than 25 years before but till now it is a most referenced book on fuzziness in Russia.

From my subjective point of view, the researches on fuzziness in Soviet Union and Russia can be divided on several periods:

1) 1965-1973: Initial period based on Zadeh's paper "Shadows of fuzzy sets" published in Russian in 1966 [37] and on earlier works on fuzziness published in English;
2) 1974-1991: A burst of works on fuzziness in Soviet Union after translation Zadeh's papers [12], [39], [40], [41] into Russian;
3) 1992- 2004: Fuzziness in new Russia;
4) 2005-2012: Fuzziness and RAFSSoftCom.

In 1965-1973 it was published only several papers on fuzziness in USSR, e.g. [18]. After publishing in 1974 Zadeh's paper [39], two other papers of Zadeh have been translated into Russian [12], [40] in 1976. The first paper [12] was published in the edited volume "Problems of Analysis and Procedures of Decision Making" containing selected papers on decision making published in foreign journals and translated into Russian. Together with paper of Bellman and Zadeh this volume contained also the papers of Bernard Roy, Ralph L. Keeney and other well known researchers in decision making. Another paper [40] was published in Russian as a separate book. Later it was published Zadeh's paper [41]. Publication of several papers of Zadeh during a short period caused in SU a burst of interest to the theory of fuzzy sets from different groups of researchers working in applied statistics, expert estimations, theory of measurements, Osgood's semantic differential, linguistics, decision making, operations research, clustering, expert systems and artificial intelligence because the concepts of fuzzy set, fuzzy relation and the problem of definition of degree of membership were closely related with the problems studied in these research areas. The theory of fuzzy sets was a topic of interests of various seminars and workshops in different universities and research institutes. The Council of Artificial Intelligence of Academy of Sciences of USSR organized two special interest groups related with applications of fuzzy concepts in control, psycholinguistics, neurophysiology and artificial intelligence. The conferences on theory and application of fuzziness have been regularly organized in different parts of Soviet Union hosted by researchers and research groups actively working in fuzziness. Many books and surveys on fuzziness and related topics [1–4],[14–16], [19–23], [25–27], [29–32], [34, 35],[42] have been published during 1974-1991. The authors of these works belonged to research groups from different cities: Moscow, Kazan, Kolomna, Sevastopol, Baku, Riga, Frunze, Taganrog, Kalinin, Tbilisi etc. Of course the list of references includes only a small part of works on fuzziness in Soviet Union before its dissolution in December of 1991.

In January of 1990 in Kazan city it was held a founding convention of Soviet Association for Fuzzy Systems (SAFS). The founding members of SAFS are presented in Fig. 5.1, first line (from left to right): Ildar Batyrshin, Alla Zaboleeva-Zotova, Dmitry Pospelov, V. Chernyaev; second line (from left to right): Valery Tarassov, Alexander Yazenin, Alexey Averkin, Askold Melikhov, Arkady Borisov; upper line (from right to left): Alexander Blishun, Alexander Shostak. The other persons were from research groups of Alexander Blishun. Askold Melikhov and later Alexey Averkin have been elected as the first and the second presidents of SAFS.

In 1989-1991, in the last years of "Perestroyka" in USSR, and in the first years of new Russia, when it was an economic collapse, research in fuzziness in Russia could survive for several reasons. In 1989-1992, research in artificial intelligence and fuzziness have been supported partially by State Research and Development Program "Perspective Information Technologies". Since 1992 these researches have been supported also by grants of Russian Foundation of Basic Research (RFBR). I used these possibilities to work actively in the area of fuzzy intelligent systems and clustering and obtained results have been included in 1996 in my Dr. Sci. (Habilitation) Dissertation [9]. It was developed, for example, a method of construction of strict monotonic conjunctions and disjunctions for processing in expert systems expert evaluations of truth measured in finite ordinal scale L. It is impossible to define such operations on L and the solution of the problem was consisted in embedding of L in the linearly ordered set of uncertainties with memory (lexicographic valuations) [11]. This problem could not be resolved in the theory of measurements [28] but was resolved in fuzzy logic. Obtained results were implemented in fuzzy expert systems shell LEXICO and in fuzzy decision support system SMOPLEX used for optimization of polymerization process in simulator of chemical reactor [11]. The last one used the knowledge-base based on the concept of fuzzy algorithm proposed by Zadeh [39].

Fig. 5.1. The founding members of Soviet Association for Fuzzy Systems, January 1990

In new Russia, in 2005, it was organized Russian Association for Fuzzy Systems and Soft Computing (RAFSSoftCom), Presidents: Ildar Batyrshin (2005-2006), Alexander Yazenin (2006-2008), Nadezhda Yarushkina (2008-2011), Sergey Kovalev (2011-2013). RAFSSoftCom is a member of IFSA, publishes the journal *Fuzzy Systems and Soft Computing* (Editor in Chief Alexander Yazenin), organizes biannual conferences on Fuzzy Systems and Soft Computing. Another biannual conference "Integrated Models and Soft Computing in Artificial Intelligence" (chair Valery Tarassov) is organized in Kolomna near Moscow by RAFSSoftCom together with Russian Association for Artificial Intelligence. Members of association actively participate in biannual conferences of Russian Association for Artificial Intelligence and in many other conferences on intelligent systems in Russia and abroad.

I met Lotfi Zadeh first time in Aachen, on EUFIT 1993, and after this we met on many conferences: in Zittau, in Antalya on ICSCCW' 2001 (Fig. 5.2), in San Francisco in 2011 etc. It was honor for me to be invited by him in Berkeley in 2002 on the meeting "State of the Art Assessment and New Directions for Research" and in 2005 on BISC-2005. I always admired his kindness, cordiality, willingness to help, activity and ability to generate new ideas.

Fig. 5.2. On ICSCCW' 2001, Antalya, Turkey, June 6-8, 2001

References

1. Averkin, A.N., Batyrshin, I.Z., Blishun, A.F., Silov, V.B., Tarassov, V.B.: Fuzzy Sets in Models of Control and Artificial Intelligence. In: Pospelov, D.A. (ed.) Fuzzy Sets in Models of Control and Artificial Intelligence. Fizmatlit, Science Publisher, Moscow (1986) (in Russian)
2. Aliev, R.A., Abdikeev, N.M., Shakhnazarov, M.M.: Industrial Systems with Artificial Intelligence. Radio and Communication Publisher, Moscow (1990) (in Russian)
3. Aliev, R.A., Zakharova, E.G., Ulyanov, S.V.: Fuzzy Control Models of Dynamic Systems. In: Results in Science and Engineering, Series Technical Cybernetics, vol. 29, pp. 127–201. VINITI AN USSR, Moscow (1990) (in Russian)

4. Aliev, R.A., Tserkovny, A.E.: Production Control with Fuzzy Initial Information. Energoizdat, Moscow (1991) (in Russian)
5. Batyrshin, I.Z.: On Some Properties of Measures of Non-probabilistic Entropy of Fuzzy Sets. In: Applied Multidimensional Statistical Analysis, pp. 345–348. Science Publisher, Moscow (1978)
6. Batyrshin, I.Z.: On Transitivity of Fuzzy Orderings. In: Operations Research and Analytical Design in Technique, pp. 67–73. KAI Publisher, Kazan (1979) (in Russian)
7. Batyrshin, I.Z.: Methods of System Analysis Based on Weighted Relations. PhD dissertation. Moscow Power Engineering Institute, Moscow (1982) (in Russian)
8. Batyrshin, I.Z.: On Fuzzinesstic Measures of Entropy on Kleene algebras. Fuzzy Sets and Systems 34, 47–60 (1990)
9. Batyrshin, I.Z.: Representation and Processing of Fuzzy Information in Intelligent Systems. Dr. Sci. (Habilitation) Dissertation. Pereslavl-Zalesskii (1996) (in Russian)
10. Batyrshin, I.Z., Rudas, T.: Invariant Hierarchical Clustering Schemes. In: Batyrshin, I., Kacprzyk, J., Sheremetov, L., Zadeh, L.A. (eds.) Perception-based Data Mining and Decision Making in Economics and Finance. SCI, vol. 36, pp. 181–206. Springer, Heidelberg (2007)
11. Batyrshin, I.Z.: Uncertainties with Memory in Construction of Strict Monotonic t-norms and t-conorms for Finite Ordinal Scales: Basic Definitions and Applications. Applied and Computational Mathematics 10(3), 498–513 (2011)
12. Bellman, R.E., Zadeh, L.A.: Decision-making in a Fuzzy Environment. Management Science 17(4), 141–164 (1970) (in Russian 1976)
13. Birkhoff, G.: Lattice Theory, 3rd edn., vol. 25. American Mathematical Society Colloquium Publications, Providence (1967)
14. Borisov, A.N., Alekseev, A.V., Krumberg, O.A., et al.: Decision Making Models Based on Linguistic Variable. Zinatne, Riga (1982) (in Russian)
15. Borisov, A.N., Alekseev, A.V., Merkurieva, G.V., Slyadz, N.N., Glushkov, V.I.: Fuzzy Information Processing in Decision Making Systems. Radio and Communication Publisher, Moscow (1989) (in Russian)
16. Borisov, A.N., Krumberg, O.A., Fedorov, I.P.: Decision Making Based on Fuzzy Models. Examples of Application. Zinatne, Riga (1990) (in Russian)
17. DeLuca, A., Termini, S.: A Definition of a Nonprobabilistic Entropy in the Setting of Fuzzy Sets Theory. Information and Control 20(4), 301–312 (1972)
18. Gusev, L.A., Smirnova, I.M.: Fuzzy Sets. Theory and Applications (Survey). Automation and Remote Control 5, 66–85 (1973) (in Russian)
19. Kafarov, V.V., Dorokhov, I.N., Markov, E.P.: System Analysis of Chemical Technology Processes. Application of the Methods of Fuzzy Sets. Science Publisher, Moscow (1986) (in Russian)
20. Kaipov, V.A., Selyugin, A.A., Dubrovsky, S.A.: Data Processing Methods in Systems with Fuzzy Information. Ilim Publisher, Frunze (1988) (in Russian)
21. Kuzmin, V.B.: Construction of Group Decisions in the Spaces of Crisp and Fuzzy Binary Relations. Science Publisher, Moscow (1982) (in Russian)
22. Malyshev, N.G., Bershtein, L.S., Bozhenyuk, A.V.: Fuzzy Models for Expert Systems in CAD. Energoatomizdat, Moscow (1991) (in Russian)
23. Melikhov, A.N., Bershtein, L.S., Korovin, S.Y.: Situational Advising Systems with Fuzzy Logic. Science Publisher, Moscow (1990) (in Russian)
24. Mirkin, B.G.: The Problem of Group Choice. Science Publisher, Moscow (1974)
25. Orlov, A.I.: Problems of Optimization and Fuzzy Variables. Knowledge Publisher, Moscow (1980) (in Russian)

26. Orlovsky, S.A.: Problems of Decision Making with Fuzzy Information. Science Publisher, Moscow (1981) (in Russian)
27. Petrov, B.N., Ulanov, G.M., Ulyanov, S.V.: Dynamic systems with random and fuzzy structures. In: Results in Science and Engineering. Series Technical Cybernetics, vol. 11, pp. 3–76. VINITI, Moscow (1979) (in Russian)
28. Pfanzagl, J.: Theory of Measurement. Physica-Verlag, Würzburg (1971)
29. Pospelov, D.A.: Logic-Linguistic Models in Control Systems. Energoizdat, Moscow (1981) (in Russian)
30. Pospelov, D.A.: Situational Control: Theory and Practice. Science, Moscow (1986) (in Russian)
31. Pospelov, D.A.: Modeling of Reasoning. Radio and Communications Publisher, Moscow (1989) (in Russian)
32. Shostak, A.P.: Two Decades of Fuzzy Topology: Basic Ideas, Concepts and Results. Russian Mathematical Surveys 44(6), 125–186 (1989)
33. Silov, V.B.: Strategic Decision Making in Fuzzy Environment in Politics, Macroeconomics, Sociology, Management, Medicine and Ecology. INPRO-RES, Moscow (1995) (in Russian)
34. Ulyanov, S.V.: Fuzzy Models of Intelligent Industrial Control Systems: Theoretical and Applied Aspects. Proceedings of the Academy of Sciences of USSR. Series Technical Cybernetics 3, 3–29 (1991) (in Russian)
35. Yazenin, A.V.: Fuzzy Mathematical Programming. Kalinin State University, Kalinin (1986) (in Russian)
36. Zadeh, L.A.: Fuzzy Sets. Information and Control 8, 338–353 (1965)
37. Zadeh, L.A.: Shadows of Fuzzy Sets. Problems of Information Transmission 2, 37–44 (1966) (in Russian)
38. Zadeh, L.A.: Similarity Relations and Fuzzy Orderings. Information Sciences 3, 177–200 (1971)
39. Zadeh, L.A.: Outline of a New Approach to the Analysis of Complex Systems and Decision Processes. IEEE Transactions on Systems, Man and Cybernetics SMS-3, 28–44 (1973) (in Russian 1974)
40. Zadeh, L.A.: The Concept of a Linguistic Variable and its Application to Approximate Reasoning. Information Sciences 8, Part I, 199–249, Part II, 301–357, 9, Part III, 43–80 (1975) (in Russian 1976)
41. Zadeh, L.A.: Fuzzy Sets and their Applications to Pattern Classification and Cluster Analysis. In: Van Ryzin, J. (ed.) Classification and Clustering, pp. 251–299. Academic Press, New York (1977) (in Russian 1980)
42. Zhukovin, V.E.: Fuzzy Multiobjective Models of Decision Making. Mezniereba, Tbilisi (1988) (in Russian)

6

The Parable of Zoltan

James C. Bezdek

I learned about fuzzy sets in 1969 when I was a graduate student in Applied Mathematics at Cornell University. Subsequently, I based my PhD thesis on Fuzzy Clustering. The notion of fuzzy sets was not only novel then, but controversial. And its basic premise - that there is a type of imprecision which cannot be adequately accounted for with probability - continues to bother many engineers and scientists today. This note is about the turbulence that is still created by this division of beliefs about mathematical models of uncertainty.

The growth of the theory and applications of fuzzy sets in the 1970s-1980s created a demand at conferences for tutorials about fuzzy models, and I sometimes gave such lectures on the use of fuzzy sets in pattern recognition. A common question then that persists to the present day was "can you give us an example that shows a real difference between fuzzy and probabilistic uncertainty?". My response to that question led me to propose an example called the "potable drinks" example. I often used this example in the late 1980s, and finally published it, first in Bezdek and Pal [1], and then again, in my introduction to fuzzy models that served as a preamble to the inaugural issue of the *IEEE Transactions on Fuzzy Systems* [2]. Here is the example:

6.1 The Potable Drinks Example (Circa 1985; cf. [1], [2]

One of the first questions asked about this scheme [fuzzy sets], and the one that is still asked most often, concerns the relationship of fuzziness to probability. Are fuzzy sets just a clever disguise for statistical models? Well, in a word, NO. Perhaps an example will help.

Let the set of all liquids be the universe of objects, and let fuzzy subset $L = \{\text{all}(\textit{potable}(= \text{"suitable for drinking"}) \text{ liquids}\}$. Suppose you had been in the desert for a week without drink and you came upon two glasses labeled A and B as in the left half of Figure 6.1 (memb = "membership", and prob = "probability").

Confronted with this pair of glasses, assuming that you will drink from the one that you choose – which one would you choose to drink from? Most readers familiar with the basic ideas of fuzzy sets, when presented with this experiment, immediately see that while A could contain, say, swamp water, it would not (discounting the possibility of a Machiavellian fuzzy modeler) contain liquids such as leaded gasoline. That is, they would know that a *membership* of 0.91 in L means that the contents of A

R. Seising et al. (Eds.): *On Fuzziness: Volume 1*, STUDFUZZ 298, pp. 39–46.
DOI: 10.1007/978-3-642-35641-4_6

are "fairly similar" to perfectly potable liquids (e.g., pure water). On the other hand, the *probability* that B is potable $= 0.91$ means that over a long run of experiments, the contents of B are expected to be potable in about 91% of the trials. And the other 9%? In these cases the contents will be unsavory (indeed, possibly deadly) – about 1 chance in 10. Thus, most observers will opt for a chance to drink swamp water, and will choose A.

Fig. 6.1. Glasses for the weary traveler – disguised and unmasked!

Another facet of this example concerns the idea of *observation*. Continuing then, suppose that we examine the contents of A and B, and discover them to be as shown in the right half of Figure 6.1 – that is, A contains beer, while B contains hydrochloric acid. After observation then, the membership value for A will be unchanged (well, this being beer, you might upgrade the membership value to 0.98 or so), whilst the probability value for B clearly drops from 0.91 to 0.0.

Finally, what would be the effect of changing the numerical information in this example? Suppose that the membership and probability values were both 0.50 – would this influence your choice? Almost certainly it would. In this case many observers would switch to a swig of the liquid in B, since it offers a 50% chance of being drinkable, whereas a membership value this low would presumably indicate a liquid unsuitable for drinking (this depends, of course, entirely on the membership function of the fuzzy set L).

In summary, my example shows that these two types of models possess philosophically different kinds of information; fuzzy memberships, which represent *similarities* of objects to imprecisely defined properties; and probabilities, which convey information about *relative frequencies*. Moreover, interpretations about and decisions based on these values also depend on the actual numerical magnitudes assigned to particular objects and events. See [3] for an amusing contrary view with lots of respondents and arguments.

Response from the probabilistic community to the potable drinks example was immediate – and predictable. Woodall and Davis [4] sent me a letter of comments on the example that was published in *IEEE TFS* 2(1). Here is their general summary from that letter:

> "We have found that many of those advocating the use of fuzzy logic have justified their methods by offering very limited views of probability. In our opinion, probability can be used to represent the information claimed to be provided only by memberships."

I published my response to them in a note titled *The Thirsty Traveler visits Gamont: A Rejoinder to "Comments on "Fuzzy Sets - What are They and Why?" "* [5]. As we approach the 50th anniversary of the first paper on fuzzy sets [6], I think it appropriate to revisit Woodall and Davis' comments and my response to them (retitled, and slightly updated here and there to account for events that have happened during the ensuing 20 years).

6.2 The Gamont Chronicles

Does the Woodall and Davis letter offer anything new? Have they finally seen through us ? I don't think so. Quoting their letter:

> "In our opinion, probability can be used to represent the information claimed to be provided only by memberships".

Woodall and Davis suggest altering my example so that a probability model behaves more like my fuzzy model of the liquids in the glasses. This sounds like – "OK, maybe fuzzy uncertainty exists, but I can still handle it at least as well, if not better, with a probability model". Let me recount for you the Gamont Chronicles, a short adventure that illustrates what I think their construction really means.

Figure 6.2 is a map of the region known as *Gamont*[1]. In the West, *Data City* is a bustling place; its members represent various populations that are distributed across the metropolitan area in many different ways. Like most big cities, it has bad neighborhoods; *Imprecise Alley* is one. You have been in neighborhoods like it – nothing there is ever really certain. At the Eastern end of Gamont is a municipality known as *Some Solution* – (it may be near Truth or Consequences, NM, but I'm not sure of this). Data City is connected to Some Solution by a modern superhighway named Statistics Parkway that passes through *Chancetown*. There is another way to get to Some Solution from Data City on a very rugged dirt path called Roughly Right Road, which passes through the mountain village of Vagueville. Finally, there's a hamlet (really, just a piglet in this tale) in Gamont – we'll get to it later.

Data City was established centuries ago, and its residents began traveling to Some Solution along the route that is now Statistics Parkway when it was still unpaved - that is, a path without much real foundation. You know progress. Both the parkway and the means for using it improved steadily, and an enterprising businessman named

[1] You can find some pretty interesting definitions for the word Gamont on the internet. I used to play the board game "DUNE" with my kids in the 1970s, and there were 3 or 4 "worthless cards" that could be drawn. One of them was titled "Trip to Gamont". Another was the "Jubba Cloak". And so on.

Mr. Probability opened a (Mercedes) dealership there about 1620. Many residents of Data City were delighted, for their families *just fit* into Probability's current models. As new models came out, inhabitants enjoyed decking one out with all the latest parameters, loading it up with little data sets, and zipping over to their favorite neighborhood in Some Solution, a down-to-earth place known as *Useful Fit*. It was a long trip, so they often stopped at the *Inferential Principle Cafe* in Chancetown. There were lots of menu choices, such as method-of-moments meatloaf, least-squares soup, maximum likelihood pie, entropy eggplant (Parmesan) and Bayesian pudding. Each family had its favorites, and sometimes they bickered a little about the confidence they should place on a particular selection, but it never caused real problems, because most choices led to pretty much the same results.

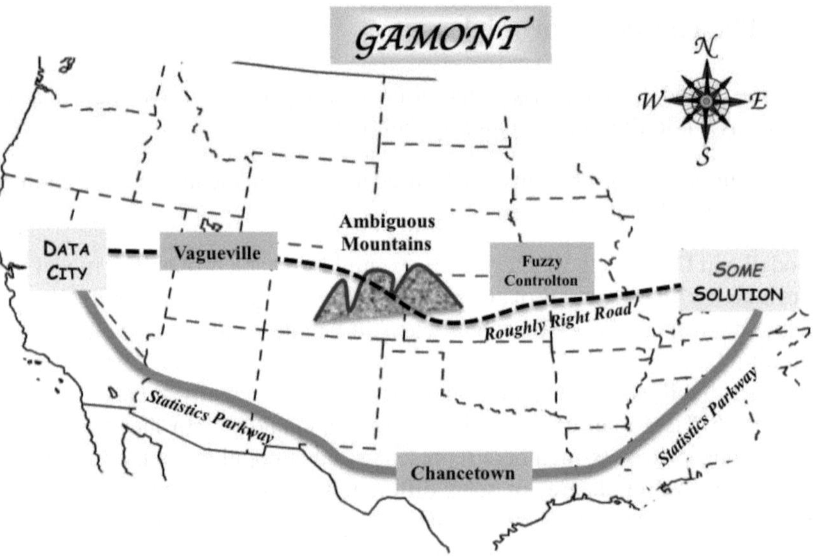

Fig. 6.2. A Trip through Gamont

But alas, some residents of Data City just *could not* fit *their* families into *any* model offered by the dealership. These were, in the main, that wretched clan that lived in Imprecise Alley. Mr. Probability felt that the Imprecise Aliens (you might have expected them to call themselves *Alleyans*, but they had very little formal training) could – and should – MAKE their families fit into one of his many models. And, since Mr. Probability had the only dealership in Data City, sometimes Imprecise Aliens did just that. But they were uncomfortable, and they spent most of their time trying to interpret the rules for traveling on Statistics Parkway, which by that time had become very complicated indeed.

Since they only rarely were able to use Mr. Probability's models, Imprecise Aliens hardly ever got out of Data City. Sometimes they packed up picnic lunches, and hiked to Some Solution by taking the mountain shortcut through Vagueville (it was quite a bit more direct, and fun too, but of course Roughly Right Road was no place for a Mercedes). These were a merry and hardy people, thick of skin and bright of eye, but they were the object of much scorn by most of the folks in Data City, who felt that anyone who really needed to get to Some Solution could always use Statistics Parkway (even if they had to hitchhike).

Then, an incredible thing happened in 1965. A *stranger* – a man from another time and place I guess, because he had an odd name like Zoltan, or something like that - moved to Data City and opened up a second dealership. Zoltan's ideas about travel were pretty radical. He sold Land Rovers. The Imprecise Aliens quickly learned that these new vehicles could easily bounce along Roughly Right Road, pass through Vagueville (taking in sights never seen by those who traveled only by Mercedes), and arrive at Some Solution with plenty of time to find Useful Fit. Families with unusually imprecise children - you know, the kind that never fitted into Probability's models at all - seemed especially comfortable in this new vehicle. Like all new models, those sold by Zoltan had some design flaws and manufacturing glitches, but after these were worked out, they became very reliable indeed. So, Land Rovers quickly multiplied, and this had some consequences.

Mr. Probability lost a little business. Not because the Mercedes was outdated; rather, it seemed more natural to sometimes take the direct route. Moreover, opening the new route to Some Solution led to the discovery of *Fuzzy Controlton*, a tiny hamlet tucked away deep in the *Ambiguous Mountains* . The residents of Fuzzy Controlton were considered inverted - their lives swung back and forth like pendulums. But they seemed to know a lot about Useful Fit, and even though they had no formal training in Cartography, they helped the Imprecise Aliens prepare a very detailed map of it.

Now the Imprecise Aliens knew that the Land Rover would run on Statistics Parkway, but they also knew it was silly to drive to Some Solution in one via the Parkway, since the Mercedes was much better suited to this task. So, they saved part of their meager incomes, and most of them finally owned both a Mercedes and a Land Rover; each was used for the trips it was most well suited to. This maximized the resale value of both vehicles, and the Imprecise Aliens were a pretty happy lot.

Mr. Probability's dealership was not in trouble; his models worked, and worked well indeed, for many families in Data City – especially for the normally distributed ones who lived in the central limit district. But he was worried. As he saw it, there were only two choices. First, he could encourage other Data Citizens to have a Land Rover *and* a Mercedes. This made sense, for then every family would have the correct model for every trip. But he felt that not every resident wanted both, or even needed both. It would be better, he thought, to show the Data Citizenry that he could modify any Mercedes so that it, too, could make the trip from Data City to Some Solution via Vagueville. He even did it, converting a 300SL so that it had an extra gas tank, four wheel drive, knobby RV tires, KC lights and the like – it was quite a sight! Prospective buyers tried it out sometimes. Oh, it made the trip alright, but it wasn't nearly as easy or comfortable to get to Useful Fit this way, and the resale value ? You decide.

There was a third choice; opening an entirely new route – one that incorporated all the good features of both routes and both vehicles. This really seemed like the best choice of all, and Zoltan happily agreed to do whatever he could to expedite it. But Mr. Probability longed for the old days - the days when the only way to Some Solution was through Chancetown. So his engineers didn't spend much time on this intriguing and eminently sensible idea.

As we leave Gamont, Zoltan and the Imprecise Aliens were last seen headed for Fuzzy Controlton – they said something about a parade. And the older, more respected dealer - really, the dealer with a much more solid foundation? Well, he was arguing with his sales force about new ways to make the Imprecise Aliens fit into his latest models. Maybe he always will - I don't know.

The Use of Mathematical Models at Cross Purposes

Woodall and Davis state that the "the fuzzy modeler has had the opportunity to sample the liquid in glass *A*, while the person evaluating glass *B* knows only that its contents were randomly selected from some population of liquids, 91% of which are potable". Then they say that, in their "alternative use of probability, the contents of glass B are examined in the same manner as those in glass *A*". The implication of this is that the evaluator (the one choosing the glass) has somehow seen *A*, but not *B*. Not so. The *evaluator* has seen *only the labels* of *A* and *B*, not (yet) sampled their contents.

If you want to know how these labels might arise, consider two bottling plants. One, run by the Imprecise Aliens, assigns a membership to every liquid they intend to bottle, based on tests, and the memberships are "tuned" until a membership function satisfying the objectives of the modelers are found. Then, whenever a run of any liquid is bottled, every bottle gets a label showing the membership value for the liquid being processed. The other plant has the same information about the liquids. However, in order to assign probabilities, it will be necessary for the second plant to decide somehow the exact boundary between the liquids that are potable, and those that are not. Why? Because probabilities refer to crisp events having hard boundaries in the sample space. Woodall and Davis may have missed this point: in probability, every liquid either is or is not potable, but in the fuzzy model, this determination is not needed, and is never made. When this second plant bottles, they let the machine randomly select, with known prior probabilities of selection (not of being potable - that is already done, and has no further bearing on the label a particular bottle receives), liquids from different storage tanks. Labels for these bottles are assigned accordingly (the manager of the plant, a Mr. Probability, was heard to say – "they can take their chances").

When I give this example, some people think I dislike probability, or don't understand it, or I am trying to discredit, or belittle it, or replace it. These people are wrong - they have missed the point. My example merely illustrates that modelers of processes may have choices. If there is no choice - fine; use what you have. But if there is a choice, pick the model that solves your problem best. The Gamont Chronicles advertise this point, and this point alone: drive nails with hammers, and screws with screwdrivers. I can drive screws with a hammer, and, with a little more effort, nails with a screwdriver. But is this the best use of these tools?

Woodall and Davis suggest a new scheme which enables probability to be used "in a way which is much more closely analogous to this use of membership", and that "this use of probability makes the argument for the usefulness of memberships much less convincing". Apparently they agree that potability is an imprecisely defined, non-random property. I fully agree with their conclusion, viz., that "probability can be used to represent the information claimed to be provided only by memberships". To me, this amounts to using probability as a means for getting good estimates of membership functions for non-random processes. I guess this is a step in the right direction.

Now it's 2012 instead of 1992. Has anything changed about this debate in the past two decades? Not really. There is still a large, loud and staunchly resistant part of our scientific community that derides the notion that fuzzy models can ever be useful. Someone asked Jim Keller recently – "aren't fuzzy sets just a cult of personality?" What would convince these folks? Nothing, I suspect. I could ask them to google up any of a hundred topics such as "fuzzy sets" (About 1,790,000 results, October 15, 2011), "fuzzy clustering" (1,620,000 results), "fuzzy control" (5,400,000 results – Fuzzy Controlton is very well populated now!), or "fuzzy patents" (3,470,000 results). Discounting the multiple hits, the false hits, and so on, this represents a substantial and reliable literature – one that guarantees us that fuzzy models are not going away anytime soon. But, if you are determined to use probability, Glenn Shafer [7] admonishes you to remember that

> "The interpretation of belief functions is controversial because the interpretation of probability is controversial"

And F. R. Moulton [8] still offers the best advice for all of us:

> "...every set of phenomena can be interpreted consistently in various ways, in fact, in infinitely many ways. It is our privilege to choose among the possible interpretations the ones that appear to us most satisfactory, whatever may be the reasons for our choice. If scientists would remember that various equally consistent interpretations of every set of observational data can be made, they would be much less dogmatic than they often are, and their beliefs in a possible ultimate finality of scientific theories would vanish."

Zen Maxim
"Great Doubt: great awakening. Little Doubt: little awakening. No Doubt: no awakening".

References

1. Bezdek, J.C., Pal, S.K. (eds.): Fuzzy Models for Pattern Recognition. IEEE Press, Piscataway (1992)
2. Bezdek, J.C.: Fuzzy Models – What are They and Why? IEEE Transations on Fuzzy Systems 1(1), 1–6 (1993)

3. Cheeseman, P.: An Inquiry into Computer Understanding. Computational Intelligence 4, 57–142 (1988) (with 22 commentaries/replies)
4. Woodall, W.H., Davis, R.E.: Comments on " Editorial: Fuzzy Models – What are They and Why?". IEEE Transactions on Fuzzy Systems 2(1), 43 (1994)
5. Bezdek, J.C.: The thirsty traveller visits Gamont: A rejoinder to "Comments on 'Editorial: Fuzzy Models – What are They and Why?'". IEEE Transactions on Fuzzy Systems 2(1), 43 (1994)
6. Zadeh, L.A.: Fuzzy Sets. Information and Control 8, 338–352 (1965)
7. Shafer, G.: Rejoinder to Comments on "Perspectives on the Theory and Practice of Belief Functions". International Journal of Approximate Reasoning 6, 445–480 (1992)
8. Moulton, F.R.: The Velocity of Light. Scientific Monthly 48, 481–484 (1939)

7

The Membership Function and Its Measurement

Taner Bilgiç

7.1 Introduction

Perhaps the most fundamental concept in fuzzy set theory is the *membership function* [24, 25]. Fuzzy sets allow for gradual degrees of membership and the membership function is a measure of that degree. The meaning of the membership function has been a question for everyone including Zadeh himself who advocated a linguistic approach [26–28] from the very early days. This idea also had support from linguists [3, 9] and psychologists [6].

In this chapter, we will take a closer look at the membership function as it relates to word meaning representation since the original idea of fuzzy sets has a strong affinity to linguistic scales and then discuss those studies that use the representational theory of measurement to provide meaning to the membership function. We argue that recent approaches to word meaning representation, a more complete theory of measurement which accounts for errors and empirical evidence that utilize brain imaging techniques can provide a much better understanding of fuzziness and a more coherent foundation to the theory. In passing, we identify potential areas of future research in this domain.

7.2 Meaning Representation in Linguistics

Since the semantics of fuzzy set theory is closely tied to the concept of a membership function one has to take a closer look at the meaning of the membership function. One possible direction is to discuss the meaning of a membership function in the context of *word meaning representation*. There are various approaches to word meaning representation[1] The classical view of meaning representation stems from antiquity [18] and has been treated in formal logic since Aristotle until Wittgenstein forcefully argued that formal logic is not adequate to represent (word) meaning [22]. This view

[1] It is widely accepted that word meaning (i.e., lexical-semantics) is grounded in conceptual knowledge. But one also needs to answer the more difficult question of whether the two are distinguishable from each other. There is empirical evidence on closeness of semantic and conceptual representations by brain imaging research. We do not delve into this debate here but note that most theories about the representation of meaning propose conceptual representations rather than semantic representations which implies that they take concepts and semantics to be the same.

R. Seising et al. (Eds.): *On Fuzziness: Volume 1*, STUDFUZZ 298, pp. 47–50.
DOI: 10.1007/978-3-642-35641-4_7

was also shared by psychologists who argued that category boundaries are fuzzy [17]. Mainly there appeared two major alternative representations [21]: the *holistic* theories based on relations among words which evaluates meanings of words in relation to other words by using differential scales or network models, and *featural* theories where the representation relies on features (or feature sets).

It appears that the meaning representation envisioned by fuzzy sets is holistic in nature trying to load all meaning for graded membership to a single mathematical function called the membership function. The featural theories also found their way into fuzzy set theory using the concepts of *conjoint measurement* [19], an area that still has room for development.

However, other models of word meaning representation (network models, semantic fields, and computational and statistical linguistic models) also require academic attention in connection with the representation of degrees of membership.

7.3 Representational Theory of Measurement

We now turn to the question of measurement of the membership function. Without delving into the practical measurement issues we discuss the implications of such a measurement. To put this question into perspective we resort to the representational theory of measurement (RTM)[8, 14]. RTM postulates axioms under which measurement (defined as a mapping from a qualitative algebraic structure to a numerical structure) is possible (representation) and *meaningful*. The meaningfulness comes as a result of the concept of a scale which is related to the uniqueness of the representation. If the representation is unique up to a positive linear transformation the resulting scale is called an interval scale where ordering and averaging are meaningful whereas if the representation is unique up to a similarity transformation the resulting scale is a ratio scale where not only ordering and averaging but taking ratios are all meaningful. This elaborate theory of measurement is the dominant view about how we think about measurement today.

The concept of meaningfulness prescribed by the RTM and the resulting scale types can shed light into the semantics of fuzzy set theory. Measurement of membership functions has been investigated in the context of RTM in quite a few studies [2, 4, 11, 12, 16, 19, 23]. What the RTM entails for measuring the membership function is an *objective* interpretation of fuzziness.

There are three main objections to the RTM [5, 13]: *(i)* RTM cannot be applied because it uses an axiomatic approach, and because it cannot give a satisfactory account of actual scientific measurement practice, *(ii)* Errors of measurement cannot be incorporated into the RTM framework, *(iii)* RTM is wrongly liberal in what it accepts as measurement. Although we do not agree with the first of these criticisms[2], the other two are largely accepted and partially defended by the proponents of RTM [10, 15].

[2] In fact, RTM carefully insists on *constructive* proofs of representation theorems. Such a constructive approach is expected to lead to practical measurement concepts like *standard sequences* [8].

Since RTM does not have room for measurement errors, what it entails for the measurement of a membership function is an *objective semantics* of fuzziness. Nevertheless, RTM provides a firm basis not only for degrees of membership but also for (concatenation) operators (usually represented as t-norms and conorms) [1, 2, 11]. The trouble of using RTM to provide semantics for fuzzy set theory is not immune to a significant deficiency of RTM as a paradigm: the impossibility of incorporating errors in measurement which is common in all practical measurements and could even be considered to be the source of fuzziness itself!

7.4 The Road Ahead

Zadeh argues that „Humans have a remarkable capability to perform a wide variety of physical and mental tasks without any measurements.“[30] and finds a logic of „... manipulating perceptions rather than measurements“more *natural* [29]. Perhaps in his wisdom Zadeh is rightly pointing to a direction where the primitive, fundamental concept in fuzzy set theory and fuzzy logic is the concept of a membership function which needs no further scrutiny of measurement. This approach evades many difficulties and criticisms raised against the vague semantics of the theory. Furthermore, one can turn the table around and build a theory of measurement on fuzzy set theory where measurement errors are explicitly accounted for[3]. This is certainly a road that is open to more improvements and developments.

On the other hand, if one considers the linguistic word meaning representation and aims to apply fuzzy sets to this domain (true to the original intentions of Zadeh) there is also much room for improvement. Particularly new theories of word meaning representation should be studied from the viewpoint of fuzziness as linguistic hedges. Recent brain imaging technologies seem to be capable of providing abundant empirical evidence for psycho-linguistic theories. Fuzzy sets and systems community should re-establish its links to the cognitive psychology and psycho-linguistics domain and re-start fruitful interaction with this community as it did in the early days of the fuzzy set theory and fuzzy logic.

References

1. Bilgiç, T., Türkşen, I.B.: Measurement-theoretic Frameworks for Fuzzy Set Theory. In: Martin, T.P., Ralescu, A.L. (eds.) IJCAI-WS 1995. LNCS, vol. 1188, pp. 252–265. Springer, Heidelberg (1997)
2. Bilgiç, T., Türkşen, I.B.: Measurement-Theoretic Justification of Fuzzy Set Connectives. Fuzzy Sets and Systems 76(3), 289–308 (1995)
3. Le Merton Bolinger, D.: Degree words. Janua linguarum. Series maior, vol. 53, The Hague Mouton (1972)
4. Bollmann-Sdorra, P., Wong, S.K.M., Yao, Y.: A measurement-theoretic axiomatization of fuzzy sets. Fuzzy Sets and Systems 60(3), 295–307 (1993)

[3] See for example [7, 20].

5. Kyburg Jr., H.E.: Measuring errors of measurement. In: Savage, C.W., Ehrlich, P. (eds.) Philosophical and Foundational Issues in Measurement Theory, ch. 5, pp. 75–92. Lawrence Erlbaum Associates Inc. Publishers, New Jersey (1992)
6. Hersh, H.M., Carmazza, A.: A Fuzzy Set Approach to Modifiers and Vagueness in Natural Language. Journal of Experimental Psychology: General 105(3), 254–276 (1976)
7. Katz, M.: Łukasiewicz Logic and the Foundations of Measurement. Studia Logica 15(3), 209–225 (1981)
8. Krantz, D.H., Duncan Luce, R., Suppes, P., Tversky, A.: Foundations of Measurement, vol. 1. Academic Press, San Diego (1971)
9. Lakoff, G.: A Study in Meaning Criteria and the Logic of Fuzzy Concepts. Journal of Philosophical Logic 2(4), 458–508 (1973)
10. Duncan Luce, R.: The Ongoing Dialog Between Empirical Science and Measurement Theory. Journal of Mathematical Psychology 40, 78–98 (1996)
11. Marchant, T.: The Measurement of Membership by Comparisons. Fuzzy Sets and Systems 148, 157–177 (2004)
12. Marchant, T.: The Measurement of Membership by Subjective Ratio Estimation. Fuzzy Sets and Systems 148, 179–199 (2004)
13. Michell, J.: An Introduction to the Logic of Psychological Measurement. Lawrence Erlbaum Associates Inc. Publishers, Hillsdale (1990)
14. Narens, L.: Abstract Measurement Theory. MIT Press, Cambridge (1985)
15. Narens, L.: A Meaningful Justification for the Representational Theory of Measurement. Journal of Mathematical Psychology 46(6), 746–768 (2002)
16. Norwich, A.M., Türkşen, I.B.: A Model for the Measurement of Membership and the Consequences of its Empirical Implementation. Fuzzy Sets and Systems 12, 1–25 (1984)
17. Rosch, E., Mervis, C.B.: Family Resemblances: Studies in the Internal Structure of Categories. Cognitive Psychology 7, 573–605 (1975)
18. Smith, E.E., Medin, D.L.: Categories and Concepts. Harvard University Press, Cambridge (1981)
19. Türkşen, I.B.: Measurement of Membership Functions and Their Assessment. Fuzzy Sets and Systems 40, 5–38 (1991)
20. Urbanski, M.K., Wasowski, J.: Fuzzy Measurement Theory. Measurement 41(4), 391–402 (2008)
21. Vigliocco, G., Vinson, D.P.: Semantic Representation. In: Gaskell, G. (ed.) Oxford Handbook of Psyholinguistics, ch. 12, pp. 195–216. Oxford University Press (2009)
22. Wittgenstein, L.: Philosophical Investigations. Blackwell Publishing, London (1953)
23. Yager, R.R.: A Measurement–informational Discussion of Fuzzy Union and Intersection. International Journal of Man–Machine Studies 11, 189–200 (1979)
24. Zadeh, L.A.: Fuzzy Sets. Information and Control 8(3), 338–353 (1965)
25. Zadeh, L.A.: Quantitative Fuzzy Semantics. Information Sciences 3(2), 159–176 (1971)
26. Zadeh, L.A.: A Fuzzy-Set-Theoretic Interpretation of Linguistic Hedges. Journal of Cybernetics 2(3), 4–34 (1972)
27. Zadeh, L.A.: PRUF a Meaning Representation Language for Natural Languages. International Journal of Man-Machine Studies 10(4), 395–460 (1978)
28. Zadeh, L.A.: Fuzzy Logic = Computing with Words. IEEE Transactions on Fuzzy Systems 4(2), 103–111 (1996)
29. Zadeh, L.A.: From Computing with Numbers to Computing with Words - from Manipulation of Measurements to Manipulation of Perceptions. In: Azvine, B., Nauck, D.D., Azarmi, N. (eds.) Intelligent Systems and Soft Computing. LNCS, vol. 1804, pp. 3–40. Springer, Heidelberg (2000)
30. Zadeh, L.A.: Is There a Need for Fuzzy Logic? Information Sciences 178(13), 2751–2779 (2008)

8

Fuzzy Models of Spatial Relations, Application to Spatial Reasoning

Isabelle Bloch

8.1 Introduction

Spatial relations are an important component of image content, that proved to be useful for recognition of individual objects and for image understanding. Indeed, spatial relations provide structural information about the scene, which is often more stable that individual object characteristics, can help disambiguating objects of similar appearance, and is often available as prior knowledge. A typical example is anatomy, where relations between anatomical structures are described in anatomical textbooks or dedicated web sites, and can be used to drive the interpretation of medical images. This will be illustrated on magnetic resonance images (MRI) of the brain, for segmenting and recognizing internal brain structures. This is a typical example where shape and appearance information may not be sufficient for recognition, in particular in pathological cases, while using structural knowledge is relevant and helps solving the problem. Similar examples can be found in understanding aerial and satellite images.

One important characteristic of spatial relations is that they often have a clear intuitive meaning in natural language, but crisp mathematical models are often to restrictive, not robust enough, and do no model the intrinsic imprecision attached to the linguistic descriptions of the relations. Fuzzy models are better suited, and allow accounting for imprecision both in the relations and in the objects. This was already mentioned in [21].

My work on fuzzy models of spatial relations was initiated while I was visiting Lotfi Zadeh's lab in Berkeley, where I spent a few months in 1995 and 1997, enjoying the stimulating environment and fruitful discussions, with researchers from different fields of fuzzy sets theory.

The main approach I proposed to model fuzzy spatial relations relies on mathematical morphology [32], because of its strong algebraic framework, which allows developing consistent models in different settings (from purely quantitative ones on sets, to purely qualitative ones in various logics), the fuzzy sets setting being a mid-way [8]. Another feature is that different types of representations can be proposed, expressing relations as numbers, fuzzy numbers, intervals, distributions, or fuzzy regions of space.

R. Seising et al. (Eds.): *On Fuzziness: Volume 1*, STUDFUZZ 298, pp. 51–58.
DOI: 10.1007/978-3-642-35641-4_8

8.2 Mathematical Morphology to Model Spatial Relations

Spatial relations include set theoretical and topological relations (inclusion, intersection, connection, adjacency...), metric ones (distances, relative direction...), and more complex ones ("between", "along", "parallel to", "aligned"...). Note that this classification extends the one of [21] and the hierarchy proposed in [25]. These relations can be binary, ternary (such as "between"), or n-ary (alignment of a series of objects for instance). Some of them can be crisply defined when objects are crisp (such as the Hausdorff distance between two well defined objects), and need to be extended to the fuzzy case (i.e. when the objects are imprecisely defined or known). Other ones are intrinsically vague ("close to", "to the right of", "between"...), and are then best modeled using fuzzy sets.

The main idea in the proposed models is to make use of morphological operations, in particular dilations, using appropriate structuring elements. This idea comes from the fact that several relations in the crisp case can be converted into algebraic expressions involving set theoretical and morphological operations, which are then easy to extend to the fuzzy case, using fuzzy mathematical morphology. Let us give a few examples:

- adjacency between two crisp objects can be expressed by the fact that the two objects do not intersect, but as soon as a dilation is applied to one of them, intersection occurs. This is translated by a conjunction (using a t-norm) of a degree of non intersection of two fuzzy sets and a degree of intersection of the dilation of one fuzzy set and the other;

- the minimal distance between two crisp objects is equivalent to the minimal size of the dilation that has to be applied to one object so that it meets the other. Again this easily extends to the fuzzy case by using fuzzy dilations.

Direct algebraic expressions have been proposed for vague relations, using similar operations. For instance the region of space which is to the right of another object is defined as the dilation of this object by a fuzzy structuring element representing the semantics of the relation (high membership functions in the horizontal direction, which decrease when going away of this direction). This is illustrated in Figure 8.1. If ν denotes the fuzzy set representing the spatial relation, and μ the reference object (fuzzy set in the spatial domain $\mathscr{S}$), then the degree of satisfaction of the relation is given by the dilation of μ by the structuring element ν: $\forall x \in \mathscr{S},\ \delta_\nu(\mu)(x) = \sup_{y\in\mathscr{S}} C[\nu(y-x), \mu(y)]$, where C denotes a fuzzy conjunction, and more specifically a t-norm. Details on mathematical morphology and the associated properties can be found in [9, 13, 27]. Assessing to which degree another object is to the right of the reference object is then performed by comparing it to the dilation result, for instance using a fuzzy pattern matching approach [19]. Details can be found in [4]. This is an example where we have a direct representation of the relation in the spatial domain, from which we can derive evaluations as numbers, intervals, distributions, etc. An example of spatial representation of the relation "between" is displayed in Figure 8.1 too. Here the reference objects are the two lungs segmented in a medical image (illustrated on one slide). The region between the lungs was used in [26] to guide the segmentation of the heart in non-contrasted 3D CT images.

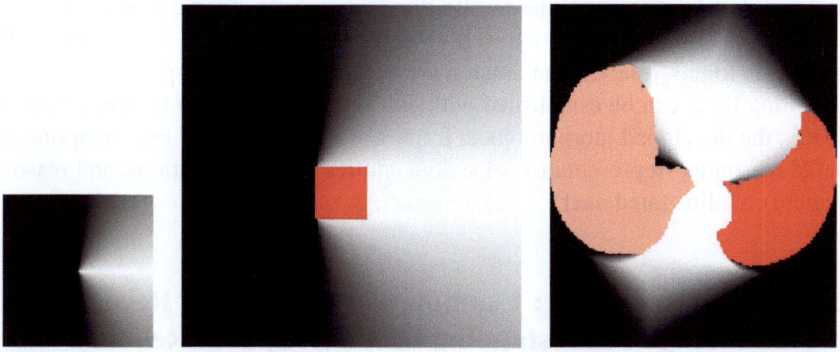

Fig. 8.1. Left: fuzzy structuring element representing the semantics of "right" in the spatial domain. Membership values are represented by grey levels (0 = black, 1 = white). Middle: region to the right of the red square. The membership at each point is the degree of satisfaction of the relation at that point. Right: region "between" the lungs.

A review on fuzzy spatial relations can be found in [7], and one on relative directions in [15], while technical details, along with examples, on several original proposals based on mathematical morphology can be found e.g. in [4, 5, 10, 14, 18, 29, 33, 34].

The main advantages of this approach is that the obtained definitions have good formal properties, provide results that fit the intuitive meaning of the relations, and are robust to the parameters defining the fuzzy structuring elements (in the sense that a fine tuning is not necessary). Moreover, having a common framework for defining several types of relations allows for their combination (see Section 8.4).

8.3 Instantiation of Spatial Relations Models in Various Settings: Towards Spatial Reasoning

Mathematical morphology, in particular its part dealing with deterministic increasing operators, relies on the algebraic framework of complete lattices [22, 31, 32]. Examples of such lattices are the powerset of a set, endowed with inclusion, functions, with the usual partial ordering, partitions, fuzzy sets, formulas in propositional logics, graphs and hypergraphs... A direct consequence is that the proposed spatial relations can be expressed in all these settings [8, 12]. In particular, it is interesting to have a symbolic expression of relations, in a logical framework (for instance by considering a formula as the logical representation of a spatial entity, whose models can be sets or fuzzy sets), to benefit from the reasoning tools inherited from the logics. This applies in propositional logics, but also in modal logics, where a structuring element gives rise to an accessibility relation, and the modalities $\Box$ and $\Diamond$ correspond to erosion and dilation, respectively [6]. Spatial relations can then be directly expressed in this logic. For instance, if φ and ψ are formulas representing two spatial entities, saying that the first one is a non tangential part of the second one can be simply expressed by: $\Diamond\varphi \rightarrow \psi$, or equivalently $\varphi \rightarrow \Box\psi$. Similarly in description logics, dilation and erosion are considered as binary predicates [24], and interesting links can be established with concept lattices [2] for reasoning purpose. Finally, the developed morphological framework deals with the two components of spatial reasoning (representation of spatial entities and their relations, and reasoning on them), as illustrated next.

8.4 Spatial Reasoning: Example of Model-Based Recognition and Image Understanding Based on Spatial Relations

Spatial reasoning can be defined as the domain of spatial knowledge representation, in particular spatial relations between spatial entities, and of reasoning on these entities and relations. This field has been largely developed in artificial intelligence, in particular using qualitative representations based on logical formalisms. In image interpretation and computer vision it is much less developed and is mainly based on quantitative representations. Our work has shown that semi-quantitative formalisms, using fuzzy sets, have many advantages. A typical example in this domain concerns model-based structure recognition in images, where the model represents spatial entities and relationships between them. For both spatial knowledge representation and reasoning, spatial relationships then constitute an important part of the knowledge we have to handle. Imprecision is often attached to spatial reasoning in images, and can occur at different levels, from knowledge to the type of question we want to answer. The reasoning component includes fusion of heterogeneous spatial knowledge, decision making, inference, recognition. Two types of questions are raised when dealing with spatial relationships:

1. given two objects (possibly fuzzy), assess the degree to which a relation is satisfied;
2. given one reference object, define the area of space in which a relation to this reference is satisfied (to some degree).

In order to answer these questions and address both representation and reasoning issues, we rely on three different frameworks and their combination: mathematical morphology [32], fuzzy set theory [35], and formal logics and the attached reasoning and inference power. The association of these three frameworks for spatial reasoning allows answering two important requirements: expressiveness and completeness with respect to the types of spatial information we want to represent [1].

As an illustration, let us consider the example where we would like to segment and recognize brain structures in a 3D MRI image, based on an anatomical model, which includes structures and their spatial relations. The usual description is given in a linguistic form, from which we can derive computational models relying on ontologies and graphs, and where spatial relations can be learned from examples [3, 11, 17, 20, 23, 24]. The model is then used for guiding the recognition. We summarize here a few approaches we have developed. Details can be found in the mentioned references.

The first approach uses a model graph and the image to segment is represented as a graph too, for instance from an over-segmentation of the image. The segmentation and recognition process is then formalized as a graph matching problem [16, 30].

In the second approach, a sequential segmentation of the internal brain structures is performed [11, 17]. The segmentation and the recognition are achieved at the same time. Each segmentation uses the spatial information encoded in the model, and more specifically the spatial relations to the previously segmented structures. This information allows restricting the search domain around the structure, in which a deformable model yields the segmentation result. The reasoning can rely on an ontology, where spatial relations are expressed based on morphological operators [23, 24], and on logical formalisms [2, 6]. In this approach, there is no initial segmentation of the image, but it raises questions on the order of segmentation of the different objects and on how to avoid the propagation of potential errors. These questions have been addressed in [20] by optimizing a segmentation path in the graph, based on saliency and structural information, and by allowing backtracking on the defined path to avoid error propagation.

Another approach was proposed in [28], which is global and uses a constraint network encoding all spatial relations that should be satisfied by the structures. Each anatomical structure is linked with a region of space which satisfies all constraints

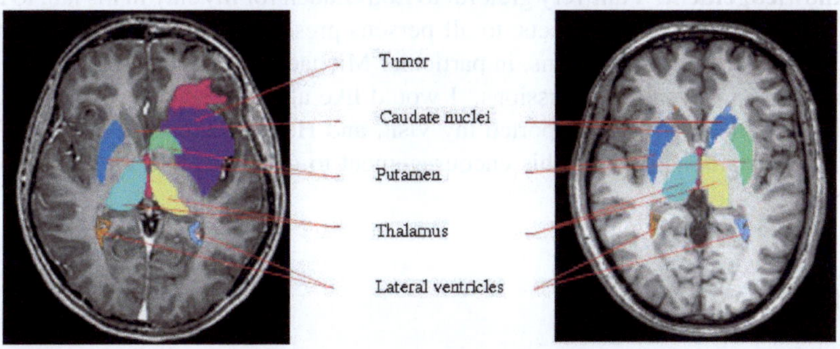

Fig. 8.2. Segmentation and recognition of a few brain structures from 3D MRI in a pathological case (left) and in a normal one (right), obtained with the sequential method. Thanks to the spatial relations, which remain stable even in presence of a pathology, the tumor does not prevent the correct segmentation of the normal structures, even if they are strongly deformed. Only one slide is displayed, but the relations are modeled and computed in 3D and the segmentation is performed in 3D too. (PhD thesis of Geoffroy Fouquier [20].)

in the network. Hence the problem is expressed as a constraint satisfaction problem (CSP). Since it is hard to solve this problem directly, only the bounds of the domain of each variable (i.e. structure to be segmented) are modified by the process and sequentially reduced using specifically designed propagators derived from the spatial constraints. In the reduced domain around the structure, the final segmentation is obtained by a minimal surface algorithm.

A typical segmentation and recognition result is shown in Figure 8.2 for a pathological case and a normal one, obtained with the sequential method. Another example is illustrated in 3D in Figure 8.3, where results have been obtained with the global CSP method.

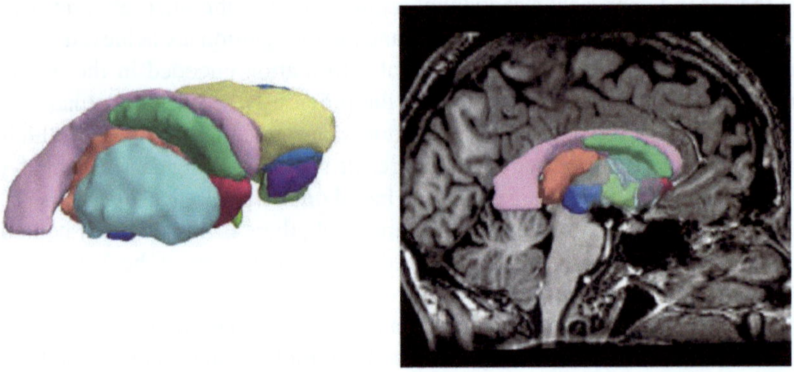

Fig. 8.3. 3D view of segmentation and recognition results obtained with the global CSP method for the following structures: caudate nuclei, putamen, lateral ventricles, thalami, third ventricle, accumbens nuclei and sub-thalami. (PhD thesis of Olivier Nempont [28].)

Acknowledgement. I am very grateful to Lotfi Zadeh for my stay in his lab, to Mori Anvari for the common projects, to all persons present in the group at the time of my visit for fruitful discussions, in particular Michael Lee and Yutaka Hata, to Eva Poinar for nice evening discussions. I would like also to thank the Fonds France-Berkeley which partially supported my visit, and Henri Maître, the head of LTCI, for his constant support and his encouragement to carry out free and independent research.

References

1. Aiello, M.: Spatial Reasoning, Theory and Practice. PhD thesis, University of Amsterdam (February 2002)
2. Atif, J., Hudelot, C., Bloch, I.: Abduction in description logics using formal concept analysis and mathematical morphology: application to image interpretation. In: 8th International Conference on Concept Lattices and Their Applications (CLA 2011), Nancy, Paris, pp. 405–408 (October 2011)

3. Atif, J., Hudelot, C., Fouquier, G., Bloch, I., Angelini, E.D.: From Generic Knowledge to Specific Reasoning for Medical Image Interpretation using Graph-based Representations. In: International Joint Conference on Artificial Intelligence, IJCAI 2007, Hyderabad, India, pp. 224–229 (January 2007)
4. Bloch, I.: Fuzzy Relative Position between Objects in Image Processing: a Morphological Approach. IEEE Transactions on Pattern Analysis and Machine Intelligence 21(7), 657–664 (1999)
5. Bloch, I.: On Fuzzy Distances and their Use in Image Processing under Imprecision. Pattern Recognition 32(11), 1873–1895 (1999)
6. Bloch, I.: Modal Logics based on Mathematical Morphology for Spatial Reasoning. Journal of Applied Non Classical Logics 12(3-4), 399–424 (2002)
7. Bloch, I.: Fuzzy Spatial Relationships for Image Processing and Interpretation: A Review. Image and Vision Computing 23(2), 89–110 (2005)
8. Bloch, I.: Spatial Reasoning under Imprecision using Fuzzy Set Theory, Formal Logics and Mathematical Morphology. International Journal of Approximate Reasoning 41, 77–95 (2006)
9. Bloch, I.: Duality vs. Adjunction for Fuzzy Mathematical Morphology and General Form of Fuzzy Erosions and Dilations. Fuzzy Sets and Systems 160, 1858–1867 (2009)
10. Bloch, I., Colliot, O., Cesar, R.M.: On the Ternary Spatial Relation Between. IEEE Transactions on Systems, Man, and Cybernetics SMC-B 36(2), 312–327 (2006)
11. Bloch, I., Géraud, T., Maître, H.: Representation and Fusion of Heterogeneous Fuzzy Information in the 3D Space for Model-Based Structural Recognition - Application to 3D Brain Imaging. Artificial Intelligence 148, 141–175 (2003)
12. Bloch, I., Heijmans, H.J.A.M., Ronse, C.: Mathematical Morphology. In: Aiello, M., Pratt-Hartman, I., van Benthem, J. (eds.) Handbook of Spatial Logics, ch. 13, pp. 857–947. Springer. Dordrecht (2007)
13. Bloch, I., Maître, H.: Fuzzy Mathematical Morphologies: A Comparative Study. Pattern Recognition 28(9), 1341–1387 (1995)
14. Bloch, I., Maître, H., Anvari, M.: Fuzzy Adjacency between Image Objects. International Journal of Uncertainty, Fuzziness and Knowledge-Based Systems 5(6), 615–653 (1997)
15. Bloch, I., Ralescu, A.: Directional Relative Position between Objects in Image Processing: A Comparison between Fuzzy Approaches. Pattern Recognition 36, 1563–1582 (2003)
16. Cesar, R.M., Bengoetxea, E., Bloch, I., Larrañaga, P.: Inexact Graph Matching for Model-Based Recognition: Evaluaton and Comparison of Optimization Algorithms. Pattern Recognition 38, 2099–2113 (2005)
17. Colliot, O., Camara, O., Bloch, I.: Integration of Fuzzy Spatial Relations in Deformable Models - Application to Brain MRI Segmentation. Pattern Recognition 39, 1401–1414 (2006)
18. Colliot, O., Tuzikov, A.V., Cesar, R.M., Bloch, I.: Approximate Reflectional Symmetries of Fuzzy Objects with an Application in Model-Based Object Recognition. Fuzzy Sets and Systems 147, 141–163 (2004)
19. Dubois, D., Prade, H., Testemale, C.: Weighted Fuzzy Pattern Matching. Fuzzy Sets and Systems 28, 313–331 (1988)
20. Fouquier, G., Atif, J., Bloch, I.: Sequential model-based segmentation and recognition of image structures driven by visual features and spatial relations. Computer Vision and Image Understanding 116(1), 146–165 (2012)
21. Freeman, J.: The Modelling of Spatial Relations. Computer Graphics and Image Processing 4(2), 156–171 (1975)

22. Heijmans, H.J.A.M., Ronse, C.: The Algebraic Basis of Mathematical Morphology – Part I: Dilations and Erosions. Computer Vision, Graphics and Image Processing 50, 245–295 (1990)
23. Hudelot, C., Atif, J., Bloch, I.: Fuzzy Spatial Relation Ontology for Image Interpretation. Fuzzy Sets and Systems 159, 1929–1951 (2008)
24. Hudelot, C., Atif, J., Bloch, I.: A Spatial Relation Ontology Using Mathematical Morphology and Description Logics for Spatial Reasoning. In: ECAI 2008 Workshop on Spatial and Temporal Reasoning, Patras, Greece, pp. 21–25 (July 2008)
25. Kuipers, B.J., Levitt, T.S.: Navigation and Mapping in Large-Scale Space. AI Magazine 9(2), 25–43 (1988)
26. Moreno, A., Takemura, C.M., Colliot, O., Camara, O., Bloch, I.: Using Anatomical Knowledge Expressed as Fuzzy Constraints to Segment the Heart in CT images. Pattern Recognition 41, 2525–2540 (2008)
27. Mike, N., Kerre, E.E.: Classical and Fuzzy Approaches towards Mathematical Morphology. In: Kerre, E.E., Nachtegael, M. (eds.) Fuzzy Techniques in Image Processing, ch. 1. STUDFUZZ, pp. 3–57. Physica-Verlag, Heidelberg (2000)
28. Nempont, O., Atif, J., Angelini, E.D., Bloch, I.: Structure Segmentation and Recognition in Images Guided by Structural Constraint Propagation. In: European Conference on Artificial Intelligence ECAI, Patras, Greece, pp. 621–625 (July 2008)
29. Nempont, O., Atif, J., Angelini, E.D., Bloch, I.: A New Fuzzy Connectivity Measure for Fuzzy Sets and Associated Fuzzy Attribute Openings. Journal of Mathematical Imaging and Vision 34, 107–136 (2009)
30. Perchant, A., Bloch, I.: Fuzzy Morphisms between Graphs. Fuzzy Sets and Systems 128(2), 149–168 (2002)
31. Ronse, C., Heijmans, H.J.A.M.: The Algebraic Basis of Mathematical Morphology – Part II: Openings and Closings. Computer Vision, Graphics and Image Processing 54, 74–97 (1991)
32. Serra, J.: Image Analysis and Mathematical Morphology. Academic Press, New-York (1982)
33. Takemura, C.M., Cesar, R.M., Bloch, I.: Modeling and measuring the spatial relation "along": regions, contours and fuzzy sets. Pattern Recognition 45, 757–766 (2011)
34. Vanegas, M.C.: Spatial relations and spatial reasoning for the interpretation of earth observation images using a structural model. PhD thesis, Télécom ParisTech (January 2011)
35. Zadeh, L.A.: Fuzzy Sets. Information and Control 8, 338–353 (1965)

9

Fuzzy Patterns for Fuzzy Modeling in Chemnitz, Germany

Steffen F. Bocklisch and Franziska Bocklisch

9.1 History

Since the foundation of the Department of Automatic Control at Chemnitz University of Technology in 1964, nonlinear systems have been a thematic priority in academic training and research. This line of research traces back to Alfred Pfeiffer (1900-1985), who was a PhD student in Berlin at the research laboratory of the famous physicist and Nobel Prize winner Walter H. Nernst (1864-1941).

During the development of technical cybernetics in the 1960s, Manfred Peschel (1932-2002) (see figure 9.1) extended the research field to include complex systems, focusing on multi-criteria optimization and decision theory. M. Peschel and Lotfi A. Zadeh met during the symposium entitled "On Systems Adaptivity and Sensitivity," held in Dubrovnik (Croatia) in 1968. Zadeh's plenary talk on "Fuzzy Sets and Systems" provided a strong impulse for further research. M. Peschel was initially reserved but at the same time, he appreciated Zadeh's profound competence in systems science as demonstrated in the book *Linear System Theory* by L.A. Zadeh and Charles A. Desoer.

In multi-criteria decisions, the concept of "quality" is very central. The insight that "quality" in complex systems is almost always defined by an insufficient number of attributes had led to the conclusion that "fuzzy" is a new concept with potential applications for research. Accordingly, the established research paradigm shifted towards using fuzzy methodology. Moreover, our efforts at the time, to describe medical diagnostic decisions (i.e., diagnosis of circulatory disorders in the human legs) using theoretical process analysis (e.g., models of flexible fluidic lines) failed, for two reasons. First, the large number of partially unknown and/or interfering objective and subjective variables (e.g., measurement technique, patient characteristics) could not be gathered. While physicians are able to solve such vague multivariate problems quite well, system analysis methods were not as easily applied. To remedy this issue, we assumed that the knowledge base and high performance of medical experts must be described by using a multidimensional space and by employing general non-classical state variables. The second reason for the departure from traditional functional modeling is due to the "medical way of reasoning." In comparison to engineers, physicians think according to "cases" (similar to case narratives in medical

R. Seising et al. (Eds.): *On Fuzziness: Volume 1*, STUDFUZZ 298, pp. 59–64.
DOI: 10.1007/978-3-642-35641-4_9

textbooks), which are defined as complex patterns/classes (e.g., circulatory disorder of a specific type). Therefore, we concluded that pattern classification methods would be an appropriate modeling tool, especially considering that clustering techniques allowed an automatically data-driven description of classes. These classes are semantically specified (e.g., as a specific type of disorder), are vague in nature, and cannot be precisely separated from each other. The conceptualization of the classes as fuzzy sets led to the development of the fuzzy pattern classification method.

Due to the technical orientation of the University in Chemnitz, fuzzy pattern classification was transferred to complex technical processes, as well. Examples include, monitoring of modules, machines and vehicles, diagnosis of errors in complex technical devices, quality management in automatic production, as well as in nonlinear dynamic control of systems with multiple inputs and outputs [1], [2]. Therefore, new fields of research to which this method were applied included environmental studies, traffic, and management studies.

(a) (b)

Fig. 9.1. (a): Professor Peschel (1983); (b): Lecture auditorium and building of the Faculty for Electrical Engineering in Chemnitz

The year 1972 marked the beginning of purposeful and systematic research in fuzzy methods in Chemnitz. First results were published in 1979 in a research report entitled "Fuzzy Classification" ("Unscharfe Klassifikation") at the *Department of Technical Cybernetics* at the *Academy of Sciences* (Akademie der Wissenschaften). Then, in 1981 and 1987, reports and textbooks of basic principles and current state-of-the-art followed [3], [4], [5], [6].

We emphasize the differences between our methodological and applied lines of research. However, to connect these lines we developed a software package that is subject to further development, and continues to serve as an important working platform for students, researchers, and users in academic training, research, and practice.

M. Peschel and a group of young researchers established a seminar in system sciences in 1975 with the two thematic foci polyoptimization and fuzzy sets. Later on, the *GDR working group for Fuzzy Sets* grew out of workshops that were marked by great enthusiasm and research dedication. The research results of the permanent and temporary members of this group are published in several scientific contributions, which demonstrate their fruitfulness and success. An early highlight in 1985, which showed evidence of growing international appreciation, was the "International Workshop on Fuzzy Sets Applications" at Wartburg in Eisenach (Germany) organized in cooperation with the *International Institute for Applied Systems Analysis* (IIASA) and the *International Atomic Energy Agency* (IAEA) (see [7]).

In 1992, a disciplinary consulting center for fuzzy technologies was established in Chemnitz. This center organized colloquiums and provided opportunities for professional development as well as initiated new research topics. Main goals were the propagation of the idea of fuzziness in education and training, and the transfer of fuzzy methods and experiences to research applications. Numerous applied and academic research projects (partly funded by public institutions) were carried out, for instance, in cooperation with Volkswagen, Siemens, and Bosch.

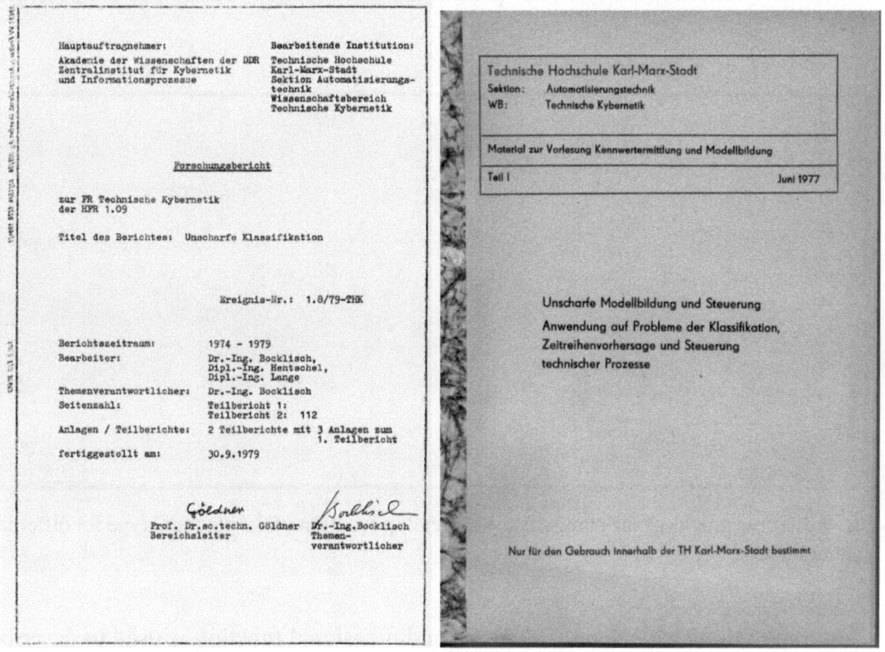

Hauptauftragnehmer:
Akademie der Wissenschaften der DDR
Zentralinstitut für Kybernetik
und Informationsprozesse

Bearbeitende Institution:
Technische Hochschule
Karl-Marx-Stadt
Sektion Automatisierungstechnik
Wissenschaftsbereich
Technische Kybernetik

Forschungsbericht

zur FR Technische Kybernetik
der HFR 1.09

Titel des Berichtes: Unscharfe Klassifikation

Ereignis-Nr.: 1.8/79-THK

Berichtszeitraum: 1974 - 1979
Bearbeiter: Dr.-Ing. Bocklisch, Dipl.-Ing. Hentschel, Dipl.-Ing. Lange
Themenverantwortlicher: Dr.-Ing. Bocklisch
Seitenzahl: Teilbericht 1: Teilbericht 2: 112
Anlagen / Teilberichte: 2 Teilberichte mit 3 Anlagen zum 1. Teilbericht
fertiggestellt am: 30.9.1979

Prof. Dr.sc.techn. Göldner
Bereichsleiter

Dr.-Ing.Bocklisch
Themenverantwortlicher

Technische Hochschule Karl-Marx-Stadt
Sektion: Automatisierungstechnik
WB: Technische Kybernetik

Material zur Vorlesung Kennwertermittlung und Modellbildung

Teil I
Juni 1977

Unscharfe Modellbildung und Steuerung
Anwendung auf Probleme der Klassifikation, Zeitreihenvorhersage und Steuerung technischer Prozesse

Nur für den Gebrauch innerhalb der TH Karl-Marx-Stadt bestimmt

(a) (b)

Fig. 9.2. (a): Report over the first reseach period in fuzzy classification at Chemnitz University of Technology (formerly Technische Hochschule Karl-Marx-Stadt); (b): First textbook for education in fuzzy modeling and control at Chemnitz University of Technology (formerly Technische Hochschule Karl-Marx-Stadt), presented by the Department of Technical Cybernetics

9.2 Fuzzy Methodology

The objective of our fuzzy methodology is the modeling of complex relations of fuzzy metric and linguistic variables. To this end, a space of generalized state variables (e.g., measured variables, quality criteria, subjective estimations) is defined. Thus, the basic idea is to identify structures, such as classes or clusters with a semantic meaning (e.g., therapeutic interventions, control strategies) in this space that can be modeled using a multimodal membership function. The structure formation can be defined by experts or data-driven algorithms. Our efficient description employs a set of unimodal class membership functions. For the purpose of class description, the following resulting parametric potential membership function type is highly efficient:

$$\mu(u) = \frac{a}{1 + \left(\frac{1}{b} - 1\right)\left(\frac{|u|}{c}\right)^{d}} \tag{9.1}$$

This is the general equation of the parametric membership function of the potential type.

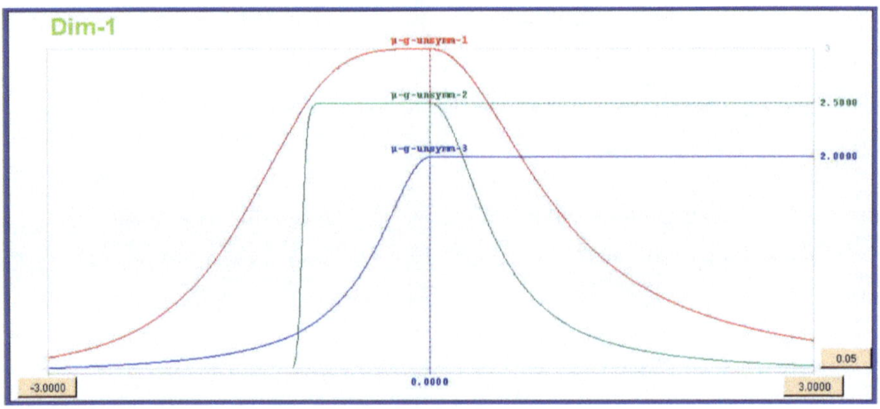

Fig. 9.3. Three graphs of the parametric membership function of the potential type for different parameter sets

In a multidimensional case, such an unidimensional function is used to describe the classes for every dimension. The function can be adapted on the basis of different specific parameters (a, b, c, and d) for every dimension. Furthermore, the distinction between left- and right handed parameters allows a flexible description of asymmetric data (see figures 9.4).

Referring to the modeling of ecological niches (Hutchinson), the parameters can be interpreted semantically (e.g., position, orientation, expansion, decline of intensity), which is of great importance for its application. Moreover, parameters can be adjusted separately depending on class structure and practical implications. This is important, for example, for the implementation of new knowledge or the optimization of modeling. In general, classification models are first designed off-line and then implemented in on-line devices through porting of the parameter set.

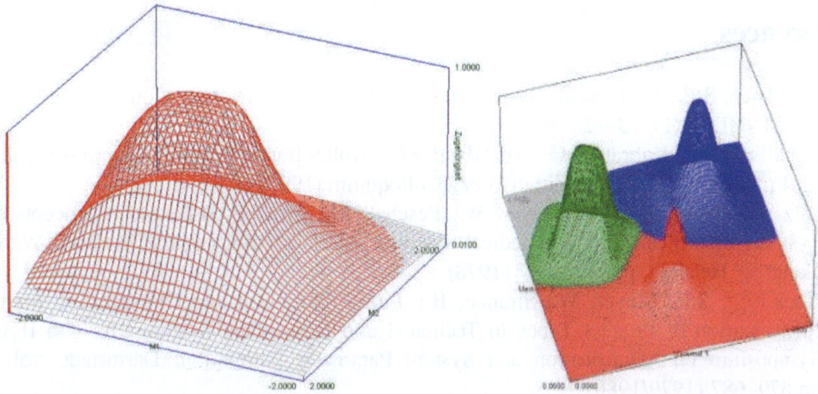

(a) (b)

Fig. 9.4. Visualization of two-dimensional non-symmetric membership functions for description of fuzzy patterns

9.3 Current State-of-the-Art and Future Perspectives

Fuzzy activities in Chemnitz are characterized by methodological and applied research activities, which are planned and ongoing. Basic topics in methodological research include:

- Modeling, analysis, and prediction of vectorial non-steady-state time series using fuzzy prototypes [8], [9]
- The description of non-convex classes by sequentially combining fuzzy classes and fuzzy anti-classes, which are all described by the same parametric potential membership function concept [10]
- Design of fuzzy classifier networks by serial and parallel connection of single fuzzy classifier models [11]

Basic research interests in applied technical and non-technical fields include:

- Multi-sensorial data analysis
- Failure diagnosis in machinery (i.e., technical devices and plants)
- In-process quality management
- Non-linear control of systems with multiple inputs and outputs
- Decision support in different fields of study, such as traffic, environment, medicine, or energy (e.g., renewable energy systems)
- Analysis and evaluation of linguistic terms in psychology and human sciences [12]

- Construction of verbal response scales for questionnaires with equidistant category labels [13]
- Fuzzy analysis of human response behavior

References

1. Bocklisch, S.F.: A Diagnosis System Based on Fuzzy Classification. Computers in Industry, CINUD4 7(1), 73–82 (1986)
2. Bocklisch, S.F., Kabranov, O.: An Adaptive Controller Based on Fuzzy Pattern Attractors. In: Proceedings of the 5th Zittau Fuzzy Colloquium (1997)
3. Bilz, F., Bocklisch, S.F., Meyer, W., Peschel, M., Straube, B.: Fuzzy Concepts for Systems Identification of Biomedical Systems. In: Proceedings of the 7th IFAC-World-Congress Helsinki, pp. 499–507 (1978)
4. Bocklisch, S.F., Meyer, W., Straube, B.: Fuzzy Clustering and Classification for the Identification of Process Types in Technical and Biomedical Systems. In: 5th IFAC-Symposium on Identification and System Parameter Estimation, Darmstadt, vol. 1, pp.679–687 (1979/1980)
5. Peschel, M., Bocklisch, S.F., Meyer, W., Straube, B., Richardt, J.: Daten- und Verhaltensanalyse mit Klassifikationsmodellen, Österreichische Studiengesellschaft für Kybernetik, Report Nr. 23 (1981)
6. Bocklisch, S.F.: Prozeßanalyse mit unscharfen Verfahren. Verlag Technik, Berlin (1987)
7. Bocklisch, S.F., Orlovski, S., Peschel, M., Nishiwaki, Y. (eds.): Fuzzy Sets Applications, Methodological Approaches, and Results (Reihe: Mathematische Forschung, Band 30). Akademie-Verlag, Berlin (1986)
8. Bocklisch, S.F., Päßler, M.: Fuzzy Time Series Analysis. In: Hampel, R., Wagenknecht, M., Chaker, N. (eds.) Fuzzy Control – Theory and Practice. AISC, pp. 331–345. Physica-Verlag, Heidelberg (2000)
9. Herbst, G., Bocklisch, S.F.: Short-Time Prediction Based on Recognition of Fuzzy Time Series Patterns. In: Hüllermeier, E., Kruse, R., Hoffmann, F. (eds.) IPMU 2010. LNCS (LNAI), vol. 6178, pp. 320–329. Springer, Heidelberg (2010)
10. Hempel, A.-J., Bocklisch, S.F.: Fuzzy Classification of Nonconvex Data-Inherent Structures. In: Hüllermeier, E., Kruse, R., Hoffmann, F. (eds.) IPMU 2010, Part I. CCIS, vol. 80, pp. 416–425. Springer, Heidelberg (2010)
11. Hempel, A.-J., Bocklisch, S.F.: A Conception for Networks of Fuzzy Pattern Classifiers. In: Proceedings of the 13th East West Fuzzy Colloquium 2006. Wissenschaftliche Berichte, Hochschule Zittau/Görlitz, Heft 89/2006, pp. 296–312 (2006)
12. Bocklisch, F., Bocklisch, S.F., Krems, J.F.: How to Translate Words into Numbers? A Fuzzy Approach for the Numerical Translation of Verbal Probabilities. In: Hüllermeier, E., Kruse, R., Hoffmann, F. (eds.) IPMU 2010. LNCS (LNAI), vol. 6178, pp. 614–623. Springer, Heidelberg (2010)
13. Bocklisch, F., Bocklisch, S.F., Krems, J.F.: Sometimes, Often, and Always – Exploring the Vague Meanings of Frequency Expression. Behavior Research Methods 44(1), 144–157 (2012)

10

Lotfi A. Zadeh, from Information Processing to Computing with Words

Bernadette Bouchon-Meunier

The first time I met Lotfi A. Zadeh was in 1977 during a colloquium, held in Cachan and entitled "Théorie de l'information: développements récents et applications", which I was co-organizing with my boss at the time, Claude-François Picard. I was slightly awed at the idea of meeting him in person. His wife, Fay, was accompanying him and I had not realized, at that time, how uncomfortable it was for them to stay in the student housings of a French institute, in the suburbs of Paris. This situation was typical of Lotfi and his eagerness to participate in workshops and conferences, whatever their size might have been, to defend fuzzy set theory and meet the community interested in his work. He has been an untiring traveler and orator and the success of fuzzy set theory certainly lies in his ability to give clear and educational lectures, always suited to the audience.

In Cachan, we had the privilege to hear him give his talk on "Possibility theory and its application to information analysis" [4], where he clearly presented the bases of possibility theory, his 1978 *Fuzzy Set and Systems* paper [5] having not yet been published. He set this introduction of the theory of possibility alongside the generalized theory of information Joseph Kampé de Fériet had been developing with Bruno Forte [6] since the end of the 60s, and on the periphery of R. Carnap and Y. Bar-Hillel's concept of semantic information [7], also explored by J. Kampé de Fériet. His later works on the computing with words paradigm [8], on the definition of precisiated natural language [9] and on a general theory of uncertainty [10] certainly all go in the same direction.

Fuzzy logic was not very popular in France at the time. A. Kaufmann's book [1] was one of the first in the world introducing fuzzy logic, but it was still recent and French scientists were not paying much attention to this new theory in a country from which famous specialists of probabilities and control engineering originated. I had discovered Lotfi Zadeh's seminal paper on fuzzy sets by chance, in the library of mathematics of the university in 1975 and was surprised to see that it fit my needs for a cooperative project I was working on with a group of sociologists exactly. They were looking for a formal model for the questionnaires they used and the human component of surveys was difficult to represent in a classic environment. I then introduced the concept of fuzzy questionnaires [2], [3], my first attempt at using fuzzy set theory in real world applications. I was still feeling rather lonely on this topic and to have the opportunity of discussing with Lotfi Zadeh was a real stroke of luck.

R. Seising et al. (Eds.): *On Fuzziness: Volume 1*, STUDFUZZ 298, pp. 65–68.
DOI: 10.1007/978-3-642-35641-4_10

With a classic background in mathematics and in the then emerging computer sciences, I was really surprised and happy to discover the "human" formalism of fuzzy set theory, at the same time rigorous and flexible, as any human being coping with the real world's complexity can be. It suited the needs of artificial intelligence so well that I soon started work on fuzzy set based knowledge representation, foreseeing its potential in automated human-like task management, such as subjective information processing or decision and diagnostics support systems.

The position of Lotfi Zadeh's research in the stream of information processing was a key element in the creation of the *International Conference on Information Processing and the Management of Uncertainty in Knowledge-based Systems* (IPMU). Lotfi Zadeh, Ronald Yager and I originally started organizing this conference in 1984, at a time when there was no regular conference dealing with fuzzy sets, since the first *IFSA World Congress* was held in 1985 in Palma de Mallorca. The first IPMU conference finally took place in 1986, in Paris, and aimed at bringing together scientists working on various methods for the management of uncertainty in intelligent systems. Our common goal and purpose was to provide a medium for the exchange of ideas between theoreticians and practitioners using different methods to address the important issue of uncertainty, all the while without isolating researchers on fuzzy sets and systems. This approach was a success because it came and answered the concerns of many researchers and the IPMU conference has been organized every two years since then in France, Spain, Italy and Germany. Lotfi Zadeh's constant and immutable support to IPMU will long be recognized.

It was, therefore, as natural for the IPMU 1992 co-presidents, Llorenç Valverde, Ronald Yager and I, to imagine and organize an award, named after J. Kampé de Fériet, to be awarded at all following IPMU conferences, as it was to present the first one to Lotfi Zadeh, who had brought such an exceptional contribution to the field of information processing and management of uncertainty. Lotfi Zadeh's visits to my laboratory were always a major scientific event and I was impressed by his willingness to come give a lecture to an audience which was the largest we had seen for any of our seminars, but still ridiculously small compared to the success his lectures garnered in Asia, for instance.

Of all his visits, I would like to mention the time he came to Paris to receive the honorary doctorate from the Université Pierre et Marie Curie-Paris 6, on January 25, 2001. Representatives of the French and Belgian fuzzy communities were present to take part in this historic moment. After the official ceremony, they were all happy to celebrate him with a friendly dinner, together with his wife Fay. The following day, he gave yet another successful lecture called "Toward an Enlargement of the Role of Natural Languages in Information Processing, Decision and Control".

Lotfi Zadeh has always been a visionary. He has laid the foundations for the main concepts behind fuzzy set theory and fuzzy logic. His most important innovation is probably the extension principle, enabling the generation of all necessary concepts by extending crisp notions to imprecise situations. He has also established the bases of various notions such as fuzzy relations, fuzzy similarities, fuzzy prototypes, fuzzy probabilities, fuzzy modifiers as well as possibility theory.

In addition, he has launched many of the application domains of fuzzy sets and systems, naturally starting from control at the beginning of the 70s, since it was his original domain of expertise. This domain has clearly been the reason for both the success and the visibility acquired by fuzzy logic in the 80s in Japan, and later in all other parts of the world. In the 90s, he realized very early on the increasing importance of natural language-like query-answering systems to interact with internet. At the same time, he had a vision of the importance of taking perceptions into account, which would soon become a component of many real world applications dealing with affective computing or, more generally, subjective information.

His various lectures on soft computing prove that he was aware of the challenging necessity to construct hybrid systems, not isolating fuzzy logic in chapels but making it work in complete synergy with neural networks, probabilistic methods and evolutionary computation to build powerful tools.

Fig. 10.1. Lotfi A. Zadeh's honorary doctorate in Paris, 2001 (from left to right: B. Bouchon-Meunier, Elie Sanchez, L. A. Zadeh, Philippe Smets)

References

1. Kaufmann, A.: Introduction à la théorie des sous-ensembles flous – éléments théoriques de base, Masson (1973)
2. Bouchon, B.: Une approche des questionnaires d'enquête. Journées Lyonnaises des Questionnaires 1, 161–169 (1975)
3. Bouchon, B.: An Application of the Theory of Questionnaires to Information. In: Topics in Information Theory, Colloquium on Information Theory, Keszthely (Hongrie), pp. 125–141. North Holland (1975, 1978)
4. Zadeh, L.A.: Possibility Theory and its Application to Information Analysis. In: Théorie de l' information, Développements Récents et Applications, Cachan, Juillet 4-8. Editions du CNRS, Paris (1977, 1978)
5. Zadeh, L.A.: Fuzzy Sets as a Basis for a Theory of Possibility. Fuzzy Sets and Systems 1(1), 3–28 (1978)
6. Kampé de Fériet, J., Forte, B.: Information et probabilité. Comptes Rendus, Academy of Sciences 265-A, 110–114 (1967)
7. Carnap, R., Bar-Hillel, Y.: Semantic Information. The British Journal for the Philosophy of Science 4(14), 147–157 (1953)
8. Zadeh, L.A.: From Computing with Numbers to Computing with Words – From Manipulation of Measurements to Manipulation of Perceptions. In: Azvine, B., Nauck, D.D., Azarmi, N. (eds.) Intelligent Systems and Soft Computing. LNCS, vol. 1804, pp. 3–40. Springer, Heidelberg (2000)
9. Zadeh, L.A.: Precisiated Natural Language (PNL). AI Magazine 25(3), 74–91 (2004)
10. Zadeh, L.A.: Generalized Theory of Uncertainty (GTU) – Principal Concepts and Ideas. Computational Statistics & Data Analysis 51(1), 15–46 (2006)

11

Fuzzy Systems at the University of Santiago de Compostela: A Personal Vision of the Last Twenty Years

Alberto J. Bugarín Diz

11.1 Preface

I suppose I should say I am at a midpoint between the newcomers and the oldest in this community, since my starting point in this field is dated (approximately) halfway between now and the seminal paper by Prof. Zadeh. [14] Almost exactly twenty years have passed between the kind invitation of the editors of this volume and my attendance to the First FUZZ-IEEE Conference at San Diego for presenting my first complete paper in this area. [4] Since round numbers as twenty have the curious property of making us stop and thinking on things, I find it a good personal coincidence for writing on our path in the field of fuzzy systems, my current vision and expectations on the topic. It is true since a number of years until now that most of we researchers are worried about publishing our work in meetings and journals that are well ranked according to citation indexes. Benefits of having objective indicators on the quality of publications is out of discussion, but sometimes has the collateral effect of mostly focusing our dedication on obtaining a quantitative "return of investment" in terms of h-index or JCR well-positioned papers. This is not bad, of course, but it has the associated risk of not considering the relevance and interest of proposals like the one by the editors of this volume, that has an indubitable and non-negligible qualitative value: helping to build, reinforce and keep united our research community. Let me firstly thank them for the initiative (and for the undeserved honor of letting me share here these reflections). One of the consequences of this is what a find one of the key values of our community: its capability to provide a warm welcome to newcomers and to allow joining new researchers with new ideas and visions. Our own experience is a good example of this, or at least it is that way how we perceive it was our (modest) arrival to the research on fuzzy systems and to the fuzzy systems community.

11.2 Some Memories of Twenty Years

I sill vividly remember the impression I got after reading the first introductory papers on fuzzy sets and systems:[7] confusion. I had started my research career a

R. Seising et al. (Eds.): *On Fuzziness: Volume 1*, STUDFUZZ 298, pp. 69–75.
DOI: 10.1007/978-3-642-35641-4_11

few months before in the topic of ECG signal processing using syntactic analysis techniques. The aim of our tiny (at that time) group was that one of us started training in new techniques that could be applied to the patients signal monitoring realm. Fuzzy sets and reasoning had attracted our attention[2] since some interesting results had been reported both in the field of medical signal processing and knowledge-based systems. [8, 10] In fact, medicine was one of the fields of application that Prof. Zadeh had foreseen, [15] by stating that „...the method described in this paper can be applied rather effectively to the formulation and approximate solution of a wide variety of practical problems, particularly in such fields as economics, management science, psychology, linguistics, taxonomy, artificial intelligence, information retrieval, medicine and biology."

All the new operators, models and techniques described in the fuzzy literature (linguistic variables, t-norms, t-conorms, implication functions, compositional rule of inference, linguistic truth values, ...) had little to do with our previous knowledge at that time that, although closely related to artificial intelligence and computer science, was still strongly tied to experimental grounds as our undergraduate education was as electronic physicists. On the other hand these new techniques were a huge shift of paradigm, since we were moving from studying syntactic (in the pure linguistic sense) techniques for medical signal processing towards focusing on the fuzzy semantics of terms.

At the end of this initial stage, the most direct connection between fuzzy sets and our research at that time did not emerge as initially expected from the field of ECG signal processing, but focused in the more general field of fuzzy knowledge representation and reasoning. Starting from the works by Godo et al. [5]and López de Mántaras, [8] we addressed the proposal of efficient mechanisms for the execution of the Mamdani type of reasoning both from the point of view of their formalization using High-Level Petri Nets and of its projection onto specific-purpose systolic hardware architectures. It is worthy remembering at this point the work by a number of relevant researchers (mostly in Japan and also in Spain) on the so-called fuzzy chips, understood as a step to the 6th generation of *fuzzy computers*. [9, 12, 13] This was a favorable context for researching on hardware implementations of fuzzy systems and their applications. In our case, the initial aim was addressing medical signals processing where real-time requirements were highly demanding, but we noticed very soon that the proposal of new models for fuzzy knowledge representation and reasoning and its efficient execution was itself a relevant topic that deserved long-term research efforts. Our first paper in this line [3] presented at the 2nd spanish conference on fuzzy logic and technologies (Madrid, 1992) allowed us to meet the fuzzy spanish community and to observe the birth of the incipient Spanish Association for Fuzzy Logic and Technologies, that was the seed of the successful and well-recognized current European Association for Fuzzy Logic and Technologies (EUSFLAT). The most relevant aspect of this little history is the warm welcome we got from widely well-known researchers that are highly recognized (and also were yet at that time) for their relevant contributions to the field. This was (and still is) a remarkable feature of our community, as we show with a couple of facts. The first one was that one of the outcomes of this conference was that we were commissioned for organizing the next Conference[1] The second was one year later, when we were also commissioned for organizing the 26th International Symposium on Multiple-Valued Logic, that allowed us to invite Prof. Zadeh as one of the keynote speakers. [6] As a novel research group in the topic of fuzzy systems it was a honor to welcoming and accompanying him in our place.

For sure, many of the readers of this volume could comment on similar experiences in many Conferences, due to the good disposition Prof. Zadeh has always shown during all his career for attending invitations throughout the world. It is fair to say here that we find it difficult to present similar examples in other scientific fields. How many people in other research areas can say they have actually met a personality of comparable scientific stature?

11.3 Personal Views on the Fuzzy Systems Area

Following the editors indications, and as a self-reflection exercise rather than having other aims, I will share here about my current personal views on this area. Note that these are unavoidably biased by my experimental education and my interests on fuzzy modeling and its applications. In the first place, I would point out to the general issue of how proposals are validated in the field of fuzzy linguistic modeling. This modeling is a central issue in Zadeh's Computational Theory of Perceptions, where all the operators that provide semantics to linguistic expressions are defined or can be framed. It is still usual that we find in the literature new proposals of fuzzy functions or operators for which only a few toy examples are provided in order to illustrate its performance. As a mature field, fuzzy approaches should increasingly provide sound evidences for their proposals, aiming to characterize them through a number of relevant properties to be confronted to thus helping other scientists to both *i*) understand the most relevant features of the models presented and their scope and *ii*) provide a framework for comparison to other approaches. A clever balance for each proposal, between simple examples that illustrate the models and a number of comprehensible characterization properties that can help to formalize it should be achieved and may be one of the (most) difficult tasks to be faced by our community nowadays in the field of fuzzy modeling. In this regard, validation methodologies for the proposals is still an open issue. There are a number of fields in the Artificial Intelligence communities that have established general procedures for validation and comparison of results, either by following a systematic methodology and/or by the existence and use of benchmark problems and data sets. The Information Retrieval or Machine Learning communities provide good examples in this regard, that could illustrate the way to be followed for fixing some standards also in our field. Building of general data sets and software tools that include the state of the art methods (as

[1] Keeping apart the wonders of the city and the quality of the keynote addresses (E. Trillas, R.R. Yager) we have for sure that almost all the attendants to this 1993 meeting still remember today the wind and rain storm that framed most of the conference days, so frequent in this part of the Atlantic coast, but so unknown for many of our visitors from other areas of Spain.

Fig. 11.1. Lotfi A. Zadeh dancing "muiñeira" (typical Galician dance) during the gala dinner at ISMVL'96 (30 May 1996)

done, for example, within the KEEL project [1] for evolutionary learning) is a need for providing feasible means for assessing the quality of new models and approaches and also its generalization capabilities.

Fig. 11.2. Lotfi A. Zadeh and the author in front of the Cathedral of Santiago de Compostela during his visit for key-note lecturing at ISMVL'96 (1 June 1996)

Finally, many of the models proposed in the literature are aimed to solve particular applied problems. This is one of the key successes of fuzzy sets, since practical applications are an inexhaustible source of both inspiration and demanding requirements for our models. Nevertheless, it is still frequent to find that the focus of many applied papers is put on the modeling tool used rather than on the quality of the results obtained and its actual relevance for the related field of application. Almost all of you surely remember Zadeh's widely and well-known "Hammer Principle," collected by R. Seising and V. Sanz ([11], p.24) as: „when the only tool you have is a hammer, everything looks like a nail."I feel that this criticism Prof. Zadeh addressed to other approaches to intelligent modeling can also be applied to fuzzy modeling itself, since it seems that a part of our community is too much "self-centered", mostly focusing to the methodological approach used (whatever it may be) rather than to the solution provided itself and the quality of the results obtained. A collective aim should be that fuzzy approaches to applied problems should be able to jump the barrier and be presented and discussed in forums (i.e., conferences, journals) that are relevant for the problem solved and also prove there its validity with the methods and tests used in such fields (that is, "playing the match away.") Reciprocally, this external forums provide in return valuable sources of inspiration like new problems to be faced or new development and validation methodologies.

11.4 My Expectations

My general expectation of the field of fuzzy systems for the near future is that it will surely become relevant (although not necessarily the most relevant one) for approaching the solution of truly complex problems and systems. Hybridization will become a key issue here, both among the different areas of soft computing (where it is currently one of the most active areas) and also with other areas and techniques from the general field of artificial intelligence. It is worthy to remember here the general topic of a relevant forum as it was the 2011 International Joint Conferences on Artificial Intelligence: *Integrated and Embedded Artificial Intelligence*. That is, crossing the discipline boundaries (fortunately not frontiers) among the different areas of AI and fostering cooperation between them (and also between AI and other disciplines) in order to face truly complex challenges and providing actually intelligent functionalities to systems. One key element here will be the enrichment of current models for fuzzy knowledge representation and reasoning for producing a repertoire of operators that capture in a more direct way the complexity of human knowledge and reasoning and provide a high-level interaction experience between systems and humans to all levels. A need here is to describe plausible models that should be integrated with other approaches and formalisms for designing systems endowed with consistent, coherent and sustained intelligent behaviors.

In a sense, I expect that the alignment of the objectives of the research in our area with this general integration challenge for AI should be easy to achieve. Keep in mind that the soft computing area is a hybrid area itself, where a number of different and very successful approaches to the management of uncertainty coexist that instead to adhere to the *hammer metaphor* previously remembered they rather resemble the new *toolbox metaphor. [11]* As a community, it is one of our tasks to contribute to fill this toolbox with sound tools, from all the theoretical, methodological and engineering points of view.

Acknowledgement. Author acknowledges all the members of the Intelligent Systems Group at the University of Santiago de Compostela involved in research in the fuzzy area during these 20 years of fruitful discussions and exchanges of opinions. Special thanks to Senén Barro, Alejandro Sobrino, Purificación Cariñena, Manuel Mucientes, Félix Díaz-Hermida and Martín Pereira-Fariña.

References

1. Alcalá-Fernández, J., Fernández-Hilario, A., Luengo Martín, J., Derrac Rus, J., García López, S., Sánchez Ramos, L., Triguero, F.H.: KEEL Data-Mining Software Tool: Data Set Repository, Integration of Algorithms and Experimental Analysis Framework. Journal of Multiple-Valued Logic and Soft Computing 17(2-3), 255–287 (2011)
2. Barro, S., Ruiz, R., Mira, J.: Fuzzy Beat Labeling for Intelligent Arrhythmia Monitoring. Computers and Biomedical Research 23(3), 240–258 (1990)

3. Bugarín, A.J., Barro, S., Ruiz, R.: Systolic Solutions in Fuzzy Control. In: Proceedings of the 2nd Spanish Conference on Fuzzy Logic and Technologies, pp. 323–338 (1992) (in Spanish)
4. Bugarín, A.J., Barro, S., Ruiz, R.: Compacting Rules for Fuzzy Production System Computation. In: Proceedings of the 1st FUZZ-IEEE International Conference, pp. 933–940 (1992)
5. Godo, L., Jacas, J., Valverde, L.: Fuzzy Values in Fuzzy Logic. International Journal of Intelligent Systems 6, 199–212 (1991)
6. Proceedings of the 26th International Symposium on Multiple-Valued Logic, Santiago de Compostela, May 29-31. IEEE Computer Society Press (1996)
7. Lee, C.C.: Fuzzy Logic in Control Systems: Fuzzy Logic Controller (Parts I and II). IEEE Transactions on Systems, Man and Cybernetics 20(2), 404–435 (1990)
8. López de Màntaras, R.: Approximate reasoning models. John Wiley and Sons (1990)
9. Moraga, C.: The ORBE Project. In: Sobrino, A., Barro, S. (eds.) Studies on Fuzzy Logic and Applications, pp. 327–334. University Santiago de Compostela (1993) (in Spanish)
10. Seising, R.: From Vagueness in Medical Thought to the Foundations of Fuzzy Reasoning in Medical Diagnosis. Artificial Intelligence in Medicine 38(3), 237–256 (2006)
11. Seising, R., Sanz González, V. (eds.): Soft Computing in Humanities and Social Sciences. STUDFUZZ, vol. 273. Springer, Heidelberg (2012)
12. Togai, M., Watanabe, H.: Expert System on a Chip: An Engine for Real-time Approximate Reasoning. IEEE Expert, 55–62 (1986)
13. Yamakawa, T., Miki, T.: The Current Mode Fuzzy Logic Integrated Circuits Fabricated by the Standard CMOS Process. IEEE Transactions Computers 35(2), 161–167 (1986)
14. Zadeh, L.A.: Fuzzy Sets. Information and Control 8(3), 338–353 (1965)
15. Zadeh, L.A.: Outline of a New Approach to the Analysis of Complex Systems and Decision Processes. IEEE Trans. Systems, Man and Cybernetics SMC-3(1), 28–44 (1973)

12

Fuzzy Sets and Their Extensions: Leitmotiv of the Research Group of Artificial Intelligence and Approximate Reasoning (GIARA)

Humberto Bustince

Once my degree on Physics by the University of Salamanca was finished, I came back to Navarra with the following goal: to carry on my Ph.D. thesis in the analysis and development of Expert Systems. For this reason, in 1989 I applied for a teaching position at the Public University of Navarra.

One of the reasons to choose that university was that, in that time, its dean was Prof. Dr. Pedro Burillo López who has already been the director of a thesis on the algebraic analysis of fuzzy relations. My idea was to use such studies to build an inference engine for an expert system.

The first Reading that Prof. Burillo proposed to me was the book by Didier Dubois and Henri Prade *Fuzzy Sets and Systems*. After studying it, I get obsessed with fuzzy sets theory and we decided to make my Ph.D. in that field. However, after checking the most widely known bibliography on that time, Pedro Burillo showed me the 1986 paper of Krassimir Atanassov entitled "Intuitionistic Fuzzy Sets". After reading it, we conclude that the simultaneous use of positive and negative information would allow us to model better the knowledge and behavior of the experts. This idea was the main objective in my thesis. I finished this thesis in 1994 with a set of theoretical developments on the concepts of entropy and intuitionistic fuzzy relations, leaving their practical application for future developments.

Looking back to that time, the most important point for me at that time was to know the concept of extension of fuzzy sets in a general way, and in particular the cases of Atanassov's intuitionistic fuzzy sets and interval-valued fuzzy sets as instances of type-2 fuzzy sets.

I consider that in my research work in fuzzy sets theory there have been three clearly differentiated periods.

12.1 First Period: 1990 – 1996

These years were characterized by carrying on an exclusively theoretical research in the field of Fuzzy Sets Extensions, particularly in Atanassov's intuitionistic fuzzy sets and interval-valued fuzzy sets. The following two papers, among others, published in the Fuzzy Sets and Systems journal, are from this time: "Vague sets are

R. Seising et al. (Eds.): *On Fuzziness: Volume 1*, STUDFUZZ 298, pp. 77–82.
DOI: 10.1007/978-3-642-35641-4_12

Intuitionistic fuzzy sets" and "Entropy on Intuitionistic Fuzzy Sets and on Interval-valued Fuzzy Sets".

Nowadays, the results of both works are being very used in fuzzy extensions classification, the first one, and in image processing, the second one (see for instance the works by H. Tizhoosh and T. Chaira).

In this period I made a stage in Bulgaria to work with Prof. K. Atanassov and it is there that I first met Prof. J. Kacprzyk from The Polish Academy of Sciences, with whom I keep a very close relation since then.

12.2 Second Period: 1997 - 2002

In this period I used for the first my previous theoretical developments in order to build a fuzzy expert system to measure the quality degree (elaboration) of rice. This project was supported by the Government of Navarra. The main obtained results were published in 2000 in the "International Journal of Approximate Reasoning". In this work I developed an inference engine for the expert system using interval-valued fuzzy sets.

At this time I realized that, in many cases, in Atanassov's intuitionistic fuzzy sets theory and in interval-valued fuzzy sets theory, what we actually were doing was an adaptation of the previously developed fuzzy sets theory. This fact led me to focus for some time only and exclusively in the basic theory of fuzzy sets (implication operators, fuzzy subsethood measures, etc.). From this work several papers arose; from my point of view, the most representative one is: "Automorphisms, Negations and Implication Operators" appeared in Fuzzy Sets and Systems in 2003.

12.3 Third Period: 2003-

This could be considered as an explosión period. In 2001 the Computer Science degree is established at the Universidad Publica de Navarra. The beginning of this new degree forced the University to hire new teachers, and this allowed me, in collaboration with Edurne Barrenechea, to create the GIARA research group: "Research Group of Artificial Intelligence and Approximate Reasoning‘" whose main research lines are:

- Extensions: Type 2 fuzzy sets, interval-valued fuzzy sets, Atanassov's intuitionistic fuzzy sets, etc...
- Fuzzy sets. Basic concepts: Aggregation functions, fuzzy implication operators, etc...
- Approximate reasoning
- Image processing
- Pattern recognition
- Decision making
- Classification.

Fig. 12.1. F.l.t.r. Javier Montero, Humberto Bustince, Lofti A. Zadeh, Tomasa Calvo and Luis Magdalena at the Technical University of Madrid (UPM) in 2007 when Lotfi A. Zadeh received the Honorary Doctorate

In this time, Edurne Barrenechea and myself started to use fuzzy sets extensions in image processing. This work was reflected in her Ph. D. thesis "Image processing with interval-valued fuzzy sets. Edge detection. Contrast" in 2005.

The third incorporation was that of Miguel Pagola in 2003. Miguel defended, in 2008, his Ph.D. thesis entitled: "Representation of uncertainty by interval-valued fuzzy sets. Application to Image Thresholding".

Later, the necessity of combining fuzzy sets theory and applications led us to incorporate to the group to the Mathematics Ph.D. Javier Fernández Fernández in 2007.

The following researchers have incorporated to GIARA: José Antonio Sanz, Mikel Galar, Daniel Paternain, Aránzazu Jurío and Carlos López-Molina. We have advised or are advising now the thesis of all of them, always in fuzzy sets theory or extensions application to several fields.

The main achievements of GIARA in this period have been:

A- To make ourselves known by the international scientific community. We collaborate with more than ten foreign universities. Moreover:
   a) We have organized three International Workshops on Artificial Intelligence, with the participation of national and international recognized researchers. As a consequence of these workshops, in 2008 we were co-editors of the

book: Fuzzy Sets and their Extensions: Representation, Aggregation and Models, Editorial Springer.

b) We have coauthored many papers with important researchers such as, for instance, B. Bedregal, G. Beliakov, B. De Baets, T. Calvo, F. Herrera, E. Hüllermeier, P. Melo-Pinto, R. Mesiar, J. Montero, A. Pradera, R. Yager, etc...

c) In 2009 we coorganize with B. De Baets y J. Fodor, the international EURO-FUSE conference. We have edited several issues of international journals, such as, for instance, Fuzzy Sets and Systems.

B- To publish, according to the Web of Knowledge database, 111 works, from which 58 are papers in JCR journals. These papers have more than 1200 references and we have a h index of 17.

a) In this period we have developed and analyzed the functional representation of the following concepts: ignorance, overlap and grouping. We have studied these representations both from the point of view of aggregations and as posible tolos to build fuzzy sets extensions. We have applied these extensions in image processing, classification, expert systems, etc...

b) We have written more than 20 book chapters and we have edited three books

C- To develop five research projects with private companies (Tracasa, Banca Cívica, Incita) and eight with public funds. In all of them we have used fuzzy sets theory.

12.4 Regarding the Future

In Dec 11, 2008 Zadeh introduced the following definition in the bisc-group list:

> "Fuzzy logic is a precise system of reasoning, deduction and computation in which the objects of discourse and analysis are associated with information which is, or is allowed to be, imperfect. Imperfect information is defined as information which in one or more respects is imprecise, uncertain, vague, incomplete, partially true or partially possible."

At the same date, Zadeh made the following remarks:

- In fuzzy logic everything is or is allowed to be a matter of degree. Degrees are allowed to be fuzzy.
- Fuzzy logic is not a replacement for bivalent logic or bivalent-logic- based probability theory. Fuzzy logic adds to bivalent logic and bivalent-logic-based probability theory a wide range of concepts and techniques for dealing with imperfect information.
- Fuzzy logic is designed to address problems in reasoning, deduction and computation with imperfect information which are beyond the reach of traditional methods based on bivalent logic and bivalent-logic - based probability theory.
- In fuzzy logic the writing instrument is a spray pen with precisely known adjustable spray pattern. In bivalent logic the writing instrument is a ballpoint pen.

- The importance of fuzzy logic derives from the fact that in much of the real world imperfect information is the norm rather than exception.

These considerations lead us to settle that in those applications in which the expert does not have enough knowledge to build an efficient membership function, it is appropriate to represent the membership degree of each element to the fuzzy set by means of an interval or by a pair of numbers, etc...; that is, it is appropriate to use some extension instead of fuzzy sets.

Fig. 12.2. GIARA Members in Eurofuse 2009. F.l.t.r. José A. Sanz, Humberto Bustince, Edurne Barrenechea, Aránzazu Jurío, Mikel Galar, Miguel Pagola, Ioritz Cía, Daniel Paternain, Carlos López-Molina and Javier Fernández.

In these conditions, if we interval-valued fuzzy sets, we will take into account the following:

1. The actual membership degree is always a numerical value inside the considered membership interval.
2. The length of the interval depends on the ignorance function that it is used to represent the lack of knowledge of the expert in the construction of the membership degree of the considered element.

Similar considerations can be made for other extensions. For instance, in the case of Atanassov's intuitionistic fuzzy sets, the ignorance function provides the intuitionistic index..

From these reflections we consider that the future of fuzzy sets extensions is very promising, since, in most of the applications, it is very difficult to build the fuzzy sets associated to that application, so it is advisable to use extensions built by means of ignorance functions, grouping functions, etc.

We also believe that in the future fuzzy sets extensions are going to be one of the elements of Soft Computing, since fuzzy sets are a particular case of them.

Nevertheless, the use of extensions of fuzzy sets or fuzzy sets should depend on the problem under consideration and the model level used to solve it. That is, we consider that it is not appropriate to impose the use of one or another type of sets for every problem. In those cases where the solution is almost the same using any of these extensions or fuzzy sets it is advisable to use fuzzy sets, since they are simpler to handle and they have been much more studied.

I would like to conclude these lines thanking professor Zadeh for his relevant achievements: Fuzzy Sets, Soft Computing, etc... Due to them I have been able of finding great colleagues, friends I have the pleasure of working with, and they are a great stimulus for my career. Thanks to all of you...

Fig. 12.3. Presentation of Eurofuse 2009. F.l.t.r. János Fodor, Javier Montero, Carlos Pérez-Nievas (Regional Minister of Education), Julio Lafuente (Rector of the Public University of Navarra, Yolanda Barcina (Mayoress of Pamplona), Bernard De Baets and Humberto Bustince.

13

On the Relevance of Fuzzy Sets in Analytics

Christer Carlsson

The title of this contribution to *On Fuzziness* has a history and a reason that bridges the past, the present and the future. The history is a paper I wrote called "On the Relevance of Fuzzy Sets in Management Science Methodology" which was published in an edited book on *Fuzzy Sets and Decision Analysis* (H. J. Zimmermann, L. A. Zadeh and B. R. Gaines (eds.), in the TIMS Studies in Management Sciences series in 1984) [2]. This was a time when we tried to make the case for fuzzy sets in management science and as a support theory for managers who plan the future, and solve management problems and make decisions in their daily activities.

If we continue the history a bit, a first version of the paper had been presented and discussed at the 11th meeting of the EURO Working Group on Fuzzy Sets at the *European Institute for Advanced Studies in Management* in Brussels on February 19-20, 1981. The EIASM is the centre for serious research on management in Europe and getting an invitation to run a workshop on fuzzy sets took some negotiation; the EURO WG was the first organized activity on the study of fuzzy sets in Europe and was organized by Hans-Jürgen Zimmermann; I was the second chairman and was running the WG for a number of years, also at the time for the workshop at the EIASM in Brussels (Hans thought that I was a natural choice as my doctoral thesis was in management science and we believed that the break-through areas for fuzzy sets in Europe would be engineering and management).

Management science methodology – and especially operations research that applied the same methodology for engineering problems and theory development – had already in 1981 been under attack for more than a decade for failing to deal with the real world problems managers have to tackle, for oversimplifying decision problems and for spending too much time with mathematically interesting but practically irrelevant solutions to problems that had been simplified to be tractable with management science theory and methodology. The message was basically that management science methodology produced theory and methods that were irrelevant for handling actual management problems. The paper in 1984 (1981) argued that fuzzy sets when properly worked into management science methodology would make the models, the algorithms and the theory more relevant and better suited to deal with management problems in practice.

Now, thirty years later, we have to admit that we were not successful in bringing it about, that fuzzy sets remained a marginal development in management science and that we have been able to get fuzzy sets based methods accepted only for limited applications, such as multiple criteria optimisation, real options valuation, logistics

R. Seising et al. (Eds.): *On Fuzziness: Volume 1*, STUDFUZZ 298, pp. 83–89.
DOI: 10.1007/978-3-642-35641-4_13

optimisation, etc. for which there have been algorithmic benefits of allowing the use of fuzzy numbers.

Management science and operations research have also changed over the decades; two major organisations in the field – TIMS and ORSA – merged and became INFORMS to combine the applications oriented research (TIMS) with the algorithms and theory oriented research (ORSA); now the annual INFORMS conferences collect 2-3000 participants; in Europe the EURO Association is a sister organisation to INFORMS and the annual EURO conferences also collect 2-3000 participants. Both organisations run major, well-established journals with high impact factors (*Management Science* and *European Journal of Operations Research*, respectively) and there are dozens of journals publishing material produced guided by the management science methodology. The field is alive and well and promotes lively research that activates thousands of researchers. The context is there, then what is needed for fuzzy sets to be relevant (again) for the research that is carried out? We will work through some arguments in the following to build the promised bridge from the history to the present and future of fuzzy sets in management theory building and applications.

Analytics is gaining support as an important business function that adds value to management; this movement, that promotes data-driven and analytical decision making, is rather recent. Analytics builds on recent software improvements in information systems that has made data, information and knowledge available in real time in ways that were not possible for managers only a few years ago [3]. Now INFORMS has found out that the new movement represents both "potential opportunities" and "challenges" to management science and operations research professionals and published a study [4] in which the INFORMS membership was asked how they want their organisation to deal with analytics. The methods and the application cases worked out in the Davenport-Harris book[3] are very close to traditional text books on management science methodology, actually so close that a manager probably fails to see any differences, which is why INFORMS finds "challenges". Then it makes sense to work out the relevance of fuzzy sets in terms of the modern analytics movement - we will do this by comparing it with the points we made on the relevance of fuzzy sets for classical management science in 1984 (1981).

Liberatore and Luo [4] state that four factors drive the analytics movement: (i) availability of data, (ii) improved analytical software, (iii) the adoption of a process orientation by organisations, and (iv) technically literate managers and executives. Compared to the early 1980's the last factor is probably the most important driver – there is a new generation of managers and executives in charge of the corporations that are using information technology as part of their daily routines. They work with data, information and knowledge on a real time basis and they continuously hunt for better and better analytical tools to help give them competitive advantages. They do not necessarily recognize the analytical tools as classical management science algorithms because analytical software (cf. (ii)) has become user-friendly through graphical user interfaces and visualisation of results, which allows them to use analytical methods without knowing too much of the mathematical background. Information technology has made data available on a real time basis – in classical management science work off line and sufficient time always had to be allocated for collecting and processing data for the models and algorithms – which allows online planning, problem solving and decision making. Maybe "allow" is not the right verb as online management work in real time now is more of a necessity to keep up with the competition. The same driver also explains the adoption of a process orientation (cf. (iii)) as management work typically is group and team work online and in real time. Davenport and Harris [3] define analytics as "the extensive use of data, statistical and quantitative analysis, explanatory and predictive models and fact-based management to drive decisions and actions". Liberatore and Luo [4] interpret this definition as representing three levels of modelling - descriptive, predictive and prescriptive – and stated that management science (they actually refer to OR, but here we do not make any difference between management science and operations research) typically would focus on advanced analytics, i.e. prescriptive modelling. Liberatore and Luo [4] also point out that analytics would focus on the managerial planning, problem solving and decision process, i.e. the transformation of data into actions through analysis and insight, which in their discussion contributes to the application cases of management science.

The modern movement of analytics appears to offer interesting possibilities and opportunities for building on fuzzy sets; the movement is data-driven which will require tools for handling imprecision; the movement is focused on managers who need to deal with real world problems, for which available data, information and knowledge are incomplete, imprecise and uncertain and should allow for fast, often intuitive conclusions; the movement builds on improved analytical software that can easily incorporate various tools using fuzzy sets (fuzzy numbers, fuzzy optimisation algorithms, linguistic modelling, etc.).

Let us then go back to the insights of 1984 (1981) and find out in what way the relevance of fuzzy sets in management science still could apply for the analytics of the 2012.

The composing of the management science methodology follows some principles, and some rules. The classical approach is to aim for either a logico-deductive or an inductive system and select the methods accordingly; the logico-deductive system is favoured in the sciences and has been favoured also for management science methodology. One of the pioneers, Russel L. Ackoff [1], outlined the system as a problem-solving methodology that handles research problems (here summarized):

i Formulate the problem; listing alternative activities that could be carried out, expected outcomes and formulating a set of criteria for comparing the outcomes;
ii Construct or select a model; describing the problem formulation in (i) with the help of a set of formal and stringent concepts;
iii Select a system of measurement; quantifying the concepts introduced in (ii) through some appropriate system of measurement and delimiting activity and solution spaces;

iv Test the model; checking the technical performance of the quantitative model in (iii) and carrying out a preliminary validation;
v Derive a solution from the model; deriving numerical values for the elements of the model in (iii); this constitutes a definite choice of a set of activities, which could be the "best" one possible;
vi Validate the model and the solution; testing and controlling both the model and the solution in order to make certain that the model is a formally valid and reliable representation of the problem and that the solution is formally correct;
vii Carry out experiments with the model, implement and control the solution; testing the applicability and relevance of both the model and the solution to the problem; continue until the model is either accepted or rejected, or modified and developed in order to better correspond to the formulation and the needs of the real world problem.

Each step in this methodology should form a deductive system: (a) a set of undefined and defined concepts to form the framework, which is developed and specified by (b) a set of assumptions, from which is deduced (c) a set of more or less formal theorems, which are confronted with (d) sets of more or less explicit facts (cf. [1]). Such a system will allow us to proceed from general and undefined concepts to specific and defined, from loose and preliminary assumptions to precise and stringent, from general theorems to specific, and from loose heaps of facts to well-structured, theory-oriented sets of facts. This corresponds quite closely to the ideal form for a scientific process and we are assured by the methodology that the solution to (for instance) a management problem will be scientifically valid, formally correct and operationally acceptable as a solution to the original real world problem.

Throughout the history of management science it has been accepted that the methodology described is – in principle – the correct and best way to find solutions to managerial problems. The methodology has been much used to explain great breakthroughs in industry and important innovations in business; it has also been useful for explaining and proving that everything necessary and relevant had been done when unexpected events have caused disasters. The typical thing is, however, that these explanations have been given *after* the processes have been carried out, not online and in real time when they would have been most needed and useful. The reasons given have been that data was not available and that there was no time to build and use the necessary decision models through the methodology summarized in (i)-(vii). Thus the end result has typically been that everybody seems to accept that management science methodology is the best way to tackle and handle management problems but that nobody uses it for practical planning, problem solving and decision making in management.

This is where we made a case for fuzzy sets in the 1984 (1981) paper [2]. First of all, the science methodology applies very well to modelling based on fuzzy sets and we do not have to make any special provisions for fuzzy modelling. The theory of fuzzy sets is developed for a domain in which descriptions of activities and observations are "fuzzy", in the sense that there are no well-defined boundaries of the set of activities or observations to which the descriptions apply.

In the context of managerial planning, problem solving and decision making we have learned over the years that "fuzziness" differs from "generality", which is the application of one description to a (well-defined) set of activities or observations; it differs also from "ambiguity", which refers to the use of several, competing (but well-defined) descriptions of a set of activities or observations. "Fuzziness" is not "uncertainty" in the sense of subjective probability theory – because it does not use its axioms – nor in the sense of classical probability theory as it does not build on the frequencies of events, activities, observations, etc. (we have of course fuzzy probability theory in order to make the world interesting). In the managerial world we quite often look for general principles to guide the development of business models and we have to deal with ambiguity and uncertainty when we work out strategic plans. In many cases management theory recommends that ambiguity and uncertainty should be handled with (subjective) probability theory even if none of the events, activities, observations, etc. is recurring. "Fuzziness" is "vagueness" or "imprecision" (but not in the sense of tolerance analysis as the tolerance interval is not sharp), which makes it a central element in human thought and perceptions, as well as in human language. Why is this important? It is a well-known fact that it is virtually impossible to give an exact description of any real physical situation; needless to say this holds also for any managerial problem situation, especially as many of these problems relate to an unknown future. A science-oriented methodology knows only exact, well-defined activities, outcomes, external activities and goals – and functional relationships. The reason for that has been the nature of and the limitations of the concepts applied in management science methodology: it has not been possible to give an adequate representation of "fuzziness" – or as Lotfi Zadeh wrote "..., *we need a radically different kind of mathematics, the mathematics of fuzzy or cloudy quantities which are not describable in terms of probability distributions*."

The theory of fuzzy sets allows us to structure and describe activities and observations which differ from each other only vaguely, to formulate them in models and to use these models for various purposes in managerial problem solving and decision making. This is ability we have as human beings but which is not supported by any science-oriented methodology. We are trained as scientists to use precise concepts and sharp definitions in order to be able to build precise and elegant models, to use mathematically well-defined algorithms so that we can give distinct descriptions, precise explanations and concise predictions; we are also trained as managers to use precise concepts and sharp definitions so that we can give distinct descriptions of business events, precise explanations of business opportunities and concise predictions of the outcomes when using resources to achieve business ends (at least we strive towards these ideals). The experience we now have (which is also supported by analytics) is that this ability is not the most important one for handling managerial problems.

The human brain is said to think and reason in imprecise, non-quantitative, vague terms which (for instance) gives us the ability to decipher sloppy handwriting and understand distorted speech. It also gives us the ability to summarize and to focus on relevant information and knowledge and to concentrate on essential aspects when handling very large amounts of data and information (a by-product of the information technology revolution). These last-mentioned three capabilities are often cited as "essential" for managers-to-be.

Over the last 30 years we have found in a number of industrial and business cases that fuzzy sets has been a very useful tool for handling imprecision, uncertainty and vagueness without undue simplifications. We have found that we can get a consistent representation of linguistically formulated knowledge that allows the use of precise operators and algorithms. We have worked with fuzzy multi-goal models, fuzzy linear programming and fuzzy multiple criteria optimization and found that fuzzy modelling offers flexibility to deal with very complex problems and the means to include subjective judgment and inexact knowledge to deal with difficult problems that include the needs to resolve future expectations. We have also found that inexactness, or fuzziness, is useful as it helps to convey sufficient information with less effort. We have found the theory of fuzzy sets helpful for carrying out the emphasis on synthesis, expansionism, adaptation and learning (in group and team work), and we have found relevance in the contention that fuzzy modelling in management will contribute to the handling of complex, real-life managerial problems.

Analytics has the same agenda as management science and is working with the same industrial and business context to support managerial planning, problem solving and decision making. Analytics has a broader scope in terms of methods - besides models and algorithms it also works with statistical methods and advanced technology for handling data, information and knowledge. The software used for analytics is several generations more advanced than the software used for management science in 1984. The managers for whom analytics support is developed is a generation more advanced in using information technology and modelling tools. Data, information and knowledge are available in real time for online use to support fast moving business operations. Davenport and Harris [3] argue that "*sophisticated quantitative and statistical analysis and predictive modelling supported by data-savvy senior leaders and powerful information technology*" are the key elements of competitive strategies that make the difference between winning and losing business.

We believe that fuzzy sets will be even more relevant for analytics than for management sci-ence as we (i) will have to deal with (often incomplete, imprecise) large data sets (now called "big data"), (ii) have much better software that can incorporate fuzzy modelling modules, (iii) have to work out support online and in real time where sufficient precision may be more im-portant than full precision, and (iv) have data-savvy users with the insight and knowledge to build on results from fuzzy modelling.

References

1. Ackoff, R.L.: Scientific Method. Optimizing Applied Research Decisions. J. Wiley & Sons, New York (1968)
2. Carlsson, C.: On the Relevance of Fuzzy Sets in Management Science Methodology. In: Zimmermann, H.-J., Zadeh, L.A., Gaines, B.R. (eds.) Fuzzy Sets and Decision Analysis. TIMS Studies in Management Sciences, pp. 11–28. North-Holland, Amsterdam (1984)
3. Davenport, T.H., Harris, J.G.: Competing on Analytics. The New Science of Winning. Harvard Business School Press, Boston (2007)
4. Liberatore, M.J., Luo, W.: INFORMS and the Analytics Movement: The View of the Membership. Interfaces 41(6), 578–589 (2011)

14

Interval Type-2 Fuzzy Logic for Hybrid Intelligent Control

Oscar Castillo

Abstract. We provide in this paper a short review of my research work on developing new methods for building intelligent control systems using type-2 fuzzy logic and soft computing techniques. Soft Computing (SC) consists of several computing paradigms, including fuzzy logic, neural networks, and genetic algorithms, which can be used to create powerful hybrid intelligent systems. Combining type-2 fuzzy logic with traditional SC techniques powerful hybrid intelligent systems can be built for solving complex control problems.

14.1 Introduction

Fuzzy logic is an area of soft computing that enables a system to reason with uncertainty [1]. A fuzzy inference system consists of a set of if-then rules defined over fuzzy sets. Fuzzy sets generalize the concept of a traditional set by allowing the membership degree to be any value between 0 and 1. This corresponds, in the real world, to many situations where it is difficult to decide in an unambiguous manner if something belongs or not to a specific class. Fuzzy expert systems, for example, have been applied with some success to problems of decision, control, diagnosis and classification, just because they can manage the complex expert reasoning involved in these areas of application. The main disadvantage of fuzzy systems is that they can't adapt to changing situations. For this reason, it is a good idea to combine fuzzy logic with neural networks or genetic algo-rithms, because either one of these last two methodologies could give adaptability to the fuzzy system. On the other hand, the knowledge that is used to build these fuzzy rules is uncertain. Such uncertainty leads to rules whose antecedents or consequents are uncertain, which translates into uncertain antecedent or consequent membership functions. Type-1 fuzzy systems, like the ones mentioned above, whose membership functions are type-1 fuzzy sets, are unable to directly handle such uncertainties. We also mention in this paper, type-2 fuzzy systems, in which the antecedent or consequent membership functions are type-2 fuzzy sets [2]. Such sets are fuzzy sets whose membership grades themselves are type-1 fuzzy sets; they are very useful in circumstances where it is difficult to determine an exact membership function for a fuzzy set.

Uncertainty is an inherent part of intelligent systems used in real-world applications. The use of new methods for handling incomplete information is of

R. Seising et al. (Eds.): *On Fuzziness: Volume 1*, STUDFUZZ 298, pp. 91–94.
DOI: 10.1007/978-3-642-35641-4_14

fundamental importance. Type-1 fuzzy sets used in conventional fuzzy systems cannot fully handle the uncertainties present in intelligent systems [3]. Type-2 fuzzy sets that are used in type-2 fuzzy systems can handle such uncertainties in a better way because they provide us with more parameters. This paper reviews the use of intelligent systems based on interval type-2 fuzzy logic for minimizing the effects of uncertainty produced by the instrumentation elements, environmental noise, etc.

14.2 Summary of Research Work on Type-2 Fuzzy Logic in Control

We have performed work on the design of type-2 fuzzy logic controllers using genetic algorithms and bio-inspired optimization methods [3]. In this section we offer a review of the work that we have done on using type-2 fuzzy logic for different control applica-tions.

As a first example, we have considered the design of type-2 fuzzy systems for the longitudinal control of an F-14 airplane using genetic algorithms [3]. The longitudinal control is carried out by controlling only the elevators of the airplane. To carry out such control it is necessary to use the stick, the rate of elevation and the angle of attack. These 3 variables are the input to the fuzzy inference system, and we obtain as output the value of the elevators. After designing the fuzzy inference system we turn to the simulation stage. Simulation results of the longitudinal control are obtained using a plant in Simulink and those results are compared against the PID controller. For optimizing the fuzzy logic control design we use a genetic algorithm (GA).

Another example is the use of an evolutionary algorithm approach for the optimization of type-2 fuzzy reactive and tracking controllers applied to a mobile robot. The algorithm optimizes the type-2 fuzzy inference systems for each of the controllers. Both the reactive and tracking controllers are needed to achieve autonomous navigation of the mobile robot. The use of new methods for handling incomplete information is of fundamental importance in engineering applications. We have also considered the simulation of the effects of uncertainty produced by the instrumentation elements in type-1 and type-2 fuzzy logic controllers to perform a comparative analysis of the systems' response, in the presence of uncertainty [3]. We have presented an innovative idea to optimize interval type-2 membership functions using an average of two type-1 systems with the human evolutionary model. We have shown comparative results of the optimized proposed method. We found that the optimized membership functions for the inputs of a type-2 system increases the performance of the system for high noise levels.

We have also considered the use of Ant Colony Optimization (ACO) for the ball and beam control problem, in particular for the problem of tuning a fuzzy controller of Sugeno type [3]. In our study case the controller has four inputs, each of them with two membership functions, and we consider the interpolation point for every pair of membership function as the main parameter and their individual shape as secondary ones in order to achieve the tuning of the fuzzy controller by using an ACO algorithm. Simulation results show that using ACO and coding the problem with just three parameters instead of six, allows us to find an optimal set of membership function parameters for the fuzzy control system with less computational effort needed.

We have also considered the application of a simple ACO as a method of optimization for membership functions' parameters of a type-2 fuzzy logic controller in order to find the optimal intelligent controller for an autonomous wheeled mobile robot [3]. Simulation results show that ACO outperforms a GA in the optimization of fuzzy logic controllers for an autonomous mobile robot.

We have also used the Particle Swarm Optimization (PSO) method to find the parameters of the membership functions of a type-2 fuzzy logic controller (Type-2 FLC) in order to minimize the state error for linear systems [3]. PSO is used to find the optimal Type-2 FLC to achieve regulation of the output and stability of the closed-loop system. For this purpose, we change the values of the cognitive, social and inertia variables in the PSO. Simulation results, with the optimal FLC implemented in Simulink, show the feasibility of the proposed approach.

In general, the above mentioned applications of type-2 fuzzy logic in intelligent control are representative of the state of art in the area. However, we also have to mention that there exist applications of type-2 fuzzy logic in pattern recognition, time series prediction, and classification, which have been successful in the real world, but are not the main concern in this paper, [5], [6]. There have also been important theoretical advances on type-2 fuzzy logic that have enable more efficient processing and type reduction, which have helped obtaining solutions to real world problems.

14.3 Motivation by Prof. Zadeh's Work

The inspiring ideas and research work of Prof. Zadeh have been fundamental in my own work [7], [8], [9]. He has always supported my research group's work and kindly accepted our invitation to offer a keynote lecture at the World IFSA 2007 Congress that was held in Cancun, Mexico in 2007, which was a very important lecture, especially for Latin America and Mexico. In Figure 1 the arrival of Prof. Zadeh to the Opening reception of IFSA 2007 is shown.

Type-2 fuzzy logic has been always supported by Prof. Zadeh as he was the first one to initially suggest the idea. In this sense he has always supported the work of the main researchers in this area, like J. M. Mendel, R. John, H. Hagras, P. Melin and others like me. We consider that type-2 fuzzy control will be one of the most important areas in the near future to consider working on as many open problems, both theoretically and application wise still remain unsolved.

14.4 Conclusions

We have described in this paper a review of the new methods for building intelligent systems using type-2 fuzzy logic and soft computing techniques. In this paper, we have considered the use of fuzzy logic to a higher order, which is called type-2 fuzzy logic. Combining type-2 fuzzy logic with traditional SC techniques, we can build powerful hybrid intelligent systems that can use the advantages that each technique offers in solving complex control problems. Finally, the application of bio-inspired optimization techniques, like GAs, PSO and ACO, has been proposed to automatically design optimal type-2 fuzzy logic controllers in different applications.

Fig. 14.1. Arrival of Prof. Zadeh to the IFSA 2007 World congress

References

1. Castillo, O., Melin, P.: Soft Computing for Control of Non-Linear Dynamical Systems. Springer, Heidelberg (2001)
2. Castillo, O., Melin, P.: Soft Computing and Fractal Theory for Intelligent Manufacturing. Physica-Verlag, Heidelberg (2003)
3. Castillo, O., Melin, P.: Type-2 Fuzzy Logic: Theory and Applications, vol. 223. Springer, Heidelberg (2008)
4. Melin, P., Castillo, O.: Modelling, Simulation and Control of Non-Linear Dynamical Systems. Springer, Heidelberg (2002)
5. Castillo, O.: Type-2 Fuzzy Logic in Intelligent Control Applications. Springer, Heidelberg (2012)
6. Melin, P., Castillo, O.: Hybrid Intelligent Systems for Pattern Recognition. Springer, Heidelberg (2005)
7. Zadeh, L.A.: Fuzzy Logic = Computing with Words. IEEE Transactions on Fuzzy Systems 4(2), 103 (1996)
8. Zadeh, L.A.: Knowledge Representation in Fuzzy Logic. IEEE Transactions on Knowledge Data Engineering 1, 89 (1989)
9. Zadeh, L.A.: Fuzzy Logic. Computer 1(4), 83–93 (1998)

15

On Fuzziness: Empiricism and Cross-Disciplinarity Unbounded

Jordi Cat

Fuzziness came to my attention through the disciplinary prism of philosophy of science, with its issues and its history. My goal is to identify the bearing of fuzziness on these abstract issues and to suggest valuable lessons.

One of my earliest interests has been physics, also its history and its philosophy. Four general ideas have stood out, I think, among others, as playing major roles: unity, relationalism -the abstract form of relativity-, symmetry and indeterminism. They share a denial of the privilege of determinate values and fixed properties of individual entities. A great deal of theoretical and experimental practices in physics, and elsewhere, are woven together by relational concepts, the description of relations or relative properties. Functions and equations are simple but clear examples in mathematics. The formalism of the calculus was introduced and applied on the notion that the proper representation of natural order is the ratio of infinitesimal differences. The differences left indeterminate, and physically insignificant, the absolute values. In the nineteenth century mathematicians explored and exploited two related forms of indeterminacy, the problem of uniqueness of solutions in potential theory and the problem of singularities. Associated with the indeterminate character of analysis were – especially by Maxwell – the thermodynamical phenomena of phase transitions and the problem of freedom of the will. That determinism of analytic functions can be a source of indeterminacy, in this case of prediction, was found manifested in the phenomenon of chaotic behavior. During the same period, the application of the calculus of probabilities was extended from the calculus of measurement errors to the statistical behavior of populations of men and molecules. Probabilities brought together much of physics under the use of the languages of indeterminacy about individual observations and entities. Quantum physics placed indeterminism on a newly fundamental level of description of physical individuals and their measurement. Finally, the emphasis on the privileged character of relational properties such as relative velocity in physical experience and theory -especially by Galileo- gave way to a history of gradually more general and abstract concepts of symmetry. The basic concept remains the invariance or indifference with respect to particular values and variations. The symmetric quantity or relation leaves indeterminate – it underdetermines – the value of another. From Galileo to Maxwell and Einstein, many physicist perceived the symmetric quantity or law are the physically objective or significant. Symmetries have become key to representing physical

R. Seising et al. (Eds.): *On Fuzziness: Volume 1*, STUDFUZZ 298, pp. 95–100.
DOI: 10.1007/978-3-642-35641-4_15

properties -think of group theory in particle physics- and thereby to unifying them, an ideal and a heuristic. Physics, then, has long thrived on indeterminacy.

My meta-scientific interests include empiricism, unification in the sciences, a priori ideas, causality, the application of mathematics, the history of philosophy of science and the relation between science and philosophy. Fuzziness bears on all of these interrelated issues. It first came to my attention looking at history. Philosophy has long engaged in the examination of empirical science; and-at least since Aristotle- it has traditionally focused on the use of language and reasoning. For instance, nineteenth-century scientist-philosophers such as William Whewell, James Clerk Maxwell and Michael Faraday considered the task of crafting language part and parcel of the construction of science as a tool in the process of the interpretation, prediction and construction of the world. What laws may relate depends on the classification we adopt of the entities involved; and that we can formulate any laws requires that we settle on criteria of classification. Equal exposure to the traditional education in rhetoric, the new romantic poetry and idealist philosophy led them to note the relation between scientific, literary and ordinary language. Maxwell introduced the notion of scientific metaphors, or metaphorical generalization of scientific terms. Whewell coined new scientific terms -alongside Faraday- and distinguished between the different naming and categorization practices in different sciences. The use of precise mathematical definitions contrasts with the use of indefinite terms outside mathematics and physics. The case in point is natural history, where he noted the operation of a mode of concept formation and naming that we can recognize; it is the one underlying modern fuzziness. The Victorian Imperial preoccupation with centers and shifting boundaries help distinguish different kinds of naming and classification strategies, boundary-based and center-based:

> "Natural Groups are best described, not by any definition which marks their boundaries, but by a Type which marks their centre. The Type of any natural group is an example which possesses in a marked degree all leading characters of the class."[1]

Indefiniteness of representation is a distinctive feature and the price of a general, empirical – possibly nominalist – strategy, centered on identifying -or declaring- concrete typical members of a group. He elaborated:

> "The apparent indefiniteness and inconsistency of the classification and definitions of Natural History belongs, in a far higher degree, to all other except mathematical speculations.' (...) 'The class is steadily fixed, though not precisely limited; it is given, though not circumscribed; it is determined, not by a boundary line without, but by a central point within; not by what it strictly excludes, but by what it eminently includes; by an example, not by a precept; in short, instead of Definition we have a Type for our director. A Type is an example of any class, for instance, a species of a genus, which is considered as eminently possessing the characters of the class. " ([10], p. 476.)

[1] Aphorism XCII, in [9], p. 16.

By the end of the nineteenth century mathematics and philosophy were beholden to symbolic notations and formal methods. Philosophy took the so-called linguistic turn and philosophy of science became the logic of science. The goal was to make sense of the ideals of intelligibility and the objectivity of scientific knowledge. In practice or in fantasy, to understand involved precise theoretical concepts and inferences. Early in the twentieth century indeterminacy or vagueness came back in the picture in attempts to represent the relation of theoretical statements to empirical data. For the physicist-philosopher Pierre Duhem, theory's symbols did not depict; and theoretical predictions were also compatible with a range of precise numerical data. As a consequence theoretical statements, even laws, couldn't be either true or false, only approximate.[2] For the philosopher-social scientist Otto Neurath, proper empirical statements, the so-called protocol statements, inevitably would include vague ordinary terms, as would social theory.[3] As a consequence, the language of science wouldn't be an idealized precise one, and the logical relation between statements wouldn't be sufficiently determining for the purpose of theory-formation and testing, and pragmatic considerations -'auxiliary motives' or 'extra-logical factors'- would be needed to settle the acceptance or rejection of any hypotheses.[4] Waismann picked up on the issue of verification and introduced the distinction between the complete or closed character of quantitative definitions and the open texture or porosity of empirical concepts [8]. Neurath was also challenging the new program of philosophy of science as precise logical analysis or logical reconstruction adopted by other so-called logical empiricists or positivists. The program sought a unified view of science – and scientific philosophy – in terms of a unified language and a unified method. Russell, however, having initiated the general program, alongside Frege, had noted the challenge of vagueness. And Lukasiewicz introduced in the Polish Aristotelian tradition of logic a multivalued logic. Around the same time, the philosopher Max Black, aware of the issues surrounding logical positivism, took up Russell's challenge and endorsed Duhem's about the value of vagueness as the condition of empirical science. [1] (Later on, he continued his analysis of language for the case of metaphors.)

I have argued that Zadeh's set-theoretic insights into the semantics of informal reasoning and ordinary language have provided the tools for a new image of empirical science and of cross-disciplinary interactions, both among the sciences and with philosophy. New models of causality play a large role. But their strength is also their weakness; and I call for improved versions [3].

In engineering, the biomedical sciences and the social sciences, for instance, the fuzzy-set theoretic framework has brought into sharper (or should I say 'fuzzier'?) relief a feature missing from traditional accounts of empirical science: it is the relevant theoretical concepts that are often fuzzy while the empirical data are precise.

[2] [6]. For Duhem, theoretical hypotheses cannot confront data in isolation, singly; data cannot definitively confirm or falsify any.

[3] [2]. For his colleague in the Vienna Circle Moriz Schlick only intuitions were indefinite, knowledge was formally structured and precise by (theoretical) construction. Duhem had illustrated this position with the difference between the concepts of warmth and temperature.

And, to move on to the other aspect of science historically emphasized and debated, this feature provides an element of partial unity; it provides a conceptual and a methodological bridge between the natural and human sciences.

The fuzzy-set theoretic framework contributed enough semantic apparatus to suggest the introduction of another conceptual tool with empirical application and connecting power: criteria of causality. From the symbolic and computational perspective that dominates the fuzzy-set paradigm, they are intended as forms of a causal calculus. The core of the different models is this: (1) the set-theoretic relational structure of subsethood relations between fuzzy sets; (2) the semantic assumption that subsethood relations can model logical relations, and, in particular, the truth conditions of material conditionals (an original motivation of Zadeh's project); and (3) the interpretive principle that causal relations are reducible to the old Aristotelian modality of necessary and sufficient conditions.

The focus of my criticism has been (3) and the failure of the derived criteria to meet important developments in causal thinking [3]. From the conceptualist approach, these models are too narrow to capture the insight that causality does not require either necessity or sufficiency. Relevant causal information takes often the form of identifying partial causes or, in J.L. Mackie's term developing J.S. Mill's account, 'INUS' conditions in complex cases: causes can be true causes while being neither necessary nor sufficient, but instead 'insufficient but necessary parts of sufficient but unnecessary causal complexes.' [7] An empiricist, extensionalist interpretation of the subsethood relations allows for its association with probabilistic relations. But this leads to limiting and problematic accounts of probabilistic, or statistical, causality. Finally, important information for the sake of explanation and intervention concerns shielding constraints and intervening mechanisms.[4] Sophisticated models of causal processes and mechanisms are available. But the fuzzy-set theoretic framework has failed to incorporate these insights.

The proliferation of causal modeling as part of scientific practice is significant in the connective role it has played and the potential it offers. This form of partial unification applies first to the natural and human sciences, and to the relation between them: cognitive maps with causal meaning were designed in political science before causal criteria were adopted and developed in computer science, neuroscience or medicine. The unification extends also to the relation between the sciences and the humanities, especially to philosophy. Causality is a philosophical category with a long and heated history, with an enduring belief in its a priori nature, that it is an intellectual product of the mind or else its investigation is. It makes empirical investigations possible and precedes experimental thinking and intervention as well as causal analyses of empirical data. Science and philosophy share this kind of conceptual investigation and creativity.[5]

[4] See, for instance [11] and [5].

[5] Both often pretend also to pursue the investigation 'scientifically' or empirically; this naturalistic stance only exposes the degree of conceptual activity in empirical research and the degree of proximity.

Zadeh's legacy of the fuzzy-set theoretic framework is most significant in a final respect: as a powerful tool for the generalization of a variety of scientific concepts, formalisms, methods and values. As a generalization of set theory, it provides, for instance, the basis for extended forms of mathematical formalisms such as the calculus, their interpretation and their application in the mathematical sciences, such as physics, biology or economics. It also entrenches the notion of approximation as conceptually accurate and methodologically acceptable. Approximate representation is not just entrenched in science; it can be an expression and source of generality, and not of limitation or compromise. Of course, embracing fuzziness uncritically at the meta-scientific level, fuzzism, risks the same sort of objections and self-referential paradoxes as relativism.

Fig. 15.1. Lotfi Zadeh and Jordi Cat at July 25 2003, 22nd International Conference of the North American Fuzzy Information Processing Society NAFIPS'2003, Chicago, Illinois, July 24-26, 2003

References

1. Black, M.: Vagueness: An Exercise in Logical Analysis. Philosophy of Science 4(4), 427–455 (1937)
2. Cat, J.: The Popper-Neurath Debate and Neurath's Attack on Scientific Method. Studies in History and Philosophy of Science 26, 219–250 (1995)
3. Cat, J.: Fuzzy Empiricism and Fuzzy-Set Causality: What is all the Fuzz About? Philosophy of Science 73, 26–41 (2006)

4. Cartwright, N., Cat, J., Fleck, L., Uebel, T.E.: Otto Neurath: Philosophy Between Science and Politics. Cambridge University Press, Cambridge (1996)
5. Craver, C.F.: When Mechanistic Models Explain. Synthese 153, 355–376 (2006)
6. Duhem, P.: Aim and Structure of Physical Theory. Atheneum, New York (1906) (1954 trans.)
7. Mackie, J.L.: The Cement of the Universe. A Study of Causation. Clarendon Press, Oxford (1975)
8. Waismann, F.: Verifiability. Proceedings of the Aristotelian Society 19(suppl.), 119–150 (1946)
9. Aphorism XCII. In: Whewell, W. (ed.) Aphorisms Concerning Ideas, Science and the Language of Science. Harrison & Co., London (1840)
10. Whewell, W.: The Philosophy of the Inductive Sciences, vol. 1, p. 476. John Parker, London (1840)
11. Woodward, J.: Making Things Happen. Oxford University Press, New York (2003)

16

Application of Fuzzy Inference Systems in Real World Scenarios

Elizabeth J. Chang, Omar K. Hussain,
and Tharam S. Dillon

16.1 Introduction

The use of computing technologies by humans for various knowledge processing and knowledge synthesis activities has grown exponentially in the recent past. One of the obvious reasons for this trend is the ease with which information is obtained, processed and computed to obtain the desired analysis. Some real-world examples of such tasks include control of a train (for example on the Sendai Subway System) [1], control of heating and cooling devices [2], signal processing [3], controlling different functions of an aircraft [4] etc. At the same time, there has also been a change in the type of users who depend on such computing technologies for information. Users range from the experienced to the intermediate or novice group of users, who are not interested in knowing the complex computations by which information is processed, but want the desired output from the computations in order to carry out various tasks. For such tasks, it is of critical importance that decision making by machines be done with the utmost precision based on an understanding of the current input scenarios. In other words, machines have to mimic the real-time decision making capability of humans so as to conduct an intelligent analysis of the underlying complex information, synthesize knowledge from it, and act on it for various activities.

This is a great challenge for machines because, unlike humans, they do not have the natural tendency to work with information that may be fuzzy or ambiguous in representation. Hence, in order to make decisions or utilize the information for further processing, they need models that extract the fuzziness from the given information so as to make the inputs quantifiable and understood by them, ensuring that the output has the required level of precision. In other words, they need models that represent the information in numeric values that exactly simulates and truly reflects the real-world scenario. This holds true for both small and large scale computing applications. One way by which the ambiguity in information is captured is by making use of a fuzzy inference system. Fuzzy systems are mathematical objects that model vagueness and uncertainty when the described phenomena do not have sharply-defined boundaries. They were developed to incorporate the concept of partial truth characterized by the fuzziness of the data which yields a more accurate mathematical representation of the perception of truth than that of crisp sets [5].

R. Seising et al. (Eds.): *On Fuzziness: Volume 1*, STUDFUZZ 298, pp. 101–107.
DOI: 10.1007/978-3-642-35641-4_16

They provide a precise approach for dealing with uncertain information by using a multi-valued logic derived from fuzzy sets.

In this chapter, we discuss three approaches where fuzzy inference systems are used to capture the fuzziness from the input values and process the information to assist the machines to perform various computations in real-life activities.

16.2 Soft Computing Approach to Measure the Usability of Software

Given the changes in the types of users, the *usability* of software systems is a crucial issue. Measuring usability is therefore an important concern [6]. Usability represents the ease with which the software can be utilized by the user to carry out a given set of actions. This is important as there are many examples of software systems that have not gained wide acceptance among users due to the poor design of the usability of the software and have been consigned to oblivion in spite of having great initial expectations. On the other hand, there are also examples of software having been modified and updated many times, only to still not be accepted by users. One of the reasons for these scenarios is the non-provision of a suitable user interface by these software and their inability to respond to the users' preferences as desired in various settings. So having a satisfactory usability measure of software is one of the important factors to be quantified and ensured during its development and testing phases.

Measuring the usability of software is a two-stage process. The first stage focuses on identifying the problems with the interface and developing solutions for these; whereas the second stage focuses on measuring the ease of use of a software on different designs in order to choose the most suitable [6]. Measuring the user friendliness of a software is a complex stage as there are several dimensions and factors that seem to impact upon it, each of which needs to be quantified and measured. Furthermore, it is possible that in some software application scenarios, several factors may play a key role in determining its ease of use compared to others in different scenarios. From the user's perspective, such variations need to be captured and considered when measuring the software's usability. In our approach, we achieve this by using a combination of physical observation and a soft computing approach. The physical observation focusses on the test layout of users, whereas the soft computing approach focusses on having a fuzzy inference system that captures the observations and the different variations in the importance of various factors in order to determine their impact and the usability of the software.The physical observation part requires a test user, a computer with the software being tested, and a test monitor user as shown in Figure 16.1 The test user performs the given task on the software that is observed by the test monitor user. This user enters a numerical score for each factor to determine their quantification for the software under observation. The factors which are used in this process are:

(a) System feedback: This factor measures the level to which the system provides feedback to the users.

(b) Consistency: This factor measures the extent to which the look, feel and behaviour of the software interface is consistent throughout and with other applications.
(c) Error prevention: This factor measures the level to which the software prevents the users from making errors.
(d) Performance/Efficiency: This factor measures the quality of the tasks completed by the users.
(e) User like/dislike: This factor measures the satisfaction of the user in using the software.
(f) Error Recovery: This factor measures the ease with which a user can exit from the situation in which he unintentionally found himself.

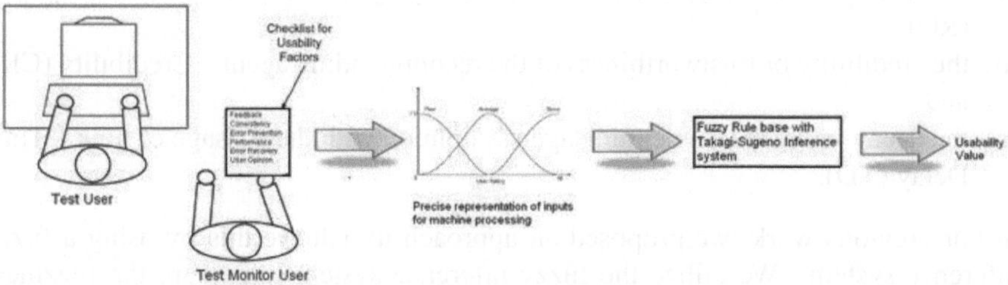

Fig. 16.1. A soft computing approach for software usability evaluation

Once the input score of each factor has been determined, they are used as inputs in the soft computing approach to determine the usability score of the software. However, before this can be done, the membership functions and the linguisitic terms which are used to transform the fuzziness in each factor are identified. In order for the machine to quantify each of the abovementioned factors, a membership function that has 3 fuzzy sets as shown in Figure 16.1 is used. Once the input variables have been fuzzified, then the Takagi-Sugeno approach for fuzzy inference is used to determine the usability of the software in crisp values. The number of homogeneous rules identified was $3^6 = 729$. The usability value when represented on a scale of 0 to 10 represents the overall usability of the software being analysed [6]. By using the fuzzy inference system, the relationship and the importance between the different input factors according to a given scenario were captured and their effect on the output usability value determined.

16.3 A Fuzzy System Approach for Reputation Calculation

Another area where the fuzzy inference system is beneficial and used is in business activities particularly over the internet. The growth of online sales has opened various avenues for businesses to maximize their profits. However, the notion of trust and reputation in such a medium is one of the important factors to be determined for maximising the user's interaction experience. While hardware systems in such a medium can be commoditized, the same does not apply to the notion of trust and reputation. In other words, they need to be measured according to specific criteria for a given context in order for an appropriate decision to be made. In today's world where human dependence on machines is ever-increasing to the extent that machines make decisions on behalf of people, intelligence needs to be introduced in them by which they capture and understand any imprecisely-defined scenarios. This automation requires a methodology that is self-adaptive and that understands semantics. In other words, a representation is needed that characterizes the important factors for reputation determination individually and combines them to ascertain the reputation of a given agent or business. The factors in question are [7]:

(a) the recommendation opinion of the third party agent – Recommendation Opinion (RO),
(b) the credibility or trustworthiness of the recommending agent – Credibility (CR), and
(c) the decay in the recommending agent's opinion with the passage of time – Time Delay (TD).

In our previous work, we proposed an approach to achieve this by using a fuzzy inference system. We utilize the fuzzy inference system to capture the fuzziness among the input variables to ascertain the trust value according to the given input conditions. The steps that we adopt in our approach are as follows and as shown in Figure 16.2:

(a) Define the universe of discourse and membership functions for the inputs.
(b) Define the fuzzy inference system by using Takagi-Sugeno approach (or the Mamdani approach).
(c) Define the fuzzy rule base. Obtain the firing rule strength associated with each rule.
(d) Compute the reputation value.

Utilizing such an approach,the fuzzy measures of each input variable can be represented and understood by the machines while ascertaining the reputation of an agent

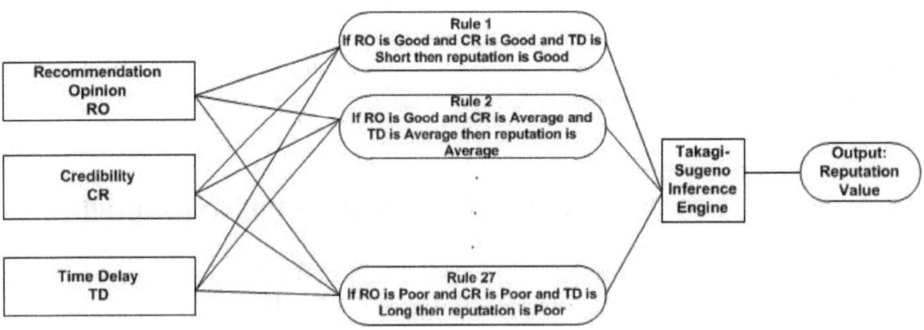

Fig. 16.2. Fuzzy approach for reputation calculation

in online business activities. This analysis can then be utilized by machines for informed decision-making in business activities.

16.4 Fuzzy Logic Based Approach for Risk Assessment in Business Activities

Risk and reward are always foremost considerations when making investment decisions and conducting business transactions. Advances in the area of Information Communication Technologies (ICT) have enabled the development of new business paradigms. Such paradigms involve transactions taking place between loosely connected parties, often totally or partially unknown to one another. One important concept required to ensure that such transactions are successful is the analysis of transactional risk that measures the likelihood of failure of the business activity and the impact or consequences as the result of failure. The importance of doing this has been demonstrated in the recent financial crisis.In our previous work, we proposed a probabilistic and convolution-based approach to ascertain the likelihood of failure and impact of failure respectively of a business activity [8]. We term these representations performance risk and financial risk in the business activity. Once these measures of risk have been determined, they need to be combined to ascertain the magnitude of transactional risk in the business activity. However during this phase, it is important to capture the variability and uncertainty that involves:

(a) determining which level/s of severity of transactional risk that might occur in the business activity (for example on a scale of 0-100, what level/s of risk occur); and
(b) determining the likelihood of occurrence of those level/s (for example, on a scale of 0-1, what is the likelihood of occurrence of the determined levels of risk).

The variability and uncertainty during transactional risk determination can be captured by numerical techniques such as possibility theory [8]. However, the broad aim of the risk assessment step is to determine and represent the level/s of transactional risk in forming a business association in a way that can be understood by the machine or risk assessing agent through a common understanding. Failure to do this may result in ambiguities or irregularities being introduced that may propagate in the risk evaluation and risk management phases. If this is the case, then the whole purpose of undertaking risk analysis is lost. For example, on the scale of $0-100$ if an interaction initiating agent 'A' determines the level of risk as 30%, 45% and 80 %, the question that arises is what does agent 'A' make of those levels of severity? Does he consider the degree of transactional risk to be 'low' or 'moderate' or 'high' during the risk assessment phase or in other words 'acceptable' or 'unacceptable' during the risk evaluation phase? Furthermore, different agents may have different interpretations of these level/s which makes it ambiguous for machines or users to deal with in the next steps of risk analysis. So in order to avoid such scenarios, the representation of transactional risk with *semantics* is important. Such representation of transactional risk defines semantic tags in the range of values between $0 - 100$ and determines the level/s of severity of occurrence of transactional risk in linguistic terms in order for machines or humans to have a common understanding.

To determine the linguistic level/s of transactional risk, an approach is needed which, based on the numeric inputs, computes the different level/s of severity of transactional risk as output on the defined semantic tags. There are different techniques to achieve this. In our approach, we utilized a fuzzy inference system to achieve this as shown in Figure 16.3 [8].

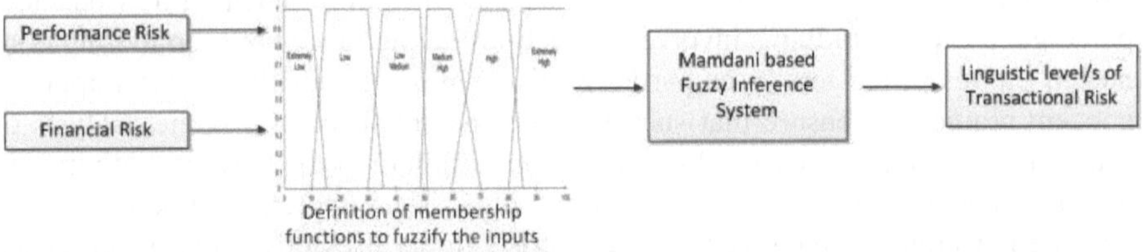

Fig. 16.3. Fuzzy Inference System for ascertaining the Linguistic representation of transactional risk

In the fuzzy inference model, there are two inputs and each input is further defined by six predicates. Hence, the total number of homogeneous rules in our system is 36. In contrast with the numeric level/s of transactional risk, the linguistic level/s represents the fuzzy sets along with their DOM that shows in semantic terms, the different level/s of transactional risk in the business activity such as $MH = 0.3$, $H = 1$ and $EH = 0.3$. This will assist the risk assessing agent to have a semantic representation of the transactional risk in the business activity. The next step after determining the level/s of transactional risk is the risk evaluation phase wherein the risk assessing agent has to make a decision about whether or not the determined level of transactional risk is 'acceptable' to it. During this phase, the linguistic representation of transactional risk with its associated semantics will be utilized better by the fuzzy inference system to evaluate it and ascertain the semantic decision output in the business activity.

16.5 Conclusion

We consider fuzziness to be one of the inherent factors that is present in our every activities. As humans, we have the ability to understand and deal with it, but in order for machines to deal with it, appropiate models and representations have to be made. The idea of fuzzy sets and fuzzy systems proposed by Lofti Zadeh have been fundamental in acheiving this. As seen from the above examples, this has enabled machines to mimic human handling of information in order for them to obtain the benefits of fast and efficient computing capabilities, which otherwise would have been difficult to achieve.

References

1. Fuzzy Logic Applications,
`http://www.calvin.edu/~pribeiro/othrlnks/Fuzzy/apps.htm`
2. Mohamad, S., Ishak, A.A., Aishah, S., Kadir, S.A.: Design of Fuzzy Logic Controller for Overdamped Temperature Response of a Process Air Heater System. In: 4th International Conference on Modeling, Simulation and Applied Optimization (ICMSAO), Kuala Lumpur, pp. 1–4 (2011)
3. Ahmad, S.A., Ishak, A.J., Ali, S.: Classification of Surface Electromyographic Signal Using Fuzzy Logic for Prosthesis Control Application. In: IEEE EMBS Conference on Biomedical Engineering and Sciences (IECBES), Kuala Lumpur, pp. 471–474 (2010)
4. Ervin, J.C., Alptekin, S.E.: Fuzzy Logic Control of a Model Airplane. In: IEEE International Conference on Systems, Man, and Cybernetics, San Diego, pp. 2320–2325 (1998)
5. Berkan, R.C., Trubatch, S.L.: Fuzzy Systems Design Principles. IEEE Press, New York (1997)
6. Chang, E.J., Dillon, T.S.: A Usability-Evaluation Metric Based on a Soft-Computing Approach. IEEE Transactions on Systems, Man and Cybernatics – Part A: Systems and Humans 36, 356–372 (2006)
7. Chang, E.J., Dillon, T., Hussain, F.K.: Trust and Reputation for Service-Oriented Environments. John Wiley & Sons Ltd., West Sussex (2006)
8. Hussain, O.K., Dillon, T.S., Hussain, F.K., Chang, E.J.: Risk Assessment and Management in the Networked Economy. Springer, Heidelberg (2012)

17

Fuzziness in Automata Theory: Why? How?

Miroslav Ćirić and Jelena Ignjatović

17.1 Introduction

The aim of this article is to explain why we study fuzzy automata and how we do it, i.e., to highlight the most efficient tools of the theory of fuzzy sets that we use in our research. In addition, we want to show how research in the theory of fuzzy automata affected our research in other areas of the theory of fuzzy sets.

We entered in the world of fuzziness when we crossed from the classical algebra and automata theory to the theory of fuzzy automata. Besides being considered as a natural generalization of ordinary automata and languages, fuzzy automata and related languages have also been studied as a means for bridging the gap between the precision of computer languages and vagueness and imprecision, which are frequently encountered in the study of natural languages (cf. [18]). During the decades, they have got a wide field of applications. However, many authors thought mainly about the properties of ordinary automata which can be transferred to fuzzy automata. We found that the theory of fuzzy automata is not only simple translation of the results from the classical automata theory to the language of fuzzy sets, but it is possible to use powerful tools of the theory of fuzzy sets in the study of fuzzy automata.

The key point is that a fuzzy automaton can be regarded as a fuzzy relational system. It can be specified by a family $\{\delta_x\}_{x\in X}$ of fuzzy transition relations on the set of states A, indexed by the input alphabet X, and fuzzy subsets σ and τ of A, the fuzzy subsets of initial and terminal states. Inductively we define the composite fuzzy transition relations $\{\delta_u\}_{u\in X^*}$ by putting that δ_ε is the crisp equality, and $\delta_{ux} = \delta_u \circ \delta_x$, for $u \in X^*$, $x \in X$. Now, the fuzzy language recognized by the fuzzy automaton $\mathscr{A}$ is defined as a fuzzy subset $L_{\mathscr{A}}$ of X^* given by $L_{\mathscr{A}}(u) = \sigma \circ \delta_u \circ \tau$, for $u \in X^*$.[1] This way of representing fuzzy automata, and fuzzy languages that they recognize, enables to study fuzzy automata using fuzzy relational calculus, and to express many problems through fuzzy relation equations and inequalities. Fuzzy relational calculus and fuzzy relation equations and inequalities have been widely used in our research.

Previously, fuzzy relational calculus and fuzzy relation equations and inequalities were used in the theory of fuzzy automata only by few authors – Peeva,

[1] Here X^* denotes the monoid of all words over X, $\varepsilon \in X^*$ is the empty word, and $\circ$ denotes the compositions of two fuzzy relations, of a fuzzy set and a fuzzy relation and two fuzzy sets, defined in the usual way over a residuated lattice or lattice-ordered monoid.

R. Seising et al. (Eds.): *On Fuzziness: Volume 1*, STUDFUZZ 298, pp. 109–114.
DOI: 10.1007/978-3-642-35641-4_17

Bělohlávek, and Li and Pedrycz (cf. [1, 19–22]). Surprisingly, such approach has not been used for ordinary nondeterministic automata, although their behavior can be expressed in terms of the calculus of two-valued relations. Probably, the reason for this is the fact that nondeterministic automata are predominantly considered from the perspective of the graph theory, and not from the perspective of the algebra of relations. A little bit similar approach has been used for weighted automata over a semiring, whose behavior is defined through the calculus of matrices with entries in the underlying semiring (cf. [8]). However, matrices over a semiring do not possess some very important properties of ordinary and fuzzy relations, and their use in the study of weighted automata is not as fruitful as the use of fuzzy relations in the study of fuzzy automata.

We will briefly explain how we used fuzzy relational calculus and fuzzy relation equations and inequalities in solving the fundamental problems of the theory of fuzzy automata: *determinization*, *equivalence* and *state reduction*.

17.2 Determinization of Fuzzy Automata

A deterministic fuzzy automaton is a fuzzy automaton having exactly one crisp initial state and a deterministic transition function, and the fuzziness is entirely concentrated in the fuzzy set of terminal states. The determinization of a fuzzy automaton is a procedure of constructing an equivalent deterministic fuzzy automaton[2]. Such procedure is usually required in most practical applications and implementation of automata. The first determinization algorithms for fuzzy automata, provided by Bělohlávek and Li and Pedrycz, generalize the well-known subset construction, and have the same shortcoming as its crisp counterpart: some states of the resulting automaton can be redundant (cf. [1, 19]). We have constructed the Nerode automaton associated with a fuzzy automaton, a deterministic fuzzy automaton which is equivalent to the original fuzzy automaton and has no redundant states. Its states are fuzzy sets of the form $\sigma_u = \sigma \circ \delta_u$, for $u \in X^*$, the single initial state is $\sigma = \sigma_\varepsilon$, the transition function δ_N is defined by $\delta_N(\sigma_u, x) = \sigma_{ux}$, for $u \in X^*, x \in X$, and the fuzzy set τ_N of terminal states is defined by $\tau_N(\sigma_u) = \sigma_u \circ \tau$, for $u \in X^*$. The Nerode automaton always has smaller number of states than automata constructed by the previous determinization methods, but nevertheless, in some cases it may be infinite. Its finiteness depends on certain local properties of the underlying structure of truth values, and necessary and sufficient conditions under which the Nerode automaton is finite have been determined. We have also provided an improved algorithm, which constructs the reduced Nerode automaton with even smaller number of states than the Nerode automaton (cf. [10, 13, 17]).

The Nerode automaton was originally constructed for fuzzy automata over a complete residuated lattice, but it was noted that the same construction can be applied to fuzzy automata over a lattice-ordered monoid, and even more, to weighted automata

[2] Two fuzzy automata are equivalent if they recognize the same fuzzy language.

over a semiring. All these structures have the multiplication which is distributive over the supremum (or addition), which ensures associativity of the composition of fuzzy relations. However, it was shown that the Nerode automaton and the reduced Nerode automaton can be constructed even if the composition is not associative, i.e., for automata with weights that are taken in a strong bimonoid, a structure which is not necessarily distributive. In particular, this includes fuzzy automata over arbitrary lattices (cf. [2, 17]).

17.3 Equivalence of Fuzzy Automata and Bisimulations

Another important problem of automata theory is to determine whether two given automata are equivalent. For deterministic automata this problem is solvable in polynomial time, but for nondeterministic and fuzzy automata it is computationally hard. It is also desirable to express the equivalence of automata as a relation between their states, if possible, or find some relation between states which implies the equivalence. The equivalence of two deterministic automata can be expressed in terms of relationships between their states, but in the case of nondeterministic and fuzzy automata the problem is more complicated, and we can only examine various relations which imply the equivalence.

It is generally agreed that the best way to model the equivalence of automata is the concept of bisimulation. They give a close enough approximation of the equivalence and are efficiently computable. Bisimulations were introduced in concurrency theory, and independently, in set theory and modal logic, and nowadays, they are successfully employed in many areas of computer science and mathematics. We have introduced two types of simulations for fuzzy automata, forward and backward simulations, and combining them, we have defined four types of bisimulations (cf. [5, 6]). Forward simulations between two fuzzy automata $\mathscr{A}$ and $\mathscr{A}'$ are defined as solutions to the system of fuzzy relation inequalities $\sigma \le \sigma' \circ \varphi^{-1}$, $\varphi^{-1} \circ \delta_x \le \delta'_x \circ \varphi^{-1}$ $(x \in X)$, $\varphi^{-1} \circ \tau \le \tau'$, backward simulations as solutions to the system $\tau \le \varphi \circ \tau'$, $\delta_x \circ \varphi \le \varphi \circ \delta'_x$ $(x \in X)$, $\sigma \circ \varphi \le \sigma'$, and bisimulations are defined by a combination of these two systems.[3] The greatest solutions to these systems, i.e., the greatest simulations and bisimulations between fuzzy automata, are computed by iterative procedures. Termination of these iterative procedures after a finite number of steps also depends on local properties of the underlying structure of truth values. Key role in the computation of the greatest simulations and bisimulations play the residuals of fuzzy relations, which we have introduced. To ensure the existence of these residuals, it is necessary that the underlying structure of truth values is also residuated, so our theory has been developed for fuzzy automata over a complete residuated lattice.[4]

---

[3] Here φ denotes an unknown fuzzy relation between the sets of states of $\mathscr{A}$ and $\mathscr{A}'$, and φ^{-1} denotes its inverse (converse, transpose) fuzzy relation.

[4] In fact, commutativity of the multiplication is not necessary, and analogous results can be obtained when the underlying structure of truth values is a quantale.

17.4 State Reduction

In contrast to deterministic automata, for which there are many fast minimization algorithms, the state minimization problem for nondeterministic and fuzzy automata is computationally hard. For these automata, a more practical problem is the *state reduction*, where we have to construct an automaton with as small as possible number of states, which is equivalent to a given automaton. This automaton need not be minimal, but must be efficiently computable.

We have reduced the state reduction problem for fuzzy automata to the problem of solving a particular system of fuzzy relation equations (cf. [7, 23]). For a given fuzzy automaton and a fuzzy equivalence on its set of states, we have defined the related factor fuzzy automaton. In general, these two fuzzy automata are not equivalent. Based on the fact that the fuzzy language recognized by a fuzzy automaton $\mathscr{A}$ can be expressed as $L_{\mathscr{A}}(u) = \sigma \circ \delta_u \circ \tau$, for $u \in X^*$, we have expressed the equivalence of a fuzzy automaton and its factor fuzzy automaton as a system of fuzzy relation equations, called the general system. Namely, we have shown that these two fuzzy automata are equivalent if and only if the fuzzy equivalence by which we perform factorization is a solution to the general system. However, the general system may consist of infinitely many equations, and finding its non-trivial solutions may be a very difficult task, so we have aimed our attention to some instances of this system which consist of finitely many equations and are easier to solve. The most interesting instances are those systems that define forward and backward bisimulations between the states of a single fuzzy automaton. We have provided effective procedures for computing the greatest forward and backward bisimulation fuzzy equivalences on a fuzzy automaton, which ensure the best reductions by fuzzy equivalences of these types. Moreover, we have shown that even better reductions can be achieved alternating reductions by forward and backward bisimulation fuzzy equivalences, and also, if we use fuzzy quasi-orders instead of fuzzy equivalences.

17.5 The Reverse Impact

As we have seen, fuzzy relational calculus and the theory of fuzzy relation inequalities and equations have had a tremendous impact on our research in the theory of fuzzy automata. However, this research has had a very strong reverse impact. Problems arising from the study of fuzzy automata have led to the launch of some new questions regarding various types of fuzzy relations. We have given many new results on fuzzy equivalences and fuzzy quasi-orders, and moreover, we have introduced a completely new concept of a uniform fuzzy relation (cf. [3, 4]). Our original intention was to introduce uniform fuzzy relations as a basis for defining such concept of a fuzzy function which would provide a correspondence between fuzzy functions and fuzzy equivalence relations, analogous to the correspondence between crisp functions and crisp equivalence relations. This was done, but also, it turned out that uniform fuzzy relations establish natural relationships between fuzzy partitions of two sets, some kind of “uniformity” between these fuzzy partitions. Roughly speaking, uniform fuzzy relations can be conceived as fuzzy equivalence relations which relate elements of two possibly different sets. They were employed to solve some systems of fuzzy relation equations that have important applications in approximate reasoning, and to define and study fuzzy homomorphisms and fuzzy relational morphisms of algebras (cf. [4, 11]). However, uniform fuzzy relations have shown their full strength in the study of equivalence between fuzzy automata, which has previously been discussed (cf. [5]).

Systems of fuzzy relation equations and inequalities that emerged from our research in the theory of fuzzy automata initiated the study of the systems of the same form from the general aspect. These systems are referred to as weakly linear systems (cf. [9, 12, 14]). There has been proved that every weakly linear system, with a complete residuated lattice as the underlying structure of truth values, has the greatest solution, and an algorithm has been provided for computing this greatest solution. This algorithm is based on the computing of the greatest post-fixed point, contained in a given fuzzy relation, of an isotone function on the lattice of fuzzy relations. The algorithm represents an iterative procedure whose each single step can be viewed as solving a particular linear system, and for this reason these systems were called weakly linear. This iterative procedure terminates in a finite number of steps whenever the underlying complete residuated lattice is locally finite, for example, when dealing with Boolean or Gödel structure. Otherwise, some sufficient conditions under which the procedure ends in a finite number of steps have been determined. If the underlying complete residuated lattice satisfies infinite distributive laws for the supremumand multiplication over infimum, for example, when dealing with a structure defined by a continuous t-norm on the real unit interval [0, 1] (an BL-algebra on [0, 1]), the greatest solution can be obtained as the infimum of fuzzy relations outputted after each single step of the iterative procedure.

It is worth noting that the methodology developed for solving weakly linear systems has been recently extended to an even broader context, and used for solving systems of inequalities and equations over partially ordered sets defined by residuated and residual functions (cf. [15]).

Acknowledgement. This research is supported by Ministry Education and Science, Republic of Serbia, Grant No. 174013.

References

1. Bělohlávek, R.: Determinism and Fuzzy Automata. Information Sciences 143, 205–209 (2002)
2. Ćirić, M., Droste, M., Ignjatović, J., Vogler, H.: Determinization of Weighted Finite Automata Over Strong Bimonoids. Information Sciences 180, 3497–3520 (2010)
3. Ćirić, M., Ignjatović, J., Bogdanović, S.: Fuzzy Equivalence Relations and Their Equivalence Classes. Fuzzy Sets and Systems 158, 1295–1313 (2007)
4. Ćirić, M., Ignjatović, J., Bogdanović, S.: Uniform fuzzy relations and fuzzy functions. Fuzzy Sets and Systems 160, 1054–1081 (2009)

5. Ćirić, M., Ignjatović, J., Damljanović, N., Bašić, M.: Bisimulations for Fuzzy Automata. Fuzzy Sets and Systems 186, 100–139 (2012)
6. Ćirić, M., Ignjatović, J., Jančić, I., Damljanović, N.: Computation of the Greatest Simulations and Bisimulations Between Fuzzy Automata. Fuzzy Sets and Systems 208, 22–42
7. Ćirić, M., Stamenković, A., Ignjatović, J., Petković, T.: Fuzzy Relation Equations and Reduction of Fuzzy Automata. Journal of Computer and Systems Sciences 76, 609–633 (2010)
8. Droste, M., Kuich, W., Vogler, H. (eds.): Handbook of Weighted Automata, Monographs in Theoretical Computer Science. EATCS Series. Springer (2009)
9. Ignjatović, J., Ćirić, M.: Weakly Linear Systems of Fuzzy Relation Inequalities and Their Applications: A Brief Survey. Filomat 26(2), 1–35 (2012)
10. Ignjatović, J., Ćirić, M., Bogdanović, S.: Determinization of Fuzzy Automata With Membership Values in Complete Residuated Lattices. Information Sciences 178, 164–180 (2008)
11. Ignjatović, J., Ćirić, M., Bogdanović, S.: Fuzzy Homomorphisms of Algebras. Fuzzy Sets and Systems 160, 2345–2365 (2009)
12. Ignjatović, J., Ćirić, M., Bogdanović, S.: On the Greatest Solutions to Weakly Linear Systems of Fuzzy Relation Inequalities and Equations. Fuzzy Sets and Systems 161, 3081–3113 (2010)
13. Ignjatović, J., Ćirić, M., Bogdanović, S., Petković, T.: Myhill-Nerode Type Theory for Fuzzy Languages and Automata. Fuzzy Sets and Systems 161, 1288–1324 (2010)
14. Ignjatović, J., Ćirić, M., Damljanović, N., Jančić, I.: Weakly Linear Systems of Fuzzy Relation Inequalities: The Heterogeneous Case. Fuzzy Sets and Systems (2011), doi:10.1016/j.fss.2011.11.011
15. Ignjatović, J., Ćirić, M., De Baets, B.: General Method for Solving Equations and Inequalities Defined by Residuated Functions (to appear)
16. Ignjatović, J., Ćirić, M., Simović, V.: Fuzzy Relation Equations and Subsystems of Fuzzy Transition Systems. Knowledge-based Systems (2012), doi:10.1016/j.knosys.2012.02.008
17. Jančić, Z., Ignjatović, J., Ćirić, M.: An Improved Algorithm for Determinization of Weighted and Fuzzy Automata. Information Sciences 181, 1358–1368 (2011)
18. Lee, E.T.-Z., Zadeh, L.A.: Note on Fuzzy Languages. Information Sciences 1, 421–434 (1969)
19. Li, Y., Pedrycz, W.: Fuzzy Finite Automata and Fuzzy Regular Expressions With Membership Values in Lattice Ordered Monoids. Fuzzy Sets and Systems 156, 68–92 (2005)
20. Peeva, K.: Behaviour, Reduction and Minimization of Finite L-automata. Fuzzy Sets and Systems 28, 171–181 (1988)
21. Peeva, K.: Finite L-fuzzy Machines. Fuzzy Sets and Systems 141, 415–437 (2004)
22. Peeva, K., Kyosev, Y.: Fuzzy Relational Calculus: Theory, Applications, and Software. World Scientific (2004)
23. Stamenković, A., Ćirić, M., Ignjatović, J.: Reduction of Fuzzy Automata by Means of Fuzzy Quasi-orders. Submitted to Information Sciences

18

Inference with Probabilistic and Fuzzy Information

Giulianella Coletti and Barbara Vantaggi

18.1 Introduction

We adopt the interpretation of fuzzy sets in terms of coherent conditional probabilities, introduced in [2–4] and presented in this issue [7] by R. Scozzafava. Aim of this chapter is to discuss (from a syntactical point of view) which concepts of fuzzy sets theory [9] are naturally obtained simply by using coherence. In particular, we focus on operations among fuzzy subsets (and relevant t-norms and t-conorms), and on Bayesian inference procedures, when statistical and fuzzy information must be taken into account. It is obvious that in the case of inference with hybrid information the proposed interpretation of the membership provides a general and well founded framework (that of coherent conditional probability) for merging and managing all the available information. For instance, in this frame the simplest inferential problem (to find the most probable element of a data base, starting from a probability distribution on the single elements and a fuzzy information expressed by a membership function defined on the elements of the data base) is referable to a Bayesian updating of an initial probability. The only remarkable question is that the Bayes formula is applied in an unusual semantic way: the distribution, which plays the role of "prior" probability, is here usually obtained by statistical data, whereas the membership function, which plays the role of "likelihood", is a subjective evaluation.

We refer (see Section 18.2) to the results about coherent conditional probability recalled in [7] to find the class of t-norms and t-conorms such that the membership of the union and intersection of two fuzzy sets obtained by them is a coherent extension of the two coherent conditional probabilities modeling the initial fuzzy sets. In Section 18.3 hybrid information is handled by maintaining consistence with the model and this gives rise to a general inferential model able to deal with different kinds of applications.

18.2 Operations among Fuzzy Sets

Let X be a (not necessarily numerical) variable, with range $\mathscr{C}_X$, and , for any $x \in \mathscr{C}_X$, let $A_x = \{X = x\}$. For any property φ related to the variable X we consider the conditional event $E_\varphi | A_x$ = {You claim (that X has the property) φ under the hypothesis that X assumes the value x }, recalled in [7]. The membership μ_φ of a

R. Seising et al. (Eds.): *On Fuzziness: Volume 1*, STUDFUZZ 298, pp. 115–119.
DOI: 10.1007/978-3-642-35641-4_18

fuzzy set $E_\varphi^* = (E_\varphi, \mu_\varphi)$ can be reinterpreted by means of the conditional probability $P(E_\varphi|\cdot)$ in fact there is complete freedom in assessing it (see [4]).

We recall that the events E_i $(i = 1,...,n)$ are logically independent if all the disjunctions $E_1^* \wedge ... \wedge E_n^*$ (where E_i^* represents the event E_i or its contrary E_i^c) are possible, that is the atoms generated by events E_i are 2^n. The following result concerns the global coherence of a set of probability assessments and it is useful for a family of fuzzy subsets such that the events E_{φ_i} are logically independent in order to show that the probability rules do not impose constrains to the membership functions (see [1]).

Theorem 1. *Let $\mathscr{C} = \{E_j|H_{j_i}\}_{i=1,...,n_j; j=1,...,n}$ be a set of conditional events such that $\mathscr{H}_j = \{H_{j_1},...,H_{j n_j}\}$ is a partition of Ω, for every j, and the events of $\mathscr{E} = \{E_j\}_{i=1,...,n}$ are logically independent. For every j, let $P_j : \mathscr{H}_j \mapsto [0,1]$ be a probability distribution and $p(E_j|\cdot) : \mathscr{H}_j \mapsto [0,1]$ a coherent conditional probability.*

If the probability distributions P_j's are "globally" coherent on $\mathscr{H}^* = \bigcup_j \mathscr{H}_j$, then the assessment $\{P_j, p(E_j|\cdot)\}_{j=1,...,n}$ is "globally" coherent in $\mathscr{E} \times \mathscr{H}^*$.

This result is important since generally events E_φ representing fuzzy sets are logically independent, even those seemingly linked: as an example we consider E_φ and E_ψ , with $\psi = \neg\varphi$ which are logical independent, since we can claim both "X has the property φ" and "X has the property $\neg\varphi''$.

Now we are going to introduce the operations between fuzzy sets by referring to [3]: the definitions of the binary operations of union and intersection and that of complementation can be obtained directly by using the rules of coherent conditional probability. For this aim let us denote by $\varphi \vee \psi$, $\varphi \wedge \psi$, respectively, the properties "φ or ψ", "φ and ψ".

As proved in [3], for any given x in the range of X, the assessment $P(E_\varphi \wedge E_\psi|A_x) = v$ is coherent if and only if it takes values in the interval

$$\max\{P(E_\varphi|A_x) + P(E_\psi|A_x) - 1, 0\} \leq v \leq \min\{P(E_\varphi|A_x), P(E_\psi|A_x)\}. \quad (18.1)$$

Now we need to define $E_{\varphi\vee\psi} = E_\varphi \vee E_\psi$, $E_{\varphi\wedge\psi} = E_\varphi \wedge E_\psi$.

Let us consider two fuzzy subsets E_φ^*, E_ψ^*, corresponding to the same variable x, with the events E_φ, E_ψ logically independent. As proved in [3], for any given X in the range of X, the assessment $P(E_\varphi \wedge E_\psi|A_x) = v$ is coherent if and only if takes values in the interval

$$\max\{P(E_\varphi|A_x) + P(E_\psi|A_x) - 1, 0\} \leq v \leq \min\{P(E_\varphi|A_x), P(E_\psi|A_x)\} \quad (18.2)$$

and moreover any choice of the values for $P(E_\varphi \wedge E_\psi|A_x)$ in the corresponding intervals is a coherent conditional probability assessment. From the probability rules, given $P(E_\varphi \wedge E_\psi|A_x)$, we get automatically also the value of $P(E_\varphi \vee E_\psi|A_x)$. Then, it is possible to put

$$E_\varphi^* \cup E_\psi^* = \{E_{\varphi\vee\psi}, \mu_{\varphi\vee\psi}\}\ ,\ E_\varphi^* \cap E_\psi^* = \{E_{\varphi\wedge\psi}, \mu_{\varphi\wedge\psi}\}\ , \quad (18.3)$$

with $\mu_{\varphi\vee\psi}(x) = P(E_\varphi \vee E_\psi|A_x)$, $\mu_{\varphi\wedge\psi}(x) = P(E_\varphi \wedge E_\psi|A_x)$.

Three possible coherent choices for the value of the conditional probability v give rise to different well-known (see, e.g., [5]) t-norms and t-conorms: in [3] (see also [2]) the choice of the so-called T_M and S_M as t-norm and t-conorm, of the Lukasiewicz t-norm and t-conorm, and, finally, of the so-called probabilistic sum S_P and product T_P is discussed also with semantic implications. As it is well known these three coherent choices correspond to the particular values $\lambda = 0$, $\lambda = 1$, $\lambda = \infty$, respectively, of the fundamental (archimedean) Frank (see [6]) t-norms T_λ and t-conorms S_λ, with $\lambda \in [0, \infty]$, which are in fact *all and only the possible coherent choices* for v and u, [3]. We notice that the condition of logical independence of events E_φ, E_ψ is crucial for proving the above assertions. So if we have a family of logically independent events E_{φ_i} and consider the algebra $\mathscr{B}$ spanned by them, we can use any Frank's t-norm and its dual t-conorm to compute any union and intersection between to relevant fuzzy sets (E_{φ_i}, μ_i); coherence rules the extension of the conditional probability $P(\cdot|A_x)$ to the other events of the algebra (for instance to the events E_φ^c), which do not support a fuzzy set. Starting from these considerations we can define the complement of a fuzzy set

$$(E_\varphi^*)' = \{E_{\neg\varphi}, \mu_{\neg\varphi}\}. \tag{18.4}$$

We recall that: $E_{\neg\varphi} \neq (E_\varphi)^c$. Then, while $E_\varphi \vee (E_\varphi)^c = \Omega$, one has $E_\varphi \vee E_{\neg\varphi} \subset \Omega$. So the membership $\mu_{\varphi\vee\neg\varphi}(x) = P(E_\varphi \vee E_\neg \varphi|A_x)$ can be different from 1 for some $x \in \mathscr{C}_X$. In other words $E^*_{\varphi\vee\neg\varphi}$ is a fuzzy set.

The case of two fuzzy subsets E^*_φ, E^*_ψ, corresponding to the random quantities X_1 and X_2, respectively, has been studied in [3] by assuming the following conditional independence condition: for every (x, x') belonging to the range of the random vector (X_1, X_2)

$$P(E_\varphi|A_x \wedge A_{x'}) = P(E_\varphi|A_x) \ , \ P(E_\psi|A_x \wedge A_{x'}) = P(E_\psi|A_{x'}). \tag{18.5}$$

In [1] the same problem has been studied without independence conditions. In both cases it is possible to conclude that the following choice for the membership of conjunction and disjunction is coherent:

$$\mu_{\varphi\vee\psi}(x,x') = P(E_\varphi \vee E_\psi|A_x \wedge A_{x'}),\ \mu_{\varphi\wedge\psi}(x,x') = P(E_\varphi \wedge E_\psi|A_x \wedge A_{x'}). \tag{18.6}$$

with the only constraints

$$\max\{\mu_\varphi(x) + \mu_\psi(x') - 1, 0\} \leq \mu_{\varphi\wedge\psi}(x,x') \leq \min\{\mu_\varphi(x) + \mu_\psi(x')\}. \tag{18.7}$$

and

$$\mu_{\varphi\vee\psi}(x,x') = \mu_\varphi(x) + \mu_\psi(x') - \mu_{\varphi\wedge\psi}(x,x'). \tag{18.8}$$

18.3 Inference with Fuzzy and Probabilistic Information

Our first aim is the following: if we have a probability distribution on the elements of $\mathscr{C}_X$ and a fuzzy information, expressed by a membership function $\mu_\varphi(\cdot) = P(E_\varphi|\cdot)$, we would choose the most probable element $x \in \mathscr{C}_X$ under the hypothesis E_φ.

By Theorem 1 the global assessment $\{P, \mu_\varphi\}$ is coherent and so we can compute, for every $x \in \mathscr{C}_X$ the value $P(A_x|E_\varphi)$. We can compute the extension by Bayes formula:

$$P(A_x|E_\varphi) = \alpha P(A_x)\mu_\varphi(x) \tag{18.9}$$

where $\alpha = (\sum_x \mu_\varphi(x)P(A_x))^{-1}$.

So, to reach our goal it is sufficient to find the events A_{x^*} with maximum posterior, i.e.

$$P(A_{x^*}|E_\varphi) = \alpha \max_x\{P(A_x)\mu_\varphi(x)\} \tag{18.10}$$

In [1] more general situations have been studied in order to give some algorithms for finding the most probable element of $\mathscr{C}_X$ also when the statistical information is related to a family different from $\{A_x\}$. Moreover, in the same paper this inferential model is applied for the virtual representation of a female avatar based also on the similarities studied in this context in [8].

A particular kind of inference is that at the basis of the "*perception based probabilistic reasoning*" introduced by Zadeh in [10]). We sketch a solution alternative to that given by Zadeh, based on our interpretation of fuzzy set, starting from a simple example.

A box contains n balls of various sizes $s_1, \ldots, s_m$, with $m \le n$, with unknown percentages. Consider an experiment consisting in drawing a ball from the box, and let E_φ be the event (referred to the drawn ball) "You claim that the size is large". *Consider also the event E_ψ=*"You claim that the size of most ball is large"*The problem is: what is the probability of $E_\varphi|E_\psi$?*

The fuzzy subset E_φ is related to the variable (the size) S and if the composition of the urn were known, then the probability of E_φ would be computed by disintegration formula. Otherwise we need to refer to the possible compositions H_k, and the probability $\alpha_k = P(E_\varphi|H_k)$ is obtained again by disintegration formula with respect to the possible values of S. Let $\mathscr{P}$ be the variable taking the possible values α_k.

Note that E_ψ can also be expressed by the sentence (referred to the ball to be drawn) "You claim that the probability of being claimed large is high", and the membership of the fuzzy subset of claiming high E_H is $P(E_H|A_{\alpha_r})$ where $A_r = \{\mathscr{P} = \alpha_r\}$.

Concerning the conditional event $E_\varphi|E_\psi$, by assuming conditional independence of E_φ and E_ψ given the possible compositions H_k, we have:

$$P(E_\varphi|E_\psi) = \sum_k P(E_\varphi|H_k)P(H_k|E_\psi), \tag{18.11}$$

where $P(H_k|E_\psi)$ is obtained by Bayes' formula.

References

1. Coletti, G., Gervasi, S., Tasso, S., Vantaggi, B.: Generalized Bayesian Inference in a Fuzzy Context: From Theory to a Virtual Reality Application. Computational Statistics & Data Analysis (2011) (in press)
2. Coletti, G., Scozzafava, R.: Probabilistic Logic in a Coherent Setting. In: Trends in Logic, vol. 15. Kluwer, Dordrecht (2002)

3. Coletti, G., Scozzafava, R.: Conditional Probability, Fuzzy Sets and Possibility: a Unifying View. Fuzzy Sets and Systems 144, 227–249 (2004)
4. Coletti, G., Scozzafava, R.: Conditional Probability and Fuzzy Information. Computational Statistics & Data Analysis 51(1), 115–132 (2006)
5. Klement, E.P., Mesiar, R., Pap, E.: Triangular Norms. Kluwer, Dordrecht (2000)
6. Frank, M.J.: On the Simultaneous Associativity of $F(x,y)$ and $x+y-F(x,y)$. Aequationes Mathematicae 19, 194–226 (1979)
7. Scozzafava, R.: The Membership of a Fuzzy Set as Coherent Conditional Probability (in this issue)
8. Scozzafava, R., Vantaggi, B.: Fuzzy Inclusion and Similarity Through Coherent Conditional Probability. Fuzzy Sets and Systems 160(3), 292–305 (2009)
9. Zadeh, L.A.: Probability measures of fuzzy events. Journal of Mathematical Analysis and Applications 23, 421–427 (1968)
10. Zadeh, L.A.: Toward a perception-based theory of probabilistic reasoning with imprecise probabilities. Journal of Statistical Planning and Inference 105, 233–264 (2002)

19

Fuzzy Conceptual Data Analysis Applied to Knowledge Management

Carmen De Maio, Giuseppe Fenza, Vincenzo Loia, and Saverio Salerno

Abstract. Conceptual data analysis has been extensively exploited to support Ontology Learning, Information Retrieval, and so on. This work emphasizes the relevant role of uncertainty in the conceptual data analysis. Specifically, Fuzzy Conceptual Data Analysis has been exploited to address two Enterprise Knowledge Management methodologies: domain ontology learning and ontology merging.

19.1 Introduction and Related Works

Nowadays Semantic Web and Web 2.0 play a crucial role in the area of Knowledge Management. Last trend is definition of ontology-based Knowledge Management platforms [3], [4]. Ontology-based Knowledge Management platform raises new requirements in terms of preparation and maintenance of domain ontologies. Specifically, domain ontologies could provide a common access point to the linked data repository. So, there is a need to define methodologies able to support life cycle of large and heterogeneous knowledge bases. Nevertheless, Semantic Web tools are still under development and ontologies maintenance (e.g., preparation, update) requires a considerable effort.

This paper addresses these challenges defining methodologies for: domain ontology learning, taking into account User Generated Content (e.g., blog, wiki, etc.), and ontology merging, to update previously extracted domain ontologies. These methodologies leverage on Fuzzy extention of Conceptual Data Analysis. Conceptual Data Analysis mainly attains with Formal Concept Analysis (FCA) [8] algorithm. In particular, this paper argues that Fuzzy Theory [7] enable us to generalise Conceptual Data Analysis introducing uncertainty management with applied to FCA (FFCA) and data analysis techniques providing support to knowledge extraction and structuring. Conceptual Data Analysis has been extensively applied to Information Retrieval because browsing lattice according to the user's query enables query augmentation and refinement [13], [14]. Specifically, (Fuzzy) Conceptual Data Analysis has been applied also for Ontology Learning (analysing text and extracting lattice with FCA) and Ontology Merging (mashing different lattices to infer new concept hierarchies).

In literature, there are many works for Ontology Learning that analises domain data by using text-mining and machine learning techniques, some of these approaches exploit FCA. Specifically, [10] introduces the L-fuzzy context, as an endeavour to combine fuzzy logic with FCA but it seems to be not practicable for

R. Seising et al. (Eds.): *On Fuzziness: Volume 1*, STUDFUZZ 298, pp. 121–127.
DOI: 10.1007/978-3-642-35641-4_19

dealing with large data sets because a human support is required to define the fuzzy linguistic variables. In [9] the Fuzzy extension of FCA theory is exploited to build hierarchical classification of the collected resources.

As for Ontology Merging, the idea of using FCA was first proposed in [5], where the FCAMerge algorithm is described. FCAMerge is a bottom-up approach to ontology merging guided by application-specific instances of the given source ontologies. Specifically, formal context has obtained analysing documents representing the two input ontologies. Instead, FCAOntMerge [11] following approach defined in [6] translates each input ontology into attributes (i.e., columns) and objects (i.e., rows) of a formal context.

On the light of described scenario, this work defines: methodology for Ontology Learning orchestrating Fuzzy C-Means (FCM) and FCA; and methodology for semi-automatic Ontology Merging extending with fuzziness FCAOntMerge approach.

The paper is organised as follows: Section 19.2 introduces theoretical background of Fuzzy Conceptual Data Analysis; Section 19.3 describe workflows of defined methodologies for Ontology Learning and Merging, then, Section 19.4 gives some experimental results of the defined methodologies. Finally conclusion close the paper.

19.2 Fuzzy Conceptual Data Analysis

This section provides most relevant notions of FFCA and FCM that are exploited to perform Fuzzy Conceptual Data Analysis. These algorithms will be orchestrated in Sections 19.3.1 and 19.3.2 to support Ontology Learning and Ontology Merging, respectively.

19.2.1 Fuzzy C-Means – FCM

FCM [1] clustering is an unsupervised process, based on c-partition [2]. It takes as input a data matrix and it tries to get an "optimal" partitioning of the feature space (composed by the data matrix). FCM aims at maximizing the homogeneity, grouping into the same cluster the patterns which are closer. Each pattern is a row of matrix. FCM recognizes spherical "clouds of points" (clusters of data) in a multi dimensional data space (i.e. data matrix) and each cluster is represented by its center point (prototype or centroid). The function minimizes the weighted sum of the distances between data points x and the centroid v, according to this formula:

$$V(U) = \sum_{i=1}^{c} \sum_{j=1}^{n} u_{i,j}^{m} \left\| \underline{x}_j - \underline{v}_i \right\|^2 \tag{19.1}$$

where $c \geq 2$ is the number of clusters, $u_{i,j} \in [0,1]$ is the membership degree of x_i in the i-th cluster and $m > 1$ controls the quantity of fuzziness in the classification process.

After the FCM execution, data partitions are returned, in a prior fixed number c of clusters.

19.2.2 Fuzzy Formal Concept Analysis – FFCA

FCA is a technique of data analysis, which exploits the ordered lattice theory. Recently, FCA and fuzzy techniques are integrated in order to deal with uncertain and vague information. In particular, this approach exploits a fuzzy extension of FCA. Fuzzy FCA (FFCA) [9] combines fuzzy logic into FCA representing the uncertainty through membership values in the range [0, 1]. Through formal contexts, FFCA enables the representation of the relationships between objects and attributes in a given domain.

Some definitions about main concepts of Formal Concept Analysis extracted from [9] and its fuzzy extension are given.

Definition 1: *A* **Fuzzy Formal Context** *is a triple $K = (G, M, I = \varphi(G \times M))$, where G is a set of objects, M is a set of attributes, and I is a fuzzy set on domain $G \times M$. Each relation $(g, m) \in I$ has a membership value $\mu(g, m)$ in [0, 1].*

Definition 2: Fuzzy Formal Concept. *Given a fuzzy formal context K=(G, M, I) and a confidence threshold T, we define $A^* = \{m \in M \mid \forall g \in A: \mu(g, m) \geq T\}$ for $A \subseteq G$ and $B^* = \{g \in G \mid \forall m \in B: \mu(g, m) \geq T\}$ for $B \subseteq M$. A fuzzy formal concept (or fuzzy concept) of a fuzzy formal context K with a confidence threshold T is a pair $(A_f = \varphi(A), B)$, where $A \subseteq G$, $B \subseteq M$, $A^*=B$ and $B^*=A$. Each object $g \in \varphi(A)$ has a membership μ_g defined as*

$$\mu_g = min_{m \in B}\, \mu(g,m)$$

where $\mu(g,m)$ is the membership value between object g and attribute m, which is defined in I. Note that if B={ } then μ_g = 1 for every g. A and B are the extent and intent of the formal concept ($\varphi(A)$,B) respectively.

The Fuzzy FCA takes into account the fuzzy formal context and performs a hierarchical arrangement of fuzzy formal concepts, so obtains fuzzy concept lattice.

Definition 3: ***A Fuzzy Concept Lattice*** *of a fuzzy formal context K with a confidence threshold T is a set F(K) of all fuzzy concepts of K with the partial order $\leq$ with the confidence threshold T .*

The fuzzy lattice evidences the membership associated to the objects and the class-subclass relationship [9]. Thanks to FCA theory, the concepts are arranged in a hierarchy, emphasizing semantic relationships like subsumption (i.e., "is-a").

19.3 Methodologies for Enterprise Knowledge Management

Following sections describe application of Fuzzy Conceptual Data Analysis algorihms to support Ontology Learning and Ontology Merging.

19.3.1 Case Study: Ontology Learning

This methodology analyses structured and unstructured (e.g., bolg, wiki, etc.) resources which are daily produced by employees in the organisation in order to extract unsupervised hiearchical conceptualisation (e.g., topics, and so on). The workflow and mapping on the exploited technological solutions is shown in Fig.19.1. It involves following phases:

- *Natural Language Processing*, that relies on several activities, such as: language detection (i.e., Apache TIKA), multiformat analysis, stopword removal, stemming and lemmatisation (i.e., Snowball), PoS tagging (i.e. Language Tool) and terms disambiguation (i.e., Wikipedia Miner), and so on. Specifically, this step exploits Wikipedia as external knowledge resource in order to enrich keywords extraction results with wikification of the main portion of text;

- *Vectorization*, that carries out feature set and term weighting of input text by mainly applying well known technique of TF-IDF;

- *Fuzzy Conceptual Data Analysis*, applies FCM in order to prune incoming data, then resulting partition is given as input to FFCA algorithm. At the end of this phase concept hierarchies have been extracted;

- *Semantic Technology Mapping*, that represents the extracted unsupervised conceptualisation (and their relationships) in a schema compliant with SemanticWeb technologies (i.e., SKOS and RDFS).

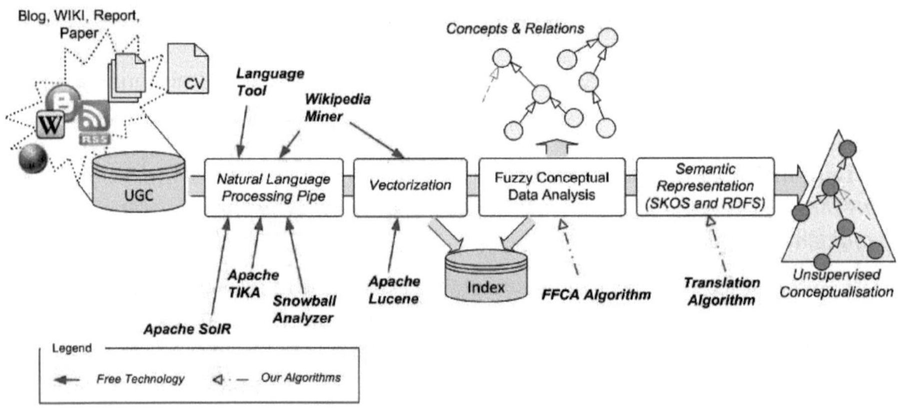

Fig. 19.1. Workflow of the Methdology for Ontology Learning

19.3.2 Case Study: Ontology Merging

This methodology is inspired to FCAOntMerge proposed in [11]. Specifically, this methodology is aimed to merge Unsupervised Conceptualisation and existing Domain Ontologies. The workflow and mapping on the exploited technological solutions is shown in Fig.19.2. It is composed of the following phases:

1. *Formal Context Creation*, input ontologies are transformed into two Formal Contexts. Each cell of the Formal Context represents instances-of relation (i.e., binary relation) between concept and individuals of input ontologies;

2. *Formal Contexts Merging*, this step exploit ontology matching results. Taking into account matching degree between different ontologies formal contexts will be merged in the union of them.

3. *Fuzzy Conceptual Data Analysis*, this step perform algorithm of FFCA on merged context in order to carry out a knowledge structure that integrates concepts of Unsupervised Conceptualisation and Domain Ontologies.

4. *Assessment and Semantic Technology Mapping*, interacting with expert user this step perform assessment in a semi-automatic manner of new knoweldge structure. After, the system translates knowledge structure into a SKOS and RDFS.

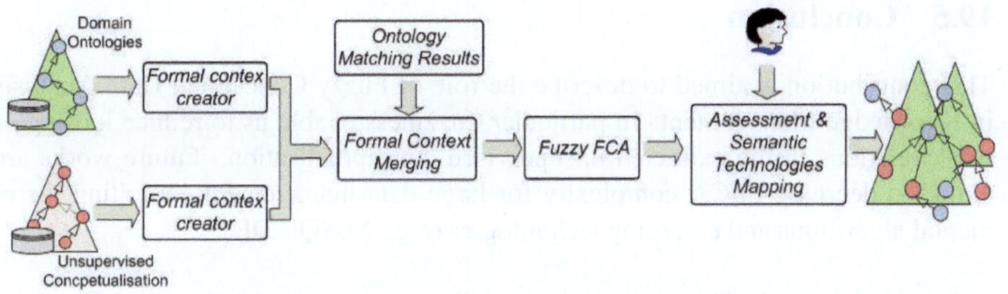

Fig. 19.2. Workflow of the Methdology for Ontology Merging

19.4 Experimental Results

In order to evaluate, the workflow of defined methodologies has been applied on subset of human classified repository of *Open Directory Project (ODP)*. Specifically, about 700 items of ODP have been analysed and classes of ODP have been used as *Domain Ontologies*(i.e., gold taxonomies). Let us note that only a brief text description of items has been exploited in the analysis process. The performances have been measured in terms of the micro-averaging of recall and precision [12]. The experimental results are shown in Fig.19.3.

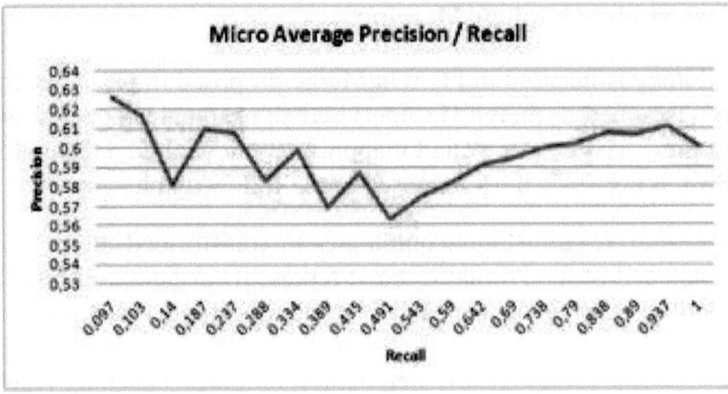

Fig. 19.3. Micro-averaging precision / recall results

From technological point of view, defined platform has been implemented exploiting existing software solutions, such as: Apache Solr, Apache Lucene, Wikipedia Miner, Sesame and OWLIM, and so on. Furthermore, services provided by the platform have been used in Ms Share Point 2010. Specifically, an existing connector framework (i.e., Manifold CF) has been instantiated in order to transparently acquire and up to date content indexes with data daily generated by the workers in Ms Share Point.

19.5 Conclusion

This contribution is aimed to describe the role of Fuzzy Conceptual Data Analysis in Knowledge Management. In particular, fuzziness enable us to reduce loosing of weak relations in the extracted unsupervised conceptualisation. Future works are aimed to decrease FFCA complexity for large data management exploiting incremental algorithms and emerging technologies (e.g., NoSQL DB).

Acknowledgement. This research is partially supported by the European Commission under the Collaborative Project ARISTOTELE "Personalised Learning & Collaborative Working Environments Fostering Social Creativity and Innovations Inside the Organisations", the Seventh Framework Programme (FP7), Theme ICT-2009.4.2 (Technology-Enhanced Learning), Grant Agreement n. 257886.

References

1. Bezdek, J.C.: Pattern Recognition and Fuzzy Objective Function Algorithms. Plenum Press, New York (1981)
2. Ruspini, E.: A new approach to clustering. Information and Control 15, 22–32 (1969)
3. Wang, Y., Guo, J., Hu, T., Wang, J.: An Ontology-Based Framework for Building Adaptable Knowledge Management Systems. In: Zhang, Z., Siekmann, J.H. (eds.) KSEM 2007. LNCS (LNAI), vol. 4798, pp. 655–660. Springer, Heidelberg (2007)

4. Buranarach, M., Supnithi, T., Chalortham, N., Khunthong, V., Varasai, P., Kawtrakul, A.: A Semantic Web Framework to Support Knowledge Management in Chronic Disease Healthcare. In: Sartori, F., Sicilia, M.Á., Manouselis, N. (eds.) MTSR 2009. CCIS, vol. 46, pp. 164–170. Springer, Heidelberg (2009)
5. Gerd, S., Maedche, A.: FCA-Merge: Bottom-Up Merging of Ontologies. In: IJCAI 2001 - Proceedings of the 17th International Joint Conference on Artificial Intelligent, Seattle, USA (2001)
6. Tao, G.: Using Formal Concept Analysis for Ontology Structuring and Building. ICIS, Nanyang Technological University (2003)
7. Zadeh, L.A.: Fuzzy Sets. Information and Control 8, 338–353 (1965)
8. Ganter, B., Wille, R.: Formal Concept Analysis: Mathematical Foundations. Springer, Heidelberg (1999)
9. De Maio, C., Fenza, G., Loia, V., Senatore, S.: Hierarchical Web Resources Retrieval by Exploiting Fuzzy Formal Concept Analysis. Information Processing & Management 48(3), 399–418 (2012)
10. Pollandt, S.: Fuzzy-Begriffe: Formale Begriffsanalyze unscharfer Daten. Springer, Heidelberg (1996)
11. Li, G.-Y., Liu, S.-P., Yan, Z.: Formal Concept Analysis Based Ontology Merging Method. In: 3rd IEEE International Conference on Computer Science and Information Technology (ICCSIT), July 9-11, vol. 8, pp. 279–282 (2010)
12. Van Rijsbergen, C.J.: "Keith" Information Retrieval, Butterworth, London (1979)
13. Godin, R., Saunders, E., Jecsei, J.: Lattice Model of Browsable Data Spaces. Journal of Information Sciences 40, 89–116 (1986)
14. Carpineto, C., Romano, G.: Concept Data Analysis: Theory and Applications. John Wiley & Sons, Ltd., Atrium Southern Gate (2004)

20

Memories of a Meeting with Professor Zadeh and His Wife Fay

Ashok Deshpande

20.1 Prologue

In the late 70's, I started working for doctoral degree on the topic: *Pumping System Reliability* which involves stochastic modeling and differential equations. What was the outcome? A Good outcome: I was awarded a PhD degree, but a Not-So-Good Outcome was – the research had litle/ no practical relevance. In those days, I was Deputy Director in Environmental Engineering Research Institute (NEERI)/CSIR India, and was asked by my engineer colleagues about its utility. Writing complex equations is good, but as an engineer, should I write complex equations all through my life? Does my brain works this way while taking real life decisions? – I was thinking almost daily about this and was getting disturbed. Is there no one who thinks this way?

My friend Dr. Chadra Mohan, Professor of Mathematics, University of Roorkee, India asked me to read a paper by Professor Lotfi Zadeh's on "Fuzzy sets" – the concept which was unknown to me and many in India in those days. Initially, I did not like it, as it was simple to understand. After deep thinking, it realized that the paper was in a new direction and was a mile stone in non probabilistic decision making. To summarize: one can say that humans take decisions with their perceptions based tacit knowledge and express them in linguistic terms. The concept itself was so fascinating that, as a research engineer, I considered to test the new idea in a real life situation and discussed with my friends. The concept of fuzzy sets and perception based modeling was considered virtually ridiculous so that very few wanted to listen, though they were comfortable with my research work based on two valued logic. Realising that their opinion would not be the final word in science, inquisitiveness about fuzzy sets and fuzzy logic continued.

20.2 Testing of the Concept of Fuzzy Sets and Fuzzy Logic

The idea of fuzzy sets was tested for the estimation of per capita water consumption based on consumer's perception. To my surprise, fuzzy operations worked very well. As a researcher, I felt it was necessary to carry out some more tests on this new fuzzy formalism! In the early 90's, a need based research proposal on Fuzzy Logic

R. Seising et al. (Eds.): *On Fuzziness: Volume 1*, STUDFUZZ 298, pp. 129–132.
DOI: 10.1007/978-3-642-35641-4_20

application in river water quality classification was presented at a research meeting in the Ministry of Environment and Forest (MoEF) of the Government of India in New Delhi. It was appreciated by Professor M. S. Swaminathan, an eminent Indian scientist who was then in the advisory board of the Prime Minister of India. Many top ranking officials did not like the presentation as it was on a non traditional approach of defining river water quality straightway in linguistic terms with a degree of certainty attached to each linguistic description of the river. No problem. The IIZUKA 92 conference was a special event in my life. I was amongst many to listen to Professor Zadeh's thought provoking speech and was keen to meet with him in person.

20.3 My Association with Professor Lotfi Zadeh and Mom Fay

In 1999, while in the bay area with my son, I sought an appointment with Professor Zadeh and the Professor immediately agreed. Though he was to go to China for a conference, he discussed with me on the application of fuzzy logic in environment management systems. I was then asked to make a presentation at the BISC after his China visit. On listening to my talk, Professor Zadeh was pleased and said that he enjoyed my describing river water quality using fuzzy logic, which was against the spirit of estimating the water quality index. While delivering a rather emotional presentation, at least three times, the words were "according to Professor Zadeh's fuzzy logic concepts. . . " He was carefully listening to me. After the seminar talk, he asked me to be in his Soda Hall office, the next day. I was afraid/ worried and thought that he might be angry with me, and perhaps scold me - may not be in the presence of any one. Once again I checked the slides and confirmed that there was nothing wrong. I could not sleep the whole night on that day! I am too small a person to meet the living legend. In our meeting the next day morning, Professor Zadeh asked me to steer the activities of a new Special Interest Group (SIG) of the BISC on Environmental Management Systems (EMS). I was overwhelmed and touched his feet, and felt blessed for my passion for fuzzy logic applications. Professor Zadeh is not only a scholar with exceptional brilliance but a great human being with excellent human qualities. Though I am not a first grade researcher in fuzzy logic, Professor Zadeh has still appreciation for my skills in implementing the concept to variety of systems: environ- informatics, medical informatics, chemo-informatics and to policy issues. Words are insufficient to narrate about his path breaking research concepts and extraordinary talent coupled with humility and affection for all. He and Mom Fay are made for each other.

Propagating Fuzzy logic via Computing with Words, world over through organising training workshops in the world, has been my passion. In this process, I learn and will continue to learn from the excellent papers written by many scholars in this important facet of soft computing.

Let me narrate a few incidents in Berkeley:

Incidence 1

Professor T. Y. Lin was the General Chair. Professor Zadeh was to Chair the discussion in Prof. T.Y. Lin's GrC2010 at SJSU San Jose. I was one of the panelists of the panel on Uncertainty. Professor Zadeh and I were scheduled to be at the venue on August 09, 2010 at 9 am. To reach from Berkeley to SJSU is not very easy. Therefore, I suggested Professor Zadeh that my son, Nikhil, would take us to San Jose. I told him that in our culture we cannot think of an elderly person travelling in a train, bus, taxi and so on. Somehow he was hesitant to accept the request but after much of persuading, he agreed.

We reached Professor Zadeh's home in Berkeley around 7:15 am. He was waiting not exactly for us - but for some one from 911, as he developed an unknown problem with the instrument (may be a pace maker). Someone from 911 (pretending as a medical doctor) arrived and asked a few questions to Professor Zadeh and concluded that it was beyond his knowledgebase; he told us to contact an adequate hospital. We all were virtually shocked. Mom Fay did not know what to do! I told her that we would take care of the great human being - Professor Lotfi Zadeh. If needed, we would take him to the hospital till the father of fuzzy logic (I consider him as my father) is brought back safe. Mom Fay was relieved after that assurance.

On seeing Professor Zadeh's pictures, the 911 people said: Oh! Is he the same person who writes on fuzzy logic? Nikhil and me looked at each other and were surprised to see their ignorance. Professor Zadeh was calm, as usual and did not say anything to them.

Nikhil then, talked with the doctor and was told that Professor Zadeh could attend the meeting and the problem was not so serious. Mom finally agreed to send Professor Zadeh to the conference when I promised her that it was our duty to bring our beloved Professor Lotfi Zadeh back home. She was happy to see our way of respecting him. We attended the panel discussion, and then Nikhil drove Professor Lotfi back home. Mom Fay was anxiously waiting for him.

Incidence 2

I had never seen Professor Zadeh annoyed on any one in my several interactions with him at the BISC office or elsewhere. If some young researcher started teaching two valued logic based probability theory he said politely in a low key "Dear, I have studied and taught this subject 60 years back."

Incidence 3

In July 2010, Dr. Vidyottama Jain (VJ) – a mathematician was unfortunately stuck and could not make headway in her research as BISC post doctoral fellow. Professor Zadeh told VJ - Let Ashok come over, he will help you. I was to be in BISC after a month. VJ had seen my profile and was happy to see me. I never met her before. On arrival, VJ requested me to assist on the topic which I had never studied in the past but she told that Professor Zadeh asked her to seek my guidance. I was happy, surprised and a little worried. VJ and I studied together and tried application of Prospect Theory (PT) and Computing with Words (CW) on India's Energy Options.

The presentation brought out the complexities inherent in PT and suggested the use of one of the facets of CW.

20.4 Impact of CW Methodology

On my return to India, the energy option issue work was discussed on phone with Mr. Prithviraj Chavan- UC Berkeley graduate in Electronics, the then Minister of State in Prime Minister's Office (presently Chief Minister of Maharashtra). He gave me a patient hearing and was in full agreement with this need based decision research. We believe that the application of CW methodology will go a long way in resolving human centric real life problems. I informed Professor Zadeh about the discussion and he was very happy.

Fig. 20.1. A. Despande with Lotfi and Fay Zadeh in Berkeley in their residence, August 2010

20.5 Epilogue

After reading the write up, if someone says: *Ashok, you are emotional.* Yes, I am and there is no secret about it. I will continue to seek guidance from Professor Lotfi – my intellectual father and affectionate love from Mom Fay for many more years.

21

Making Large Information Sources Better Accessible Using Fuzzy Set Theory

Guy De Tré

21.1 Introduction

Nowadays our society still witnesses an ever growing amount of digital information sources that are made publicly accessible via the internet. Along with the availability of the huge quantity of data comes the need for query engines and tools to efficiently explore and access these data and provide users with the facilities to retrieve exactly what they are looking for. As users most efficiently express their retrieval preferences using natural language and as matching in information retrieval and query processing in such cases often becomes a matter of degree or in some cases even a matter of uncertainty, fuzzy set theory and its related possibility theory offer an excellent mathematical basis for the development of advanced data access methods. These observations are the rationales behind the research of the Database, Document and Content Management group at Ghent University. In what follows we briefly describe the evolution of our research in the recent twenty years. Herewith we also try to give insight in the developments in fuzzy set theory and information management that formed the inspiration for this evolution. Furthermore, we present our vision on some trends for future developments.

21.2 Early Personal Research Experiences

The first time I came in contact with fuzzy set theory was at the beginning of the 90's during a math course taught by Etienne Kerre. Etienne has just finished his book on 'Introduction to the Basic Principles of Fuzzy Set Theory and some of its Applications' and was talking with such an enthusiasm about its topics and their potential applications that without any doubts this theory must have been something really beautiful. At that time I could not even imagine how fuzzy sets would have an impact on my future work and life.

I was lucky to start my professional career as knowledge engineer in a small spinoff company of the Artificial Intelligence lab of the Vrije Universiteit Brussel. Here I had to study the problem of efficient time modelling for complex train schedules. After two years I received an opportunity from Rita De Caluwe to join the Computer Science Laboratory and start PhD studies at Ghent University. This

R. Seising et al. (Eds.): *On Fuzziness: Volume 1*, STUDFUZZ 298, pp. 133–137.
DOI: 10.1007/978-3-642-35641-4_21

was the start of my academic career and at the same time for me a re-initiation to fuzzy set theory. My first research topic was the definition of a flexible, fuzzy time model able to handle complex time indications, as often encountered in natural language expressions, and the incorporation of this time model in a temporal database framework.

My first scientific research results were presented at the EUFIT'97 conference in Aachen [3] where I also attended the plenary talk by Lotfi Zadeh on the usefulness of generalized constraints. I was impressed by Lotfi's talk, but even more impressed by his openness and willingness to briefly discuss some of my questions and comments during the coffee breaks. The concept of a generalized constraint started intriguing me and was later on the basis of my PhD work. In this work I studied the use of generalized constraints in fuzzy object-oriented database modelling [4]. Hereby, generalized constraints, as proposed by Zadeh, were used for semantic data integrity modelling purposes, as well as for query formulation purposes. Uncertainty in the stored data was modelled using possibility distributions as initially proposed in [10], whereas for the evaluation of generalized constraints, an multiple-valued possibilistic logic, based on possibilistic truth values [9], but extended to explicitly cope with missing information has been developed [5]. I obtained my PhD, entitled 'A formal generalized object oriented database model, appropriate for the exploitation of crisp and non-crisp information', in applied sciences at Ghent University in June 2000.

After obtaining my PhD, I continued specialising myself in fuzziness and soft computing in database management and information retrieval. Hereby, investigating among others, the application of generalized constraints for the modelling of fuzzy and uncertain spatio-temporal information in databases [6], the use of level-2 fuzzy sets for dealing with concurrent, orthogonal occurrences of fuzziness and uncertainty in fuzzy database modelling [7], and the handling of null values in fuzzy databases [8].

21.3 Research at the DDCM Lab

In 2004, I obtained a tenured professor position at the Faculty of Engineering and Architecture of Ghent University with research area 'fuzzy information processing'. In that year we also established the Database, Document and Content Management (DDCM) research group by restructuring and renaming the former Computer Science Laboratory, now explicitly focussing its research, education and service activities on the handling and management of (imperfect) information.

The research mission of the DDCM group is to search for new soft computing techniques allowing to make large, heterogeneous data collections better accessible. The rationale behind this mission is in essence to find solutions for the demand of our society to handle the ever growing amount of digital information sources more efficiently. Additionally, the envisioned research offers better potentials for industrial and practical applications than pure fuzzy database modelling research offers, what is an important consideration in Ghent University's engineering faculty. The main research topics of the group include:

- **Coreference detection.** An important issue in database querying and information retrieval encompasses the task to find out whether two pieces of information refer to the same real world entity or not. If this is the case, we call the pieces coreferent. Beside being important for data access purposes, coreference detection is also useful to help guaranteeing data quality. Our group studies coreference detection of atomic data, collections, structured data, multimedia, texts and more general unstructured data. Among the applications under development is the ear identification application which is established in close cooperation with the Medical Imaging Center of the KULeuven and the Disaster Victim Identification team of the Belgian Federal Police. Coreference detection can be done both on the data (e.g., database) and on the metadata (e.g., database schema).

- **Information fusion.** Once detected, coreferent data can be further processed. For example, in database context, storage of coreferent data should be avoided as this would imply the storage of duplicated and often inconsistent information. As a solution the coreferent data can be merged or fused. The fusion of coreferent data is being studied by the group. Special research focus goes to the fusion of texts in the context of multiple document summarization.

- **Spatio-temporal information modelling.** Spatial and temporal data are generally recognised as special characteristics of information which deserve special care in database and information system contexts. Indeed, a lot of facts are registered in a database in a given spatial and or temporal context. Our research studies the handling of imperfect (imprecise, uncertain, incomplete) spatio-temporal information. For the handling of imperfect temporal information we closely cooperate with Olga Pons of the University of Granada. An application dealing with imperfect time indications in a database of medieval charters is under development. For the spatial data processing we cooperate with Nico Van de Weghe of the Geography department of Ghent University and with Jörg Verstraete who is currently employed at the Systems Research Institute of the Polish Academy of Sciences. Current research includes the efficient handling and analysis of moving objects, a research topic where also Bernard De Baets of Ghent University is involved in.

- **Bipolar information handling.** The DDCM research group also has an excellent cooperation with Sławomir Zadrożny and Janusz Kacprzyk of the Systems Research Institute of the Polish Academy of Sciences. This joint research can be best categorised under fuzzy querying and fuzzy databases and comprises the handling of bipolarity in both database querying and database modelling. Bipolarity hereby refers to the fact that information, as communicated by humans, often has positive and negative components which do not necessarily have to complement each other. Extensions of fuzzy set theory, based on interval-valued fuzzy sets, Atanassov's intuitionistic fuzzy sets or twofold fuzzy sets form an excellent basis for further research on bipolarity.

- **Decision support.** Multiple criteria decision support systems have many things in common with flexible querying systems. In both systems, the user has to specify criteria which have to be evaluated (for each case under consideration, resp. for each relevant database record) and in both systems evaluation results have to be aggregated to an overall degree of suitability or degree of satisfaction. The DDCM research group closely collaborates with Jozo Dujmović of San Francisco State University to study the applicability of the general logical scoring of preference (LSP) method for geographical suitability map construction and more recently for efficiently dealing with the opinions of multiple decision makers. Another research aspect concerns the further improvement of fuzzy querying techniques with LSP facilities.

Fig. 21.1. From left to right: Jozo Dujmović, Lotfi Zadeh and Guy De Tré at the WConSC'11 conference diner, San Francisco, 2011

21.4 Some Future Trends

There is currently a high demand from industry to manage the tremendous amount of unstructured data like texts as easy as structured database data. This implies that there is a need for semantic richer text interpretation and text analysis algorithms. For that reason, we foresee in the near future a growing importance of semantic rich text parsing mechanisms which allow to extract essential information and context from texts. Such mechanisms would also allow for smarter indexing and information retrieval techniques and will hopefully bring us a step closer to automatic ontology generation. Protoforms, as proposed by Lotfi Zadeh, could play an important role in such developments. Beside textual information, the content-based retrieval of multimedia documents like photographs, audio and video is in our opinion another challenge for future research.

By considering texts as sequences of words which on their turn are sequences of characters it is worth to investigate whether symbol sequences obtained from the annotation of sensor data or biomedical data (DNA, proteins, peptides, etc.) could be meaningfully processed as texts as described above. This, in order to obtain a semantic richer interpretation of these data.

Further research in the above mentioned areas is required and planned by the DDCM research group. The future will learn us whether this research could bring us a little bit closer to tools for the efficient and full exploration of the huge, ever growing quantity of data that is and still becomes available through the internet and the information systems of organisations, companies, societies, etc. We at least are enthusiastic and well motivated to tackle these challenges.

References

1. Buckles, B.P., Petry, F.E.: A Fuzzy Representation of Data for Relational Databases. Fuzzy Sets and Systems 7, 213–226 (1982)
2. Cavallo, R., Pittarelli, M.: The Theory of Probabilistic Databases. In: Stocker, P.M., Kent, W., Hammersley, P. (eds.) Proceedings of 13th International Conference on Very Large Data Base, September 1-4, pp. 71–81. Morgan Kaufmann, Brighton (1987)
3. De Tré, G., De Caluwe, R., Van der Cruyssen, B.: Dealing with Time in Fuzzy and Uncertain Object-Oriented Database Models. In: Proceedings of the EUFIT 1997 Conference, Aachen, Germany, September 9-11, pp. 1157–1161 (1997)
4. De Tré, G., De Caluwe, R., Van der Cruyssen, B.: A Generalised Object-Oriented Database Model. In: Bordogna, G., Pasi, G. (eds.) Recent Issues on Fuzzy Databases. STUDFUZZ, pp. 155–182. Physica-Verlag, Springer, Heidelberg, Germany (2000)
5. De Tré, G.: Extended Possibilistic Truth Values. International Journal of Intelligent Systems 17(4), 427–446 (2002)
6. De Tré, G., De Caluwe, R., Hallez, A., Verstraete, J.: Modelling of Fuzzy and Uncertain Spatio-Temporal Information in Databases: A Constraint-based Approach. In: Bouchon-Meunier, B., Foulloy, L., Yager, R.R. (eds.) Intelligent Systems for Information Processing: From Representation to Applications, pp. 117–128. Elsevier Science B.V., Amsterdam (2003)
7. De Tré, G., De Caluwe, R.: Level-2 fuzzy sets and their usefulness in object-oriented database modelling. Fuzzy Sets and Systems 140(1), 29–49 (2003)
8. De Tré, G., de Caluwe, R., Prade, H.: Null Values Revisited in Prospect of Data Integration. In: Bouzeghoub, M., Goble, C.A., Kashyap, V., Spaccapietra, S. (eds.) ICSNW 2004. LNCS, vol. 3226, pp. 79–90. Springer, Heidelberg (2004)
9. Prade, H.: Possibility Sets, Fuzzy Sets and Their Relation to Lukasiewicz Logic. In: Proceedings of the 12th International Symposium on Multiple-Valued Logic, Los Angeles, USA, pp. 223–227 (1982)
10. Prade, H., Testemale, C.: Generalizing Database Relational Algebra for the Treatment of Incomplete or Uncertain Information and Vague Queries. Information Sciences 34, 115–143 (1984)

22

Fuzzy Transform for Coding/Decoding Images: A Short Description of Methods and Techniques

Ferdinando Di Martino, Vincenzo Loia, Irina Perfilieva, and Salvatore Sessa

Abstract. Fuzzy transform (for short, F-Transform) is a new fuzzy image compression method that improves the quality of the images coded/decoded using fuzzy relations and provides results comparable with the ones obtained using the well known JPEG method for any compression rate. In this paper the F-transform method is described and its employability is discussed by reporting test results.

22.1 Introduction

Since 1965 when Lofti Zadeh proposed the theory of fuzzy sets [13] to formalize imprecision and vagueness mathematically, fuzzy logic has been applied in numerous fields to solve problems with the help of methods of approximate reasoning. Substantially, these methods are based on a grained view of a real system whose description is formalized by fuzzy logic tools. In image analysis, we apply data compression algorithms to digital images in order to transmit image data more efficiently. In applications (where a minor loss of fidelity is acceptable), a loss of fine details of an image for the sake of a storage space and fast transmission is appreciated. The known lossy image compression algorithms are: fractal [6] and wavelet [12] The most widely used image compression method is the JPEG algorithm [11], which is adopted for many image file formats.

Fuzzy relation calculus is also used in image processing because a normalized image can be considered as a fuzzy relation. A known application is the (lossy) Image Compression Fuzzy method (for short, ICF) – a method based on fuzzy relation equations [1], [2], [7],[8]. Comparison of qualities of reconstructed images coded/decoded with the help of ICF and JPEG shows the advantage of the latter method which increases exponentially with an increase of the compression rate. With the objective to improve the performances of the ICF technique, a new (lossy) image compression method has been developed on the basis of the concept of fuzzy (F-) transform [9,10]. In [3],[4], the authors have shown that this method gives better results in comparison with the ICF and comparable results with respect to standard JPEG image compression method, even if one increases the compression rate. This result has been confirmed in the case of color images provided that the YUV space is used instead of the RGB [5]. In Section 22.3, we describe briefly the F-transform method. In Section 22.4, the comparison results are discussed.

R. Seising et al. (Eds.): *On Fuzziness: Volume 1*, STUDFUZZ 298, pp. 139–146.
DOI: 10.1007/978-3-642-35641-4_22

22.2 F-Transform Image Compression Method

Let $[a,b]$ be a closed interval, $n \geq 2$ and $x_1, x_2, \ldots, x_n$ be points of $[a,b]$, called *nodes*, such that $x_1 = a < x_2 < \ldots < x_n = b$. We say that an assigned family of fuzzy sets $A_1, \ldots, A_n : [a,b] \to [0,1]$ is a *fuzzy partition* of $[a,b]$ if the following conditions hold:

1. $A_i(x_i) = 1$ for every $i = 1, 2, \ldots, n$;
2. $A_i(x) = 0$ if $x \notin (x_{i-1}, x_{i+1})$, where we assume $x_0, x_1 = a$ and $x_n, x_{n+1} = b$;
3. $A_i(x)$ is a continuous function on $[a,b]$;
4. $A_i(x)$ strictly increases on $[x_{i-1}, x_i]$ for $i = 2, \ldots, n$ and strictly decreases on $[x_i, x_{i+1}]$ for $i = 1, \ldots, n-1$;
5. $\forall x \in [a,b], \sum_{i=1}^{n} A_i(x) = 1$.

The fuzzy sets $A_1, \ldots, A_n$ are called *basic functions*. Moreover, we say that they form an *uniform fuzzy partition* if

6 $n \geq 3$ and $x_i = a + h \cdot (i-1)$, where $h = (b-a)/(n-1)$ and $i = 1, 2, \ldots, n$ (the nodes are equidistant);
7 $A_i(x_i - x) = A_i(x_i + x)$ for every $x \in [0,h]$ and $i = 2, \ldots, n-1$;
8 $A_{i+1}(x) = A_i(x-h)$ for every $x \in [x_i, x_{i+1}]$ and $i = 1, 2, \ldots, n-1$.

Let $A_1, \ldots, A_n$ be a fixed fuzzy partition of $[a,b]$ and f be a continuous function on $[a,b]$. In the discrete case, the function f is assumed to be defined at points $p_1, \ldots, p_m$ of $[a,b]$. We assume that the set $P = \{p_1, \ldots, p_m\}$ is *sufficiently dense with respect to the fixed partition*, i.e. for each $i = 1, \ldots, n$ there exists $j \in \{1, \ldots, m\}$ such that $A_i(p_j) > 0$. Then we say that the n-tuple $[F_1, \ldots, F_n]$ is the discrete *F-transform* of f with respect to $\{A_1, \ldots, A_n\}$, if each component F_i is given by

$$F_i = \frac{\sum_{j=1}^{m} f(p_j) A_i(p_j)}{\sum_{j=1}^{m} A_i(p_j)}, i = 1, \ldots, n. \tag{22.1}$$

The function $f_{F,n} : P \to [a,b]$ is called the *inverse discrete F-transform of f with respect to $A_1, \ldots, A_n$*:

$$f_{F,n}(p_j) = \sum_{i=1}^{n} F_i A_i(p_j). \tag{22.2}$$

By an appropriate choice of partition, the inverse F-transform can approximate the original function with an arbitrarily chosen precision.

We can easily extend the above concepts to functions in two variables. Assume that our universe of discourse is a rectangle $[a,b] \times [c,d]$, and let $n, m \geq 2$, $x_1, x_2, \ldots, x_n \in [a,b]$ and $y_1, y_2, \ldots, y_m \in [c,d]$ be assigned points, called *nodes*, such that $x_1 = a < x_2 < \ldots < x_n = b$ and $y_1 = c < y_2 < \ldots < y_m = d$. Furthermore, let $A_1, \ldots, A_n : [a,b] \to [0,1]$ and $B_1, \ldots, B_m : [c,d] \to [0,1]$ be respective fuzzy partitions and f be a continuous function on $[a,b] \times [c,d]$. In the discrete case, we assume that the function f is defined at points $(p_i, q_j) \in [a,b] \times [c,d]$, where $i = 1, \ldots, N$ and $j = 1, \ldots, M$. Moreover, the sets $P = \{p_1, \ldots, p_N\}$ and $Q = \{q_1, \ldots, q_M\}$ of these points are sufficiently dense with respect to the chosen partitions. In the discrete case, the matrix $[F_{kl}]$ is the discrete F-transform of f with respect to $\{A_1, \ldots, A_n\}$ and $\{B_1, \ldots, B_m\}$ with components

$$F_{kl} = \frac{\sum_{j=1}^{M} \sum_{i=1}^{N} f(p_i, q_j) A_k(p_i) B_l(q_j)}{\sum_{j=1}^{M} \sum_{i=1}^{N} A_k(p_i) B_l(q_j)}, k = 1, \ldots, n, l = 1, \ldots, m. \tag{22.3}$$

By extending 22.2 to the case of two variables, we define the inverse discrete F-transform of f with respect to $\{A_1, A_2, \ldots, A_n\}$ and $\{B_1, \ldots, B_m\}$ as follows:

$$f_{nm}^F(p_i, q_j) = \sum_{k=1}^{n} \sum_{l=1}^{m} F_{kl} A_k(p_i) B_l(q_j), i \in \{1, \ldots, N\}, j \in \{1, \ldots, M\}. \tag{22.4}$$

22.3 Coding/Decoding Images Using the F-Transforms

Let R be a gray image of the sizes $N \times M$, which we consider as a fuzzy relation $R : \{1, \ldots\ldots, N\} \times \{1, \ldots, M\} \to [0, 1]$. $R(i, j)$ is a normalized value of gray color intensity at pixel (i, j). We assume that fuzzy sets $\{A_1, \ldots, A_n\}$ and $\{B_1, \ldots, B_m\}$, with $n < N$ and $m < M$, form fuzzy partitions of the real intervals $[1, N]$ and $[1, M]$, respectively. The relation R is divided into *blocks* R_B, which are subrelations of $N(B) \times M(B)$ sizes. Each block is compressed by the discrete F-transform to a smaller block $[F_{kl}^B]$ of $n(B) \times m(B)$ sizes where $n(B) < N(B)$ and $m(B) < M(B)$) and

$$F_{kl}^B = \frac{\sum_{j=1}^{M(B)} \sum_{i=1}^{N(B)} R_B(i, j) A_k(i) B_l(j)}{\sum_{j=1}^{M(B)} \sum_{i=1}^{N(B)} A_k(i) B_l(j)}, k = 1, \ldots, n(B), l = 1, \ldots, m(B). \tag{22.5}$$

The compressed blocks are reconstructed by the inverse discrete F-transform to $R_{n(B)m(B)}^F$:

$$R_{n(B)m(B)}^F(i, j) = \sum_{k=1}^{n(B)} \sum_{l=1}^{m(B)} F_{kl}^B A_k(i) B_l(j), (i, j) \in \{1, \ldots, N(B)\} \times \{1, \ldots, M(B)\}. \tag{22.6}$$

A quality of the reconstructed image is evaluated by the Peak Signal to Noise Ratio (shortly, *PSNR*) given by

$$PSNR = 20 log_{10} \frac{255}{RMSE}, \tag{22.7}$$

where the value of $RMSE$ (Root Mean Square Error) is given by

$$RMSE = \sqrt{\frac{\sum_{i=1}^{N}\sum_{J=1}^{m}(R(i,j) - R_{NM}^{F}(i,j))^2}{N \times M}} \tag{22.8}$$

Note that R_{NM}^{F} in formula 22.8 represents the whole reconstructed image obtained from the blocks $R_{n(B)m(B)}^{F}(i,j)$.

In our experiments, we used color images in the YUV space. The reason is that the resolution of an image in the Y band (brightness) is more visible to a human eye than in the U or V bands (chrominance). Thus, the same image can be compressed in U and V bands with a higher rate than in the Y band. This trick is a trade-off between a desired quality of a reconstructed image and a compression rate. The theoretical explanation can be found in [10]. Below, we justify this by showing better results for images in the YUV space in comparison with the same images in the RGB space. Then both methods are compared with the classical JPEG method.

22.4 Test Results

Numerous tests were performed with images from the well-known databases. For the sake of completeness, we limit ourselves to two examples: the 256 × 256 gray image "Bridge" from the database Corel Gallery (Arizona Directory) and the 256 × 256 color image "Tree" extracted from the Image Database of the University of Southern California (http://sipi.usc.edu/database/).

In Table 22.1, the $PSNR$ values obtained for the image "Bridge" by using three compression methods: the F-transform, ICF and JPEG are presented with various

(a) (b)

Fig. 22.1. (a): "Bridge" (b): "Tree"

compression ratios, and the % of gain indexes. The latter are computed by using the following schemas:

$$\% \text{ Gain FTR over ICF} = \frac{(PSNR\ FTR) - (PSNR\ ICF)] \times 100}{(PSNR\ FTR)}$$

Table 22.1. *PSNR* and % gain for "Bridge"

$\rho(B)$	*PSNR* in FTR	*PSNR* in ICF	*PSNR* in JPEG	% Gain FTR over ICF	% Gain JPEG over FTR
0.035	20.7262	11.0283	22.6985	87.9364	9.5159
0.062	21.4833	14.2812	24.7253	50.4306	15.0907
0.140	23.2101	16.4632	28.1149	65.5220	21.1321
0.250	24.6975	19.7759	31.2148	24.8868	26.3885
0.444	27.0960	23.7349	37.2367	14.1610	37.4250

Figure 22.2 shows reconstructed images which are obtained after compression made by three methods: ICF, FTR and JPEG. Compression ratios are: ρ = 0.444 and ρ = 0.25.

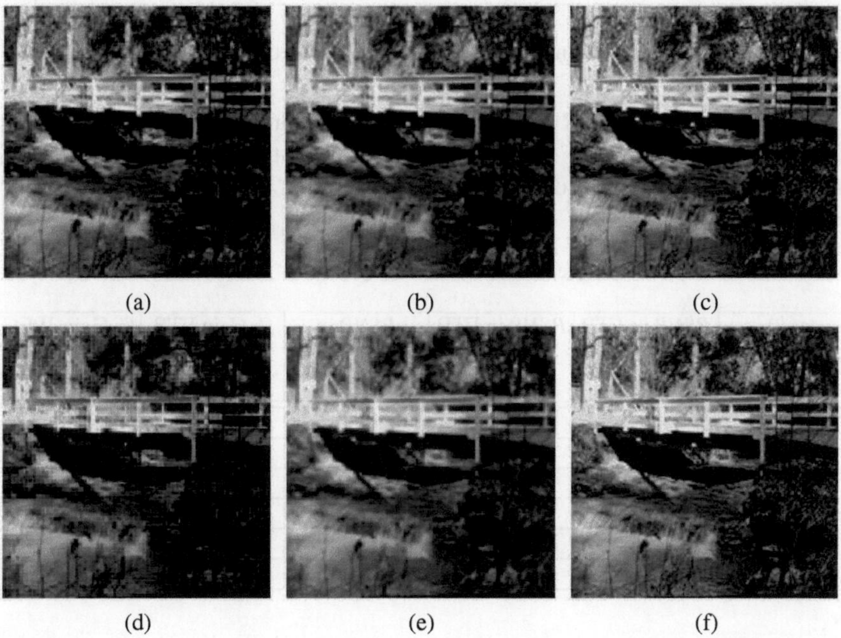

Fig. 22.2. (a): ICF, $\rho = 0.444$ (b): FTR, $\rho = 0.444$ (c): JPEG, $\rho = 0.444$ (d): FTR, $\rho = 0.444$ (e): FTR, $\rho = 0.25$ (f): JPEG, $\rho = 0.25$

In Figure 22.3 the trends of the *PSNR* index are shown for the image "Bridge" and three compression methods with various ratios.

The above given results demonstrate that the *PSNR* value of the F-Transform method is better than the one of the ICF method and is close to the *PSNR* obtained by using the JPEG method. The gain of FTR over ICF is more visible if a compression ratio is low (strong compressions).

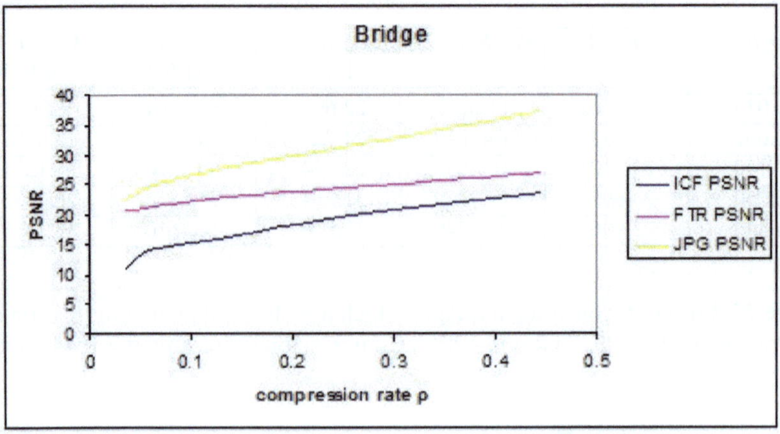

Fig. 22.3. *PSNR* in the FTR, ICF and, JPEG methods for "Bridge"

In Table 22.2, the *PSNR* values obtained for the image "Tree" by using three compression methods: the F-transform in YUB, the F-transform in RGB and JPEG are presented with various compression ratios, and the % of gain indexes.

Table 22.2. *PSNR* and % gain for "Tree"

$\rho(B)$	*PSNR* in FTR YUV	*PSNR* in FTR RGB	*PSNR* in JPEG	% Gain FTR YUV over FTR RGB	% Gain JPEG over FTR YUV
0.06375	24.2024	21.6114	29.329533	10.70555	21.1844
0.033508	21.6241	19.9084	27.169767	7.934203	25.645769
0.140625	27.0294	23.4003	32.9186	13.426491	21.788127
0.1979167	28.1302	25.432	34.9764	9.5918266	24.337545
0.2978317	29.7371	27.3001	35.650067	8.1951502	19.88414
0.4444444	32.676	28.6888	35.9372	12.202228	9.9804138

Figure 22.4 shows the decoded image obtained using the three methods for ρ = 0.444 and ρ = 0.197.

Figure 22.5 shows the trends of the *PSNR* index for three methods mentioned above.

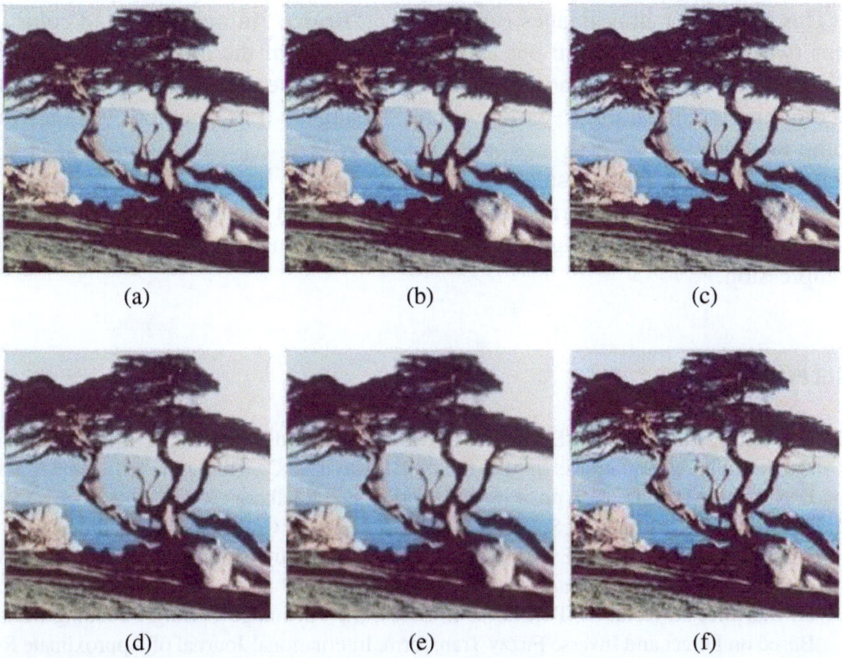

Fig. 22.4. (a): FTR YUV, $\rho = 0.444$ (b): FTR RGB, $\rho = 0..444$ (c): JPEG, $\rho = 0.444$ (d): FTR YUV, $\rho = 0.197$ (e): FTR RGB, $\rho = 0..197$ (f): JPEG, $\rho = 0.197$

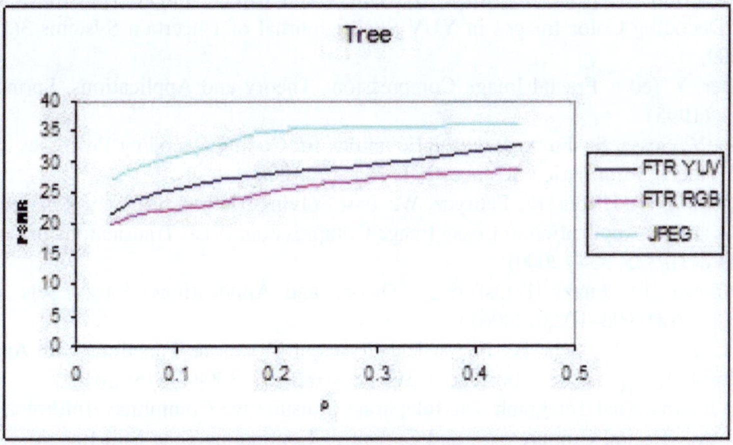

Fig. 22.5. *PSNR* in the FTR YUV, FTR RGB and JPEG methods for "Tree"

Figure 22.5 shows that for any compression ratio, the quality of the image compression by the F-transform method in the YUV space is better than the quality of the same method in the RGB space, and it is comparable to the JPEG method.

This is true for gray images (we used the "Bridge" in our tests) and color images (we used the "Tree" in our tests). The quality of the images coded/decoded using the F-Transform method is always better than the one obtained using the ICF method and is comparable to the one obtained using the JPEG method even for strong compression.

Encouraged by these results, a future research in image analysis will be performed using the F-transform method in multispectral high resolution image compression, image fusion, image information retrieval, image watermarking and video compression.

References

1. Di Martino, F., Loia, V., Sessa, S.: A Method for Coding/Decoding Images by Using Fuzzy Relation Equations. In: De Baets, B., Kaynak, O., Bilgiç, T. (eds.) IFSA 2003. LNCS (LNAI), vol. 2715, pp. 436–441. Springer, Heidelberg (2003)
2. Di Martino, F., Loia, V., Sessa, S.: A Method in the Compression/Decompression of Images by Using Fuzzy Equations and Fuzzy Similarities. In: Proceedings of the 10th IFSA World Congress, Istanbul, pp. 524–527 (2003)
3. Di Martino, F., Loia, V., Perfilieva, I., Sessa, S.: An Image Coding/Decoding Method Based on Direct and Inverse Fuzzy Transform. International Journal of Approximate Reasoning 48(1), 110–131 (2008)
4. Di Martino, F., Sessa, S.: Compression and Decompression Images with Discrete Fuzzy Transforms. Information Sciences 117(11), 2349–2362 (2007)
5. Di Martino, F., Loia, V., Sessa, S.: Direct and Inverse Fuzzy Transforms for Coding/Decoding Color Images in YUV Space. Journal of Uncertain Systems 3(1), 11–30 (2009)
6. Fisher, Y. (ed.): Fractal Image Compression: Theory and Applications. Springer, New York (1995)
7. Loia, V., Sessa, S.: Fuzzy Relation Equations for Coding/Decoding Processes of Images and Videos. Information Sciences 171, 145–172 (2005)
8. Nobuhara, H., Hirota, K., Pedrycz, W.: Fast Solving Method of Fuzzy Relational Equations and its Application to Lossy Image Compression. IEEE Transactions of Fuzzy Systems 8(3), 325–334 (2000)
9. Perfilieva, I.: Fuzzy Transforms: Theory and Applications. Fuzzy Sets and Systems 157(8), 993–1023 (2006)
10. Perfilieva, I., De Baets, B.: Fuzzy Transforms of Monotone Functions with Application to Image Compression. Information Sciences 180(17), 3304–3315 (2010)
11. The International Telegraph and Telephone Consultative Committee: Information Technology – Digital Compression and Coding of Continuous-Tone Still Images – Requirements and Guidelines, Recommendation T81 (1992), `http://www.w3.org/Graphics/JPEG/itu-t81.pdf`
12. Topiwala, P.N.: Wavelet Image and Video Compression. International Series in Engineering and Computer Science. Springer, New York (2004)
13. Zadeh, L.A.: Fuzzy Sets. Information and Control 8, 338–353 (1965)

23

From Aristotle to Lotfi

Patrik Eklund, M. Ángeles Galán, Robert Helgesson,
and Jari Kortelainen

At the IFSA Congress in Seattle 1989, Lotfi Zadeh spoke about "The Coming Age of Fuzzy Logic". Now decades after, can we say it has arrived? Is it already here? Was it there already at that time? Some say it was, referring back e.g. to Jan Łukasiewicz three-valued logic from the early 20th century. Some say it never came, and some ask the question: Will it ever?

Fuzzy sets quickly became fuzzy set theory, and relation matters drew much attention. Sets are easy to handle, and in particular standard sets. Structured sets like powersets are still easy, but moving e.g. to filters starts to reveal non-triviality. The simple identification of relations being morpisms in the Kleisli category of the powerset monad didn't have much impact, even if that opens up avenues to quite fascinating aspects of 'generalized relations'.

Fuzzy extension of traditional relational composition became the 'logic engine' for fuzzy control, and 'logic' on top of fuzzy set theory was well received by numerous researchers around the world. This eventually became 'fuzzy logic in the wider sense' and logicians started to investigate the roots of logical modelling of uncertainty. However, this never really went out of the realm of fuzzy sets, since the set of truth values was the only object of investigation. Joseph Goguen saw the set of truth values as a lattice, and the next decades of 'fuzzy logic in the narrow sense' were spent on looking at algebraic structures of that set of truth values. Everything was still according to the predicativist's view, and even worse, nobody did anything about the language structures. The 'fuzzy logic in the narrow sense' community soon adopted category theory for further deepening of theory. At this point there were some first interactions with computer science, and 'internalizing logic' by topos theoretic means was seen as a breakthrough. It was not a breakthrough for fuzzy language, but still just an extension to views on the set of truth values. In a topos this is the subobject classifier and predicativist's connectives are derived from categorical limits. A topos is a cartesian closed category so the categorical product underlies conjunction in that intuitionistic logic 'within' the topos. Further abstractions using enriched categories has been seen to be very thrilling, but all that still focuses only on sophistication and abstract manipulation of truth values.

Logic indeed remained manifested by predicates, and there was very little efforts to identify and investigate fuzzy as appearing in the language of logic. Hurdles are many, and a first step there is to realize that the fuzzy set of truth values resides on the semantic side. We could also say that fuzzy didn't have much impact on syntax.

R. Seising et al. (Eds.): *On Fuzziness: Volume 1*, STUDFUZZ 298, pp. 147–152.
DOI: 10.1007/978-3-642-35641-4_23

Looking at *Logic*, it is a structure containing *signatures, terms, sentences, theoremata (structured sets of sentences), entailments, algebras, satisfactions, axioms, theories and proof calculi*[1], so shouldn't we then say that *Fuzzy Logic*, again as a structure, contains *fuzzy signatures, fuzzy terms, fuzzy sentences, fuzzy theoremata, fuzzy entailments, fuzzy algebras, fuzzy satisfactions, fuzzy axioms, fuzzy theories and fuzzy proof calculi*, i.e., fuzzy "distributes" over the "operator" that glues substructures in logic into a whole.

Yes, of course, we should[2]. At this point, note how the signature and terms are ingredients for information (as "data"), where further inclusions of sentence and theoremata is more epistemic and are indeed ingredients for knowledge ("representation"). Entailment and inference rules are then finally needed in order to "compute" or "infer" with knowledge (sentences and theoremata), using data syntactically as terms and in substitutions, or semantically as terms as specified by assignments.

Lotfi Zadeh started fuzzy logic in 1965 by his paper on fuzzy sets. In 1962, Lotfi Zadeh wrote "... we need a radically different kind of mathematics, the mathematics of fuzzy or cloudy quantities which are not describable in terms of probability distributions ...". Mathematics is formalism, so this requires the use of ambiguous formalism for these "quantities". "Cloudy" is in our view modeled by monads and partially ordered monads. Yes, they are not "describable in terms of probability distributions", since probability theory is at most part of some "semantic logic". We would also suggest speaking not only about "quantity" in the sense of terms (syntactically) and term values (semantically), but also about sentence and how to calculate (logically) with sentences. Thus, we speak not only about physical quantity, but metaphysical quantity or quantity of knowledge.

---

[1] It is interesting to note how early 20th century logicians didn't all that much respect this order of producing building blocks for logic. Self-referentiality was allowed like appearing in Gödel's numbering. Gödel starts off at a very flexible view of sentences, and then doesn't close that door to sentences, when he moves over to theory and inference calculi. Gödel then uses proof sequences and indexing to produce 'new' sentences, and goes back through that door still left open to allow for adding these 'new' sentences. We should therefore be at least a bit careful when we aim at drawing conclusions from such consistency and completeness matters where "all doors are always wide open".

[2] We have produced a number of papers over the last decade in this area, but the purpose of this paper is not to refer to them explicitly. These papers deal with strict management of terms and sentences, and so on and so forth, using monads and partially order monads, where substitution is a key concept. Once a functor has been produced, the "door cannot be left open" to came back and modify that functor. Modifying some of or something in those functors changes the logic, so the possibility for self-referentiality is very much left out in the cold. We should also underline that we do not necessarily have a single logic covering reasoning within all applications and all human thinking including mental, social and intelligent behaviour. We must be allowed to use different logics, and the important property in these respects is that there are mappings between these logics so that knowledge, represented in a particular logic, can be carried over to be represented in another logic, understood by respective *logic stakeholder*. The categorical and monadic approach for such approaches to logic is critical in particular for these homomorphic transformations in the realm of what we call *substitution logic*.

It is easy to say, as is often done, that this is "quality", so that quantity and quality are distinguished. If we then speak about 'quantifiers' and 'qualifiers', we may touch upon intuition that tries to make a distinction between physical value and metaphysical value. Quantifiers bring reality into logic like qualifiers bring logic into reality. This is perhaps the reason why it is so difficult to *come to terms*[3] with quantifiers.

An important point entirely forgotten is the distinction between "computing with fuzzy" and "fuzzy computing". We can explain this distinction e.g. using relational composition, which has been successfully used e.g. in fuzzy control. Relational composition is not inference calculus in the *narrow sense* of fuzzy logic, but it can indeed be seen as 'logic' in the *wider sense*. From monadic point of view, relations are "morphisms in the Kleisli category of the powerset monad over the category of sets", and fuzzy relations are "morphisms in the Kleisli category of the fuzzy powerset monad over the category of sets". Very few seem to have considered the option "morphisms in the Kleisli category of the powerset monad over the Goguen category of fuzzy sets". We have, and there is indeed the distinction to be made between "computing with fuzzy" and "fuzzy computing", and really in that order, as the latter is related to uncertainty of the operator, i.e., at large related to fuzzy languages. Term monads over the Goguen category are here important, and term monads in general over selected categories opens up entirely new aspects concerning our view and management of variables. It may be that "computing with fuzzy" is where the larger engineering impact comes from, and doing subtleties with "fuzzy computing" leads to less impact. It may also be that the impact of "fuzzy computing" will be in entirely different application areas, such as involving social aspects and human behaviour.

Łukasiewicz didn't consider any of this, as he was looking only at the set of truth values. Notably, when we talk about the early 20th century, in our view we are talking all too little about early 20th century more broadly. It's mostly Łukasiewicz for 'fuzzy', and Tarski for 'modal'. In that Lwów-Warsaw school of logic established by Kazimierz Twardowski[4], the third musketeer, or maybe the first one, was Stanislaw Lesniewski. Lesniewski is mostly forgotten even if he did some interesting things on quantifiers. Tarski was Lesniewski's student.

[3] This "*come to terms*" actually has a very promising logical meaning. We have shown e.g. how the existential quantifier used in description logic is nothing but modeled by the powerset type constructor on the so called "superceding level" related to the original and background signature. The relation between quantifiers and type constructors must be explored in much more detail. A fascinating question is also whether or not everything is about terms and term manipulations, so that sentences in the end also are describable as terms in a broad sense, and thus *everything comes down to terms*.

[4] Twardowski was a student of Franz Brentano (1838-1917), who was working on his theory of intentionality, a concept used also by Jeremy Bentham (1748-1832) in his doctrine of consciousness. Bentham worked with utility closely e.g. with his secretary James Mill, and Mill's son, John Stuart, who eventually provided significant contributions to the theory of utilitarianism. Intentionality dates back even to Anselm of Canterbury (ca. 1033-1109) and the ontological argument in his *Proslogium* for the existence of God.

We allow ourselves here provide a remark that the traditional meaning of ontology, such as in web ontology, either completely neglects or is intentionally informal in particular about the sentential and inferential parts of logic. This view of ontology, in particular as seen in the case of a health terminology like SNOMED CT, is more like a mereology since the mereonomic type hierarchies in SNOMED CT still seek to find a proper inclusion of deductive elements, and therefore in some sense disqualifies to be called ontology. Another way of speaking is to say that nomenclature is not sufficient, since we need a *calculus of nomenclature*, not at all enabled by description logic and its derivatives. Lesniewski was interested in those directions in his *calculus of names* [1]. Lesniewski's work on protothetics, ontology and mereology, as nearby concepts and theories to set theory and logic, deserve more attention in the fuzzy logic community.

Haskell Curry acknowledged the work of Lesniewski in his *Foundations of Mathematical Logic* [2]. Speaking of Curry, the impact of Göttingen for fuzzy is also not very visible. Schönfinkel's combinators [3] were in progress before 1920, i.e., at the time of Hilbert's lectures 1917-1922, eventually leading to his and Ackermann's *Grundzüge* [4]. This was at that time a culmination, where Frege's *Begriffsschrift* [5] was the kick-off. Frege's logic was originally intended as logic only for mathematical reasoning, i.e., logic for mathematics. Later Hilbert pointed out the difficulties concerning natural numbers and logic, and the question about which comes first. There is obviously no metalanguage for this *fons et origo* first-order logic, which makes it absurd from mathematical point of view. Of course, Foundations of Mathematics is in debate, always was, always will be, but as long as self-referentially is avoided, we can live with it. Indeed, the process of finding difficulties within this logic, and set theory supported by it, there was at that time kind of a step-by-step incorporated rendering of these difficulties, and this rendering indeed continued for decades. Between Frege's *Begriffsschrift* and Hilbert's and Ackermann's *Grundzüge*, there is Peano, there is Russell, and there are many many others. Many things happen in the discussion on logic in that period. Moving to the 1930's, Curry had been a student of Hilbert and Ackermann in Göttingen just before Alonzo Church moves on to creating λ-calculus [6]. After the war, Tarski, Curry, Church, Kleene, and many many others, came together and worked together in the US, and it's therefore quite surprising that the impact of all those foundations of computing didn't reach out more than it did e.g. for the early years of fuzzy sets and logic. It still doesn't, and we think it should.

Having said all that, we still have quite a long way to Aristotle and syllogism. Some say Aristotle came up with much more than we can imagine, while others, like Szabó [7], say Aristotle actually didn't. The dialectics between Aristotle's writings and modern logics as we know it today is obviously not happening between Aristotle and modern logicians as we know them today, but this dialectic is apodeictically presented by Aristotle's fans. Szabo further rightfully points out that Aristotle didn't start logic, but Zeno and pre-Socratic time did. Logic started surely even before that, as humans were reasonable way before any pre-time imaginable. Uncertainty in human thinking was always there, and it was there in *terms and sentences*, it was there in *interpretations*, and it was there in *inference*. It was never just about sets of truth values.

In his Categoriae[5], Aristotle deals with perception in the sense of it ceasing to exist once the perceptible is annihilated. Aristotle also says that the annihilation of perception does not cancel the existence of the perceptible, and makes the distinction between a body perceived and a body in which perception takes place. This is the same as making the distinction between the observer (operator in the underlying signature) and the observation (syntactic term or semantic value of term), as the observer is the recorder of the perception, and the observation is the recording of the perception. Aristotle does not speak about such recordings, as he does not consider the third party, namely the computer (database, in a broader sense) into which the perception is registered and stored for later retrieval.

Plato's dialogues are unfortunately not seen as logical. They mostly underpin theater and drama, sociology and theology. Dialogues are at most seen as philosophical, and indeed the philosophy of dialogue is evident e.g. in Martin Buber's *Ich und Du* (1923), where Buber makes a distinction between "I" and "It" (Ich und Es) and "I" and "Thou" (Ich und Du), the former being more of the intralogical, and the mostly missing analytics about the relation between "it in me" and "it in you", corresponding to the interlogical adventure formalized by morphisms between logics in substitution logic. Note indeed that these morphisms e.g. map observers to observers and observations to observations.

Aristotle does not reach out to such logic morphisms, neither in *Analytica Priora*, nor in *Analytica Posteriora*, the former covering syllogism and induction, the latter speaking of knowledge that is 'prior' and 'better known'. In this context, Aristotle also speaks about these things being "closer to sense", like a closer or stronger "I-It" relation. However, and as related to his logical thinking, Aristotle never clearly goes into discussions on the logical nature of communication, not even in his ethical or aesthetical writings.

We can also but only speculate on how Plato and Aristotle may have covered this without having provided such academic discussions into print, and obviously not having the necessary formal and mathematical apparatus to do that. Aristotle's discussions with Alexander the Great may also have touched upon these questions in dealing with strategy and negotiation in one form or another, but clear imprints do not exist to support historical development of logic morphisms, as compared to the path from Aristotle's syllogisms and predications to Frege's logic and his *Bedeutung*.

Here is clearly an area that can be investigated more thoroughly, namely, finding the roots of logic more broadly in writings including politics and art, passing through historical milestones of all kind, such as the French revolution, e.g. including logical analysis of what Marquis de Condorcet really wrote. There are so many important historical periods involving active academic and logic writing, which open up discussions on still mostly non-existent areas like the logic of political science, logic of care and its related ethos, or the logic of music, theatre and drama. It is almost

[5] Part of Aristotle's Organon.

tempting to extend the word *dialogic*[6] for the substitution logic we are presently developing. Substitution logic is substitution based on object level in the category of substitution logics, but more dialogic on morphism level.

References

1. Lesniewski, S.: Grundzüge eines Neuen Systems der Grundlagen der Mathematik. Fundamenta Mathematicae 14, 1–81 (1929)
2. Curry, H.B.: Foundations of Mathematical Logic. McGraw-Hill, New York (1963)
3. Schönfinkel, M.: Über die Bausteine der mathematischen Logik. Mathematische Annalen 92, 305–316 (1924)
4. Hilbert, D., Ackermann, W.: Grundzüge der theoretischen Logik. Springer, Göttingen (1928)
5. Frege, G.: Begriffsschrift: eine der arithmetischen nachgebildete Formelsprache des reinen Denkens. Louis Nebert, Halle (1879)
6. Church, A.: A formulation of the simple type theory. The Journal of Symbolic Logic 5, 56–68 (1940)
7. Szabó, Á.: The Beginnings of Greek Mathematics. Reidel, Dordrecht (1978)

[6] Dialogic or dialogism has also been used by Mikhail Bakhtin (1895-1975) in his literary theory. Bakhtin goes further than Buber, as Bakhtin includes the *Other for Me* aspect, i.e., an opposite arrow for the logic morphism representing his *I for the Other.* This comes already quite close to the intuition behind morphisms in the category of substitution logic. Also Peirce's semiotic logic is intralogical only, so again we can say that our discussion clearly invites to open up a study e.g. of interlogical semiotics.

24

Fuzzy Set-Based Approximate Reasoning and Mathematical Fuzzy Logic

Francesc Esteva and Lluís Godo

24.1 Introduction

Zadeh proposed and developed the theory of approximate reasoning in a long series of papers in the 1970's (see e.g. [28–32, 34, 35]), at the same time when he introduced possibility theory [33] as a new approach to uncertainty modeling. His original approach is based on a fuzzy set-based representation of the contents of factual statements (expressing elastic restrictions on the possible values of some parameters) and of if-then rules relating such fuzzy statements. Zadeh himself wrote in [37] that *fuzzy logic in narrow sense*

> "(...) is a logical system which aims a formalization of approximate reasoning. In this sense it is an extension of many-valued logic. However the agenda of fuzzy logic (FL) is quite different from that of traditional many-valued logic. Such key concepts in FL as the concept of linguistic variable, fuzzy if-then rule, fuzzy quantification and defuzzification, truth qualification, the extension principle, the compositional rule of inference and interpolative reasoning, among others, are not addressed in traditional systems."

Thus, according to Zadeh, fuzzy logic is something more than a system of many-valued logic, in particular it clearly departs at first glance from the standard view of (many-valued) logic where inference does not depend on the contents of propositions.

On the other hand, the study of the so-called *t-norm based fuzzy logics* corresponding to formal many-valued calculi with truth-values in the real unit interval $[0,1]$ defined by a conjunction and an implication interpreted respectively by a (left-) continuous t-norm and its residuum[1], has had since the mid nineties a great development from many points of view (logical, algebraic, proof-theoretical, functional representation, and computational complexity). This has been witnessed by a number of important monographs that have appeared in the literature since then, see e.g. [17, 18, 25], and has very recently culminated with the handbook [3]. It is worth noticing that, although formal, all these systems originated as an attempt to provide

[1] Thus, including e.g. the well-known Łukasiewicz and Gödel infinitely-valued logics, defined much before fuzzy logic was born, and corresponding to the calculi defined by Łukasiewicz and min t-norms respectively.

R. Seising et al. (Eds.): *On Fuzziness: Volume 1*, STUDFUZZ 298, pp. 153–163.
DOI: 10.1007/978-3-642-35641-4_24

sound logical foundations for fuzzy set theory as well as to address computational problems related to vagueness and imprecision. Indeed, Hájek, in the introduction of his celebrated monograph [18] makes the following comment to Zadeh's quotation:

> "Even if I agree with Zadeh's distinction (...) I consider formal calculi of many-valued logic to be the kernel of fuzzy logic in the narrow sense and the task of explaining things Zadeh mentions by means of this calculi to be a very promising task."

Following this line of thought, a main part of our research efforts in the last years have been devoted to the study and definition of different t-norm based fuzzy logic systems (see e.g. [1, 2, 7–13]), but having in mind that a main task was to address as much as possible the different aspects of the agenda of the fuzzy logic in narrow sense not in principle directly covered by them, e.g. the approximate reasoning machinery (flexible constraints propagation, generalized modus ponens, compositional rule of inference, etc.).

In this short note, as our modest homage to Prof. Lotfi Zadeh and his great contributions, we revisit an approach (c.f. [4, 16, 18]) to understand some of Zadeh's approximate reasoning principles as sound deductions within a formal system of mathematical fuzzy logic, so trying to bridge the gap between both areas.

24.2 Propositional and Predicate T-Norm Based Fuzzy Logics

T-norm based (propositional) logics correspond to logical calculi with the real interval $[0,1]$ as set of truth-values and defined by a conjunction & and an implication $\rightarrow$ interpreted respectively by a left-continuous t-norm $*$ and its residuum $\Rightarrow$, and where negation is defined as $\neg\varphi = \varphi \rightarrow \overline{0}$, with $\overline{0}$ being the truth-constant for falsity. In this framework, each left continuous t-norm $*$ uniquely determines a semantical (propositional) calculus $PC(*)$ over formulas defined in the usual way from a countable set of propositional variables, connectives $\wedge$, & and $\rightarrow$ and truth-constant $\overline{0}$ [18]. Further connectives are defined as follows:

$\varphi \vee \psi$ is $((\varphi \rightarrow \psi) \rightarrow \psi) \wedge ((\psi \rightarrow \varphi) \rightarrow \varphi)$,
$\neg\varphi$ is $\varphi \rightarrow \bar{0}$,
$\varphi \equiv \psi$ is $(\varphi \rightarrow \psi)\&(\psi \rightarrow \varphi)$.

Evaluations of propositional variables are mappings e assigning to each propositional variable p a truth-value $e(p) \in [0,1]$, which extend univocally to compound formulas as follows:

$$\begin{aligned}
e(\overline{0}) &= 0 \\
e(\varphi \wedge \psi) &= \min(e(\varphi), e(\psi)) \\
e(\varphi \& \psi) &= e(\varphi) * e(\psi) \\
e(\varphi \rightarrow \psi) &= e(\varphi) \Rightarrow e(\psi)
\end{aligned}$$

Note that, from the above defintions, $e(\varphi \vee \psi) = \max(e(\varphi), e(\psi))$, $\neg\varphi = e(\varphi) \Rightarrow 0$ and $e(\varphi \equiv \psi) = e(\varphi \rightarrow \psi) * e(\psi \rightarrow \varphi)$. A formula φ is a said to be a 1-tautology of $PC(*)$ if $e(\varphi) = 1$ for each evaluation e. The set of all 1-tautologies of $PC(*)$ will be denoted as $TAUT(*)$.

Well-known axiomatic systems, like Łukasiewicz logic (Ł), Gödel logic (G), Product logic (Π), Basic Fuzzy logic (BL) and Monoidal t-norm logic (MTL) syntactically capture different sets of $TAUT(*)$ for different choices of the t-norm $*$, see e.g. [3, 17, 18]. In other words, the following condiitons hold true, where $*_{Ł}$, $*_G$ and $*_\Pi$ respectively denote the Łukasiewicz t-norm, the min t-norm and the product t-norm:

φ is provable in Ł iff $\varphi \in TAUT(*_{Ł})$
φ is provable in G iff $\varphi \in TAUT(*_G)$
φ is provable in Π iff $\varphi \in TAUT(*_\Pi)$
φ is provable in BL iff $\varphi \in TAUT(*)$ for all cont. t-norm $*$
φ is provable in MTL iff $\varphi \in TAUT(*)$ for all left-cont. t-norm $*$

These completeness results also extend to deductions from a finite set of premises but, in general, they do not extend to deductions from an infinite set of premises. Prominent exceptions are the case of Gödel logic and MTL.

Predicate logic versions of propositional t-norm based logics have also been defined and studied in the literature. Following [20] we provide below a general definition of the predicate logic $L_*\forall$ for any propositional logic L_* of a t-norm $*$. As usual, the propositional language of L_* is enlarged with a set of predicates *Pred*, a set of object variables *Var* and a set of object constants *Const*, together with the two classical quantifiers $\forall$ and $\exists$. An $[0,1]$-valued L-interpretation for a predicate language $\mathscr{PL} = (Pred, Const)$ of $L_*\forall$ is a structure

$$\mathbf{M} = (M, (r_P)_{P \in Pred}, (m_c)_{c \in Const})$$

where $M \neq \emptyset$, $r_P : M^{ar(P)} \rightarrow [0,1]$ and $m_c \in M$ for each $P \in Pred$ and $c \in Const$. For each evaluation of variables $v : Var \rightarrow M$, the truth-value $\|\varphi\|_{\mathbf{M},v}$ of a formula (where $v(x) \in M$ for each variable x) is defined inductively from

$$\|P(x, \cdots, c, \cdots)\|_{\mathbf{M},v} = r_P(v(x), \cdots, m_c \cdots),$$

taking into account that the value commutes with connectives (according to the above rules for the propositional case), and defining

$$\|(\forall x)\varphi\|_{\mathbf{M},v} = \inf\{\|\varphi\|_{\mathbf{M},v'} \mid v(y) = v'(y) \text{ for all variables, except } x\}$$
$$\|(\exists x)\varphi\|_{\mathbf{M},v} = \sup\{\|\varphi\|_{\mathbf{M},v'} \mid v(y) = v'(y) \text{ for all variables, except } x\}$$

From a syntactical point of view, the additional axioms on quantifiers for $L_*\forall$ are the following ones:

($\forall$1) $(\forall x)\varphi(x) \to \varphi(t)$ (t substitutable for x in $\varphi(x)$)
($\exists$1) $\varphi(t) \to (\exists x)\varphi(x)$ (t substitutable for x in $\varphi(x)$)
($\forall$2) $(\forall x)(\nu \to \varphi) \to (\nu \to (\forall x)\varphi)$ (x not free in ν)
($\exists$2) $(\forall x)(\varphi \to \nu) \to ((\exists x)\varphi \to \nu)$ (x not free in ν)
($\forall$3) $(\forall x)(\varphi \vee \nu) \to ((\forall x)\varphi \vee \nu)$ (x not free in ν)

Rules of inference of $L_*\forall$ are modus ponens and generalization: from φ infer $(\forall x)\varphi$.

The above mentioned propositional completeness results do not easily generalize to the first order case, MTL$\forall$ and G$\forall$ being remarkable exceptions. For more details on predicate fuzzy logics, including completeness and complexity results and model theory, the interested reader is referred to [3, 20].

24.3 T-Norm Based Fuzzy Logic Modelling of Approximate Reasoning

In the literature one can find several approaches to cast main Zadeh's approximate reasoning constructs in a formal logical framework. In particular, Novák and colleagues have done much in this direction, using the model of fuzzy logic with evaluated syntax, fully elaborated in the monograph [25] (see the references therein and also [6]), and more recently he has developed a very powerful and sophisticated model of fuzzy type theory [21, 24]. In his monograph, Hájek [18] also has a part devoted to this task.

In what follows, we show a simple way of how to capture at a syntactical level, namely in a many-sorted version of predicate fuzzy logic calculus, say MTL$\forall$, some of the basic Zadeh's approximate reasoning patterns, basically from ideas in [16, 18]. It turns out that the logical structure becomes rather simple and the fact that fuzzy inference is in fact a (crisp) deduction becomes rather apparent.

Consider the simplest and most usual expressions in Zadeh's fuzzy logic of the form

$$\text{“}\mathbf{x} \text{ is } A\text{"},$$

with the intended meaning the variable x takes the value in A, represented by a fuzzy set μ_A on a certain domain U. The representation of this statement in the frame of possibility theory is the constraint

$$(\forall u)(\pi_{\mathbf{x}}(u) \leq \mu_A(u)),$$

where $\pi_{\mathbf{x}}$ stands for the possibility distribution for the variable $\mathbf{x}$. But such a constraint is very easy to represent in MTL$\forall$ as the formula[2]

[2] Caution: do not confuse the logical variable x in this logical expression from the linguistic (extra-logical) variable $\mathbf{x}$ in “$\mathbf{x}$ is A".

$$(\forall x)(X(x) \rightarrow A(x))$$

where A and X are many-valued *predicates* of the same sort in each particular model **M**. Their interpretations (as fuzzy relations on their common domain) can be understood as the membership function $\mu_A : U \longrightarrow [0,1]$ and the possibility distribution $\pi_{\mathbf{x}}$ respectively. Indeed, one can easily observe that $\|(\forall x)(X(x) \rightarrow A(x))\|_M = 1$ if and only if $\|X(x)\|_{M,e} \leq \|A(x)\|_{M,e}$, for all x and any evaluation e. From now on, variables ranging over universes will be $\mathbf{x}, \mathbf{y}, \mathbf{z}$; "**x** is A" becomes $(\forall x)(X(x) \rightarrow A(x))$ or just $X \subseteq A$; if **z** is 2-dimensional variable $(\mathbf{x}, \mathbf{y})$, then an expression "**z** is R" becomes $(\forall x, y)(Z(x,y) \rightarrow R(x,y))$ or just $Z \subseteq R$.

In what follows, only two (linguistic) variables will be involved $\mathbf{x}, \mathbf{y}$ and $\mathbf{z} = (\mathbf{x}, \mathbf{y})$. Therefore we assume that X, Y (corresponding to the possibility distributions $\pi_{\mathbf{x}}$ and $\pi_{\mathbf{y}}$) are projections of a binary fuzzy predicate Z (corresponding to the joint possibility distribution $\pi_{\mathbf{x},\mathbf{y}}$). The axioms we need to state in order to formalize this assumption are:

$$\Pi 1 : (\forall x, y)(Z(x,y) \rightarrow X(x)) \;\&\; (\forall x, y)(Z(x,y) \rightarrow Y(y))$$
$$\Pi 2 : (\forall x)(X(x) \rightarrow (\exists y) Z(x,y)) \;\&\; (\forall y)(Y(y) \rightarrow (\exists x) Z(x,y))$$

Condition $\Pi 1$ expresses the monotonicity conditions $\pi_{\mathbf{x},\mathbf{y}}(u,v) \leq \pi_{\mathbf{x}}(u)$ and $\pi_{\mathbf{x},\mathbf{y}}(u,v) \leq \pi_{\mathbf{y}}(v)$, whereas both conditions $\Pi 1$ and $\Pi 2$ used together express the marginalization conditions $\pi_{\mathbf{x}}(u) = \sup_v \pi_{\mathbf{x},\mathbf{y}}(u,v)$ and $\pi_{\mathbf{y}}(v) = \sup_v \pi_{\mathbf{x},\mathbf{y}}(u,v)$. These can be equivalently presented as the only one condition $Proj$, as follows:

$$\textit{Proj:}\; (\forall x)(X(x) \equiv (\exists y) Z(x,y)) \;\&\; (\forall y)(Y(y) \equiv (\exists x) Z(x,y))$$

Next we shall consider several approximate reasoning patterns, and for each pattern we shall present a provable tautology (in MTL$\forall$) and its corresponding derived deduction rule, which will automatically be sound.

1. *Entailment Principle:* From "**x** is A" infer "**x** is A^*", whenever $\mu_A(u) \leq \mu_{A^*}(u)$ for all u.

   Provable tautology:

   $$(A \subseteq A^*) \rightarrow (X \subseteq A \rightarrow X \subseteq A^*)$$

   Sound rule:

   $$\frac{A \subseteq A^*, X \subseteq A}{X \subseteq A^*}$$

2. *Truth-qualification:* From "**x** is A" infer that "(**x** is A^*) is α-true", where $\alpha = \inf_u \mu_A(u) \Rightarrow \mu_{A^*}(u)$.

   Provable tautology:

   $$(X \subseteq A) \rightarrow (A \subseteq A^* \rightarrow X \subseteq A^*)$$

   Sound rule:

   $$\frac{X \subseteq A}{A \subseteq A^* \rightarrow X \subseteq A^*}$$

3. *Truth-modification:* From "(**x** is A) is α-true" infer that "**x** is A^*", where $\mu_{A^*}(u) = \alpha \Rightarrow \mu_A(u)$.
   Provable tautology (where $\overline{\alpha}$ denotes a truth-constant):
   $$(\overline{\alpha} \to (X \subseteq A)) \to (X \subseteq (\overline{\alpha} \to A)$$
   Sound rule:
   $$\frac{(\overline{\alpha} \to (X \subseteq A)}{X \subseteq (\overline{\alpha} \to A)}$$
4. *Cylindrical extension:* From "**x** is A" infer "(**x**,**y**) is A^+", where $\mu_{A^+}(u,v) = \mu_A(u)$ for each v.
   Provable tautology:
   $$\Pi 1 \to [(X \subseteq A) \to ((\forall xy)(A^+(x,y) \equiv A(x)) \to (Z \subseteq A^+))]$$
   Sound rule:
   $$\frac{\Pi 1, X \subseteq A, (\forall xy)(A^+(x,y) \equiv A(x))}{Z \subseteq A^+}$$
5. *Projection*: From "(**x**,**y**) is R" infer "**y** is R_Y", where $\mu_{R_Y}(y) = \sup_u \mu_R(u,v)$ for each v.
   Provable tautology:
   $$\Pi 2 \to ((Z \subseteq R) \to (\forall y)(Y(y) \to (\exists x)R(x,y)))$$
   Sound rule:
   $$\frac{\Pi 2, Z \subseteq R}{(\forall y)(Y(y) \to (\exists x)R(x,y))}$$
6. *min–Combination*: From "**x** is A_1" and "**x** is A_2" infer "**x** is $A_1 \cap A_2$", where $\mu_{A_1 \cap A_2}(u) = \min(\mu_{A_1}(u), \mu_{A_2}(u))$.
   Provable tautology:
   $$(X \subseteq A_1) \to ((X \subseteq A_2) \to (X \subseteq (A_1 \wedge A_2)))$$
   Sound rule:
   $$\frac{X \subseteq A_1, X \subseteq A_2}{X \subseteq (A_1 \wedge A_2)}$$
   where $(A_1 \wedge A_2)(x)$ is an abbreviation for $A_1(x) \wedge A_2(x)$.
7. *Compositional rule of inference*: From "(**x**,**y**) is R_1" and "(**y**,**z**) is R_2" infer "(**x**,**z**) is $R_1 \circ R_2$", where $\mu_{R_1 \circ R_2}(u,w) = \sup_v \min(\mu_{R_1}(u,v), \mu_{R_2}(v,w))$.
   Provable tautology:
   $$(Z_1 \subseteq R_1) \to ((Z_2 \subseteq R_2) \to (Z_3 \subseteq (R_1 \circ R_2)))$$
   Sound rule:
   $$\frac{Z_1 \subseteq R_1, Z_2 \subseteq R_2}{Z_3 \subseteq (R_1 \circ R_2)}$$
   where $(R_1 \circ R_2)(x,z)$ is an abbreviation for $(\exists y)(R_1(x,y) \wedge R_2(y,z))$.

Note that the following rule

$$\frac{Cond, Proj, X \subseteq A,\ Z \subseteq R}{Y \subseteq B},$$

where *Cond* is the formula $(\forall y)(B(y) \equiv (\exists x)(A(x) \wedge R(x,y)))$, formalizing the particular instance of *max–min composition rule*, from "**x** is A" and "$(\mathbf{x},\mathbf{y})$ is R" infer "**y** is B", where $\mu_B(y) = \sup_u \min(\mu_A(u), \mu_R(u,v))$, is indeed a derived rule from the above ones.

More complex patterns like those related to inference with fuzzy if-then rules "if **x** is A then **y** is B" can also be formalized. As it has been discussed elsewhere (see e.g. [4, 5]), there are several semantics for the fuzzy if-then rules in terms of the different types constraints on the joint possibility distribution $\pi_{\mathbf{x},\mathbf{y}}$ it may induce. Each particular semantics will obviously have a different representation. We will describe just a couple of them.

Within the implicative interpretations of fuzzy rules, gradual rules are interpreted by the constraint $\pi_{\mathbf{x},\mathbf{y}}(u,v) \leq \mu_A(u) \Rightarrow \mu_B(v)$, for some residuated implication $\Rightarrow$. According to this interpretation, the folllowing is a derivable (sound) rule

$$\frac{Cond, Proj, X \subseteq A^*, Z \subseteq A \to B}{Y \subseteq B^*},$$

where $(A \to B)(x,y)$ stands for $A(x) \to B(y)$ and *Cond* is $(\forall y)[B^*(y) \equiv (\exists x)(A^*(x) \wedge (A(x) \to B(y)))]$.

Finally, within the conjunctive model of fuzzy rules (i.e. Mamdani fuzzy rules), where a rule "if **x** is A then **y** is B" is interpreted by the constraint $\pi_{\mathbf{x},\mathbf{y}}(u,v) \geq \mu_A(u) \wedge \mu_B(v)$, and an observation "**x** is A^*" by a positive constraint $\pi_{\mathbf{x}}(u) \geq A^*(u)$, one can easily derive the Mamdani model (here with just one rule)

$$\frac{Cond, Proj, X \supseteq A^*, Z \supseteq A \wedge B}{Y \supseteq B^*},$$

where *Cond* is $(\forall y)[B^*(y) \equiv (\exists x)(A^*(x) \wedge A(x)) \wedge B(y)]$.

24.4 Conclusions

In this short note we have put forward our thesis that Mathematical Fuzzy logic is not only the basic kernel of fuzzy logic in narrow sense (with which Zadeh and many fuzzy logicians agree) but also a logical framework where many of the well known concepts and approximate reasoning inference rules of fuzzy logic in narrow sense can be properly formalized. Obviously there are some others fuzzy concepts that are more difficult to be fully interpretable in Mathematical Fuzzy logic. Among them, we can cite:

- Linguistic modifiers. They have been partially interpreted as unary connectives in the logical framework, in particular the so-called fuzzy truth hedges (*very true*, *slightly true*, etc.). These are usually classified in two classes: truth-stressers, that modify the truth-value of an expression by decreasing it, and truth-depressers, that modify the truth-value of an expression by increasing it. The formalization in this kind of connectives has been within the framework of t-norm based fuzzy logics has been addressed in several papers, e.g. by Hájek [19], Vichodyl [26] and Esteva et al. [15].

- Fuzzy quantifiers. There is a nice chapter in Hájek's book [18] devoted to this topic where he axiomatizes the quantifier "*many*", but it is only a first step in the work needed to do (there are many non-answered questions). Also Nóvak has done very interesting work (see e.g. [22]) on formalizing linguistic quantifiers. In fact in first-order fuzzy logics like the ones mentioned in Section 24.2, the only formalized quantifiers are the classical ones $\forall$ and $\exists$, interpreted as inf and sup respectively.

Nevertheless we believet that, in the near future, new developments in mathematical fuzzy logic[3] will make possible the non-trivial task of defining formal systems of fuzzy logic closer and closer to the "logic" of human approximate reasoning as envisaged and proposed by Lotfi A. Zadeh since long ago.

Fig. 24.1. Llorenç Valverde, Enric Trillas, Francesc Esteva and Joan Jacas when L. A. Zadeh was awarded the PhD Honoris Causa by the University of Oviedo (Spain) in 1995

[3] See e.g. [23] for a list of possible future tasks in the study of mathematical fuzzy logic and its applications.

Fig. 24.2. Participants of the Workshop "The logic of Soft Computing" held in Gargnano (Italy), Nov 19-24, 2001. Standing row (from left to right): Beata Konikowska, Tanja Kisielova, Lotfi Zadeh, Daniele Mundici, Norbert Preining, Peter Vojtas, Lluís Godo, NN, Petr Cintula, Paolo Farina, Petr Hájek, NN, Ryszard Wójcicki, Matthias Baaz. Front row (from left to right): Arnon Avron, Brunella Gerla, Agata Ciabbattoni, Francesc Esteva, Antonio Di Nola, Nicola Olivetti, Paolo Amato.

References

1. Cignoli, R., Esteva, F., Godo, L., Torrens, A.: Basic Fuzzy Logic is the Logic of Continuous t-norms and their Residua. Soft Computing 4(2), 106–112 (2000)
2. Cintula, P., Esteva, F., Gispert, J., Godo, L., Montagna, F., Noguera, C.: Distinguished Algebraic Semantics for t-norm Based Fuzzy Logics: Methods and Algebraic Equivalences. Annals of Pure and Applied Logic 160(1), 53–81 (2009)
3. Cintula, P., Hájek, P., Noguera, C. (eds.): Handbook of Mathematical Fuzzy Logic (in 2 volumes). Studies in Logic, Mathematical Logic and Foundations, vol. 37, 38. College Publications, London (2011)
4. Dubois, D., Esteva, F., Godo, L., Prade, H.:Fuzzy Set-based Logics – A History-oriented Presentation of Their Main Developments. In: Gabbay, D.M., Woods, J. (eds.) Handbook of the History of Logic. The Many Valued and Non-monotonic Turn in Logic, vol. 8, pp. 325–449. Elsevier (2007)
5. Dubois, D., Prade, H.: What are Fuzzy Rules and How to Use Them. Fuzzy Sets and Systems 84, 169–185 (1996)
6. Dvořák, A., Novák, V.: Formal Theories and Linguistic Descriptions. Fuzzy Sets and Systems 143(1), 169–188 (2004)

7. Esteva, F., Gispert, J., Godo, L., Montagna, F.: On the Standard and Rational Completeness of Some Axiomatic Extensions of the Monoidal t-norm Logic. Studia Logica 71(2), 199–226 (2002)
8. Esteva, F., Gispert, J., Godo, L., Noguera, C.: Adding Truth-constants to Continuous t-norm Based Logics: Axiomatization and Completeness Results. Fuzzy Sets and Systems 158, 597–618 (2007)
9. Esteva, F., Godo, L.: Monoidal t-norm Based Logic: Towards a Logic for Left-continuous t-norms. Fuzzy Sets and Systems 124(3), 271–288 (2001)
10. Esteva, F., Godo, L., García-Cerdaña, A.: On the Hierarchy of t-norm Based Residuated Fuzzy Logics. In: Fitting, M., Orlowska, E.S. (eds.) Beyond Two: Theory and Applications of Multiple-Valued Logic, pp. 251–272. Springer (2003)
11. Esteva, F., Godo, L., Hájek, P., Navara, M.: Residuated Fuzzy Logics With an Involutive Negation. Archive for Mathematical Logic 39(2), 103–124 (2000)
12. Esteva, F., Godo, L., Hájek, P., Montagna, F.: Hoops and Fuzzy Logic. Journal of Logic and Computation 13, 531–555 (2003)
13. Esteva, F., Godo, L., Montagna, F.: The ŁΠ and Ł$\Pi\frac{1}{2}$ Logics: Two complete Fuzzy Systems Joining Łukasiewicz and Product Logics. Archive for Mathematical Logic 40(1), 39–67 (2001)
14. Esteva, F., Godo, L., Montagna, F.: Equational Characterization of the Subvarieties of BL Generated by t-norm Algebras. Studia Logica 76, 161–200 (2004)
15. Esteva, F., Godo, L., Noguera, C.:Fuzzy Logics With Truth Hedges Revisited. In: Proc. of EUSFLAT - LFA 2011, pp. 146–152 (2011)
16. Godo, L., Hájek, P.: Fuzzy Inference as Deduction. Journal of Applied Non-Classical Logics 9(1), 37–60 (1999)
17. Gottwald, S.: A Treatise on Many-valued Logics. Studies in Logic and Computation, vol. 9. Research Studies Press Ltd., Baldock (2001)
18. Hájek, P.: Metamathematics of Fuzzy Logic. Trends in Logic, vol. 4. Kluwer, Dordercht (1998)
19. Hájek, P.: On Very True. Fuzzy Sets and Systems 124(3), 329–333 (2001)
20. Hájek, P., Cintula, P.: Triangular Norm Predicate Fuzzy Logics. Fuzzy Sets and Systems 161, 311–346 (2010)
21. Novák, V.: On Fuzzy Type Theory. Fuzzy Sets and Systems 149(2), 235–273 (2005)
22. Novák, V.: A Formal Theory of Intermediate Quantifiers. Fuzzy Sets and Systems 159, 1229–1246 (2008)
23. Novák, V.: Reasoning About Mathematical Fuzzy Logic and its Future. Fuzzy Sets and Systems 192, 25–44 (2012)
24. Novák, V., Lehmke, S.: Logical Structure of Fuzzy IF-THEN Rules. Fuzzy Sets and Systems 157(15), 2003–2029 (2006)
25. Novák, V., Perfilieva, I., Močkoř, J.: Mathematical Principles of Fuzzy Logic. Kluwer, Dordrecht (1999)
26. Vychodil, V.: Truth-depressing Hedges and BL-Logic. Fuzzy Sets and Systems 157(15), 2074–2090 (2006)
27. Zadeh, L.A.: Fuzzy sets. Information and Control 8, 338–353 (1965)
28. Zadeh, L.A.: Outline of a New Approach to the Analysis of Complex Systems and Decision Processes. IEEE Transactions on Systems, Man and Cybernetics 3, 28–44 (1973)
29. Zadeh, L.A.: Fuzzy Logic and Approximate Reasoning (In memory of Grigore Moisil. Synthese 30, 407–428 (1975)
30. Zadeh, L.A.: Calculus of Fuzzy Restrictions. In: Zadeh, L.A., Fu, K.S., Tanaka, K., Shimura, M. (eds.) Fuzzy Sets and their Applications to Cognitive and Decision Processes, pp. 1–39. Academic Press, New York (1975)

31. Zadeh, L.A.: The concept of a linguistic variable and its application to approximate reasoning. Information Sciences Part 1: 8, 199–249, Part 2: 8, 301–357, Part 3: 9, 43–80 (1975)
32. Zadeh, L.A.: A Fuzzy-algorithmic Approach to the Definition of omplex or Imprecise Concepts. International Journal of Man-Machine Studies 8, 249–291 (1976)
33. Zadeh, L.A.: Fuzzy Sets as a Basis for a Theory of Possibility. Fuzzy Sets and Systems 1, 3–28 (1978)
34. Zadeh, L.A.: PRUF – A Meaning Representation Language for Natural Languages. International Journal of Man-Machine Studies 10, 395–460 (1978)
35. Zadeh, L.A.: A Theory of Approximate Reasoning. In: Hayes, J.E., Michie, D., Mikulich, L.I. (eds.) Machine Intelligence, vol. 9, pp. 149–194. Elsevier (1979)
36. Zadeh, L.A.: Fuzzy Logic. IEEE Computer 21(4), 83–93 (1988)
37. Zadeh, L.A.: Preface. In: Marks-II, R.J. (ed.) Fuzzy Logic Technology and Applications, IEEE Technical Activities Board, Piscataway (1994)

25

Consensus Modelling in Group Decision Making: A Dynamical Approach Based on Zadeh's Fuzzy Preferences

Mario Fedrizzi, Michele Fedrizzi, and R.A. Marques Pereira

25.1 Introduction

The notion of consensus plays an important role in the theory of group decisions, particularly when the collective preference structure is generated by a dynamical process of aggregation of the single individual preference structures. In this process of preference aggregation each single decision maker gradually transforms his/her preference structure by combining it, through iterative weighted averaging, with the preference structures of the other decision makers. In this way the collective decision emerges dynamically as a result of the consensual interaction among the various decision makers in the group. From the point of view of applied mathematics, the models of consensual dynamics stand in the context of multi-agent complex systems, with interactive and non linear dynamics. The consensual interaction among the various decision makers acts on their preferences in order to optimize an appropriate measure of consensus, which can be of type 'hard' (full agreement within the group of decision makers) or 'soft' (partial agreement within the group of decision makers). The 'soft' approach to consensus modeling is based on the premise that the key elements in decision making, which is based on human thinking, are not numbers but labels of fuzzy sets since, as L.A. Zadeh [25] pointed out, „... much of the logic behind human reasoning is not the traditional two-valued or even multivalued logic, but a logic with fuzzy truths, fuzzy connectives, and fuzzy rules of inference". In this paper we review some results we have obtained in the modelling of consensus reaching in a 'soft' environment, i.e., when the individual opinions are assumed to be expressed as fuzzy preference relations. Here consensus is meant as the degree to which most of the predominant experts agree on the preferences associated to most relevant alternatives. First of all we derive a degree of disagreement based on linguistic quantifiers and then we introduced a form of network dynamics in which the quantifiers are represented by scaling functions. Assuming that the decision makers can express their preferences in a more flexible way, i. e., by using triangular fuzzy numbers, we propose an iterative process of opinion transformation towards consensus via the gradient dynamics of a cost function expressed as a convex combination of a disagreement cost and an inertial cost.

R. Seising et al. (Eds.): *On Fuzziness: Volume 1*, STUDFUZZ 298, pp. 165–170.
DOI: 10.1007/978-3-642-35641-4_25

25.2 A Historical Overview

A classical problem addressed by political and social science concerning the theory and practice of political behavior reagrds the design of models for making decisions when a group of two or more individuals must aggregate their opinions (preferences) in order to obtain a social choice. The first systematic approaches to the problem were developed around the time of the French Revolution with the fundamental work of Borda [5] (1781) and Condorcet [7] (1785), who initiated the formal discipline of voting theory in social choice.

The subject of social choice was revived in the twentieth century by Arrow (1951), who was concerned with the difficulties of group decisions and the inconsistencies they can generate, leading to the well-known Impossibility Theorem. Along with two other works in economics, von Neumann and Morgenstern's *Theory of Games and Economic Behavior*, and Black's *The Theory of Committees and Elections*, Arrow's book demonstrated the applicability of formalized theoretical argument to the understanding of a broad range of political phenomena. In so doing, the three works set the stage for a partial reintegration of political science and economy by showing that one paradigm could perform yeoman service in two disciplines. For an up to date review see Nurmi [22].

Over the last two decades of the twentieth century a number of authors have extended the theory of social choice (group decision making) in various ways in order to encompass fuzziness in individual and group preferences. Barret, Pattanaik and Salles [2] investigated the structure of fuzzy aggregation rules which, for every feasible profile of individual preferences, specify a fuzzy social ordering. Dutta [8] (1987), permitting that both individual and social preferences of the Arrowian framework be fuzzy, showed that, under a weaker transitivity condition, the fuzzy counterpart of Arrow's assumptions result in oligarchic and not dictatorial aggregation rules. Montero [20] introduced rationality as a fuzzy property by suggesting a definition of fuzzy opinion different from the classical fuzzy preference relation, and showed how to escape from impossibility theorems through the idea of fuzzy rationality. More details and useful references can be found in Nurmi and Kacprzyk [21].

Consensus is traditionally meant as a strict and unanimous agreement. However, since various decision makers have different more or less conflicting opinions, the traditional strict meaning of consensus is unrealistic. The human perception of consensus is much 'softer', and people are willing to accept that a consensus has been reached when most or the more predominant actors agree on the preferences associated to the most relevant alternatives.

As a formal tool for deriving the degree of consensus Zadeh's fuzzy logic based calculus of linguistically quantified propositions has been used [26]. This calculus is a prerequisite for Zadeh's representation of commonsense knowledge as a collection of dispositions, i.e., propositions with implicit linguistic quantifiers [27]. The use of this calculus in the development of the degree of consensus may thus be viewed as an attempt at introducing commonsense into the essence of consensus.

The problem of consensus reaching modeling in a fuzzy environment was addressed at first in Ragade [23], Bezdek, Spillman, and Spillman [4], Spillman,

Bezdek, and Spillman [24] , and then developed in Kacprzyk and Fedrizzi [17], Kacprzyk, Fedrizzi, and Nurmi [18], and Kacprzyk, Nurmi, and Fedrizzi [19]. Some authors addressed the problem introducing linguistically-based preference relations, see for instance, among others, Herrera-Viedma et al. [16] and Ben-Arieh and Chen [3]. For an up to date analysis and discussion of the advantages and drawbacks of different consensus approaches in fuzzy group decision making problems see Cabrerizo, Moreno, Perez, and Herrera-Viedma [6].

25.3 A Dynamical Approach to Consensus Reaching

The dynamical approach to consensus under fuzziness has been developed in Fedrizzi, Fedrizzi, and Marques Pereira [11] on the basis of the soft consensus paradigm introduced in Kacprzyk and Fedrizzi [17] in the framework of reciprocal fuzzy preferences. The basic model was then extended in Fedrizzi, Fedrizzi, and Marques Pereira [12] and in Fedrizzi, Fedrizzi, Marques Pereira, and Brunelli [13], [14]. In this model, instead of degree of consensus, a measure of the degree of disagreement has been introduced and derived in three steps. Firstly, for each pair of individuals a degree of disagreement as to their preferences between a pair of alternatives is derived. Then, these degrees are aggregated to obtain a degree of disagreement of each pair of individuals as to their preferences between Q_1 (a linguistic quantifier as, e.g., 'most', 'almost all', 'more than 50%',...) pairs of relevant alternatives. Finally, these degrees were pooled to obtain a degree of disagreement of Q_2 (a linguistic quantifier similar to Q_1) pairs of individuals as to their preferences between Q_1 pairs of relevant alternatives.

The consensus reaching dynamics is generated starting from a combination of a soft measure of collective disagreement with an inertial mechanism of opinion changing aversion. It acts on the network of single preference structures by a combination of a collective process of (nonlinear) diffusion and an individual mechanism of (nonlinear) inertia. The overall effect of the dynamics is to outline and enhance the natural segmentation of the decision makers group into homogeneous preference subgroups.

In the soft consensus model each decision maker $i = 1,...,n$ is represented by a pair of connected nodes, a primary node (dynamic) and a secondary node (static). The n primary nodes form a fully connected sub network and each of them encodes the individual opinion of a single decision maker. The n secondary nodes, on the other hand, encode the individual opinions originally declared by the decision makers and each of them is connected only with the associated primary node.

The iterative process of opinion transformation corresponds to the gradient dynamics of a cost function W, depending on both the present and the original network configurations. The value of W combines a measure V of the overall disagreement in the present network configuration and a measure U of the overall change from the original network configuration.

The various interactions involving node x_i are mediated by interaction coefficients whose role is to quantify the strength of the interaction. The diffusive interaction between primary nodes x_i and x_j is mediated by the interaction coefficient $v_{ij} \in (0,1)$, whereas the inertial interaction between the primary node x_i and the associated secondary node s_i is mediated by the interaction coefficient $u_i \in (0,1)$. It turns out that the values of these interaction coefficients are given by the derivative f' of the scaling function.

The diffusive component of the network dynamics results from the consensual interaction between each node x_i and the remaining $n-1$ nodes $x_{j \neq i}$ in the network. The aggregated effect of these $n-1$ interactions can be represented as a single consensual interaction between node x_i and a virtual node $\bar{x}_i$ containing a particular weighted average of the remaining opinion values (for details see Fedrizzi, Fedrizzi and Marques Pereira [12]).

25.4 Concluding Remarks

The computer simulations of our model, particularly those in the fuzzy preference framework illustrated in [13] and [14], provide clear evidence that the fuzzy soft consensus model exhibits interesting non standard opinion changing behaviour in relation to the original crisp version of the model. Future research should explore the particular features of the fuzzy soft consensus model and demonstrate the potential of the methodology as an effective support for the modelling of consensus reaching in multicriteria and multiagent decision making.

Fig. 25.1. At the conference Iizuka 1994 in Japan, f.l.t.r.: Gabriella Pasi, Lotfi A. Zadeh, Margit Kovacs and Mario Fedrizzi

References

1. Arrow, K.J.: Social Choice and Individual Values, 2nd edn. Cowles Foundation Monograph, vol. 12. Yale University Press, New Haven (1951)
2. Barrett, C.R., Pattanaik, P.K., Salles, M.: On the Structure of Fuzzy Social Welfare Functions. IEEE Transactions on Fuzzy Sets and Systems 19, 1–10 (1986)
3. Ben-Arieh, D., Chen, Z.: Linguistic Labels Aggregation and Consensus Measure for Autocratic Decision Making Using Group Recommendations. IEEE Transanctions on Systems, Man, and Cybernetics (A) 36, 558–568 (2006)
4. Bezdek, J.C., Spillman, B., Spillman, R.: Fuzzy Measures of Preferences and Consensus in Group Decision Making. In: Fu, K.S. (ed.) Proceedings of the IEEE Conference on Decision and Control, pp. 1303–1309 (1977)
5. de Borda, J.C.: Memoire sur les elections au scrutin. In: Histoire de L'Academie Royale des Sciences, Paris (1781); Translated by de Grazia, A.: Mathematical Derivation of an Election System. Isis 44(1-2), 42–51 (1953)
6. Cabrerizo, F.J., Moreno, J.M., Pérez, I.J., Herrera-Viedma, E.: Analyzing Consensus Approaches in Fuzzy Group Decision Making: Advantages and Drawbacks. Soft Computing 14, 451–463 (2010)
7. Condorcet, Marie-Jean-Antoine Nicolas de Caritat, Marquis de: Essai sur l'application de l'analyse a la probabilite des decisions rendues a la pluralite des voix. L'Impremerie Royale, Paris (1785)
8. Dutta, B.: Fuzzy Preferences and Social Choice. Mathematical Social Sciences 13, 215–229 (1987)
9. Fedrizzi, M., Fedrizzi, M., Marques Pereira, R.A.: Consensual dynamics: an unsupervised learning model in group decision making. In: Proceedings of the 3rd International Conference on Fuzzy Logic, Neural Nets and Soft Computing, IIZUKA 1994, pp. 99–100. Fuzzy Logic Systems Institute FLSI, Iizuka (1994)
10. Fedrizzi, M., Fedrizzi, M., Marques Pereira, R.A.: Consensual Dynamics in Group Decision Making. In: Proceedings 6th International Fuzzy Systems Association World Congress, IFSA 1995, vol. II, pp. 145–148. International Fuzzy Systems Association IFSA and INPE-Brazil, Sao Paulo (1995)
11. Fedrizzi, M., Fedrizzi, M., Marques Pereira, R.A.: Soft Consensus and Network Dynamics in Group Decision Making. International Journal of Intelligent Systems 14, 63–77 (1999)
12. Fedrizzi, M., Fedrizzi, M., Marques Pereira, R.A.: Consensus Modelling in Group Decision Making: A Dynamical Approach Based on Fuzzy Preferences. New Mathematics and Natural Computation 3, 219–237 (2007)
13. Fedrizzi, M., Fedrizzi, M., Marques Pereira, R.A., Brunelli, M.: Consensual Dynamics in Group Decision Making with Triangular Fuzzy Numbers. In: Proceedings of the 41st Annual International Conference on System Sciences HICSS 2008, Waikoloa, Big Island, Hawaii USA, January 7-10, pp. 70–78. IEEE Computer Society, Los Alamitos (2008)
14. Fedrizzi, M., Fedrizzi, M., Pereira, R.A.M., Brunelli, M.: The Dynamics of Consensus in Group Decision Making: Investigating the Pairwise Interactions between Fuzzy Preferences. In: Greco, S., Pereira, R.A.M., Squillante, M., Yager, R.R., Kacprzyk, J. (eds.) Preferences and Decisions. STUDFUZZ, vol. 257, pp. 159–182. Springer, Heidelberg (2010)
15. Fedrizzi, M., Kacprzyk, J., Nurmi, H.: Consensus Degrees Under Fuzzy Majorities and Fuzzy Preferences Using OWA (Ordered Weighted Averaging) Operators. Control and Cybernetics 22, 77–86 (1993)

16. Herrera-Viedma, E., Martinez, L., Mata, F., Chiclana, F.: A Consensus Support System Model for Group Decision Making Problems with Multigranular Linguistic Preference Relations. IEEE Transanctions on Fuzzy Systems 13, 644–658 (2005)
17. Kacprzyk, J., Fedrizzi, M.: A 'Soft' Measure of Consensus in the Setting of Partial (Fuzzy) Preferences. European Journal of Operational Research 34, 316–325 (1988)
18. Kacprzyk, J., Fedrizzi, M., Nurmi, H.: Group Decision Making and Consensus Under Fuzzy Preferences and Fuzzy Majority. Fuzzy Sets and Systems 49, 21–31 (1992)
19. Kacprzyk, J., Nurmi, H., Fedrizzi, M. (eds.): Consensus Under Fuzziness. International Series in Intelligent Technologies. Kluwer Academic Publishers, Dordrecht (1997)
20. Montero, J.: Arrow's Theorem Under Fuzzy Rationality. Behavioral Science 32, 267–273 (1987)
21. Nurmi, H., Kacprzyk, J.: Social Choice Under Fuzziness: a Perspective. In: Fodor, J., De Baets, B., Perny, P. (eds.) Preferences and Decisions under Incomplete Knowledge, pp. 107–130. Physica, Heidelberg (2000)
22. Nurmi, H.: Voting Procedures under Uncertainty. Springer, Heidelberg (2002)
23. Ragade, R.K.: Fuzzy Sets in Communication Systems and in Consensus Formation Systems. Journal of Cybernetics 6, 21–38 (1976)
24. Spillman, B., Bezdek, J.C., Spillman, R.: Coalition Analysis with Fuzzy Sets. Kybernetes 8, 203–211 (1979)
25. Zadeh, L.A.: Outline of a New Approach to the Analysis of Complex Systems and Decision Processes. IEEE Transactions on Systems, Man, and Cybernetics 3, 28–44 (1973)
26. Zadeh, L.A.: A Computational Approach to Fuzzy Quantifiers in Natural Languages. Computer and Mathematics with Applications 9, 149–183 (1983)
27. Zadeh, L.A.: A Theory of Commonsense Knowledge. In: Skala, H.J., Termini, S., Trillas, E. (eds.) Computer and Mathematics with Applications, vol. 9, pp. 149–183 (1983)

26

How Did I Come to Fuzzy Logic and to Fuzzy Decision Making – A Personal View

Rudolf Felix

In the year of 1975 after having finished the eight-year Zofia Naukowska primary school I entered the Mikolaj Kopernik secondary school in the Silesian city of Opole in Poland at the age of 15. After a four-year term at this school one used to get a diploma that qualified for the examination for university admission. Already in the first hour of the classes of mathematics I realized that it would not be as easy as it was in the primary school where I spent the first eight years of education. But let me tell the story step by step.

The first topic we had in our math course was propositional logic. Basically we had to learn how to find truth values of expressions using truth value tables. I studied as hard as I could but in the first test we had to pass, our teacher surprised us with linguistically formulated sentences to which we had to find the corresponding truth value. And of course, the sentences were like "If 5=10 then the earth is a plate". Somehow, instead of applying what we had learned, I started to intuitively interpret the sentences. And of course, the truth value I decided to choose was 0 instead of 1 because the earth was not a plate. And of course, I did not pass the test and, of course, for quite a while I was very afraid that I would not be able to understand this kind of stuff at all and to finish the secondary school with a diploma. Fortunately, our teacher in mathematics, Mrs. Maria Opiela, to whom I am to this day grateful for her excellent way of teaching, helped me to get out of my mental dead end quite soon. And with her help I accepted that science is not necessarily intuitive. But the remembrance of how surprised I was that my intuition did not lead to a correct result has remained. Retrospectively I would say that this was the first time I felt that formal systems should not be too strict and should better match with the intuition.

Twelve years later when I already was a PhD student in computer science at the University Dortmund, Germany, something similar happened. I was working there in the computer science department in a working group belonging to the chair of Bernd Reusch and I got in contact with Claudio Moraga who was also a professor there. The topic of the working group was the exploitation of Artificial Intelligence techniques for the so called high level VLSI synthesis. High level VLSI synthesis was, roughly speaking, a process of automatic generation of very large scale integrated circuits from behavioral descriptions. The descriptions were given in a formal language similar to high level programming languages like Pascal, for instance, and there were different subtasks defined by the working group, one of which was the

R. Seising et al. (Eds.): *On Fuzziness: Volume 1*, STUDFUZZ 298, pp. 171–176.
DOI: 10.1007/978-3-642-35641-4_26

so called VLSI chip-architecture selection. Here we had to solve a kind of classification problem and there was a set of characteristics defined based on which a chip architecture scheme had to be selected out of a set of about ten predefined schemes. The problem was that the values of the characteristics where linguistically expressed terms and that, in some cases, some of the information about the values was not complete. Intuitively and looking at the values at a glance, it was quite clear which chip-architecture scheme was the proper one. But the rule based approaches we tried to apply were not promising because of the exploding number of rules one had to manage. Again - as at the time when I was in the secondary school – my inner feeling was that intuition and formalisms were not matching. And I became afraid that there would be no efficient solution at all for the selection problem. And, of course, that this might be the end of my PhD ambitions.

Fig. 26.1. Blanes, Spain, 1989. A Workshop on Fuzzy Logic. From left to right: Rudolf Felix, Didier Dubois, Bernadette Bouchon-Meuneir, Lotfi A. Zadeh, Lluis Godo (almost not to see), Francesc Esteva, Juan Luis Castro (Granada), Josep Aguilar (Toulouse), Eduard Bonet, Christian Freksa, NN, Amparo Vila, Ramon Lopez de Mantaras, NN, Llorenç Valverde, Claudi Alsina, Enric Trillas, Jose Luis Verdegay (alias Curro), Joan Jacas, NN, Henri Prade, Enrique Ruspini.

Fortunately, just at this point of time it was Claudio Moraga who proposed me to participate in a scientific exchange project founded by the German Academic Exchange Service (DAAD). And more or less because he knew that my hobby-horse was the Spanish language – Claudio Moraga as a native speaker in Spanish would talk with me in Spanish from time to time - he offered me to take part in the project - as I also believe - because the project partner should be a Spanish institute belonging to CSIC council. The institute was called Centro de Estudios Avanzados de Blanes

(CEAB) and was located close to Barcelona. By the way, now the institute is called IIIA – Instituto de Investigación en Inteligencia Artificial and is located directly in Barcelona.

So, from Claudio Moraga I received the first fuzzy paper in my life as preparation material for the first visit at CEAB. And reading this paper was my break-through experience with respect to my further career. The paper was entitled "Managing Linguistically Expressed Uncertainty in MILORD-Application to Medical Diagnosis" by Lluis Godo, Ramon Lopez de Mantaras, Carles Sierra and Albert Verdaguer. Already after reading the first pages of this paper, it was clear to me that the chip-architecture selection problem had an efficient solution. The solution was a classification approach based on fuzzy if-then rules. Together with my colleague Peter Grabienski we quickly solved the problem still during our first stay at CEAB using the MILORD environment designed by Carles Sierra whom we met at CEAB. Lluis Godo and Ramon Lopez de Mantaras were working there, too. During some subsequent stays in Blanes, my second colleague Achim Höffman and I finished the work on the chip-architecture selection system. My future working direction was now clear. Fuzzy techniques would accompany my career. And I made new friends in Spain.

Fig. 26.2. Dortmund, Germany, 1996: EFDAN ´96 European Workshop on Fuzzy Decision Analysis for Management Planning an Optimization, A Panel Discussion: From left to right: Hans Hellendoorn, Lotfi A. Zadeh, Rudolf Felix (Chairman and Organizer), Janusz Kacprzyk, Didier Dubois.

CSIC´s CEAB was also the place where I met Lotfi Zadeh for the first time. It was in Blanes during a workshop on fuzzy and related techniques that took place there in 1989 (see Photo). During this time I was still working in the field of VLSI synthesis. My topic was modeling of design processes. Inspired by a paper of Jack Mostow about modeling of design decisions and the importance of goal conflicts I wanted to dedicate myself to the field of decision analysis in my PhD thesis. My intuition was that decision making and balancing of conflicts between multiple decision goals was something fuzzy by nature. During a coffee break of the workshop I asked Lotfi Zadeh about his opinion on the fuzziness of decision making saying something like "I plan to dedicate myself to the field of decision making because my impression is that almost everything could be viewed as a decision making process with fuzzy conflicts between the goals and many applications could be expected in future. What do you think?" The answer I got was a typical Zadehan one: "Yes, to a certain extent. But go on". I went on.

Fig. 26.3. Dortmund, Germany, 1996: EFDAN ́96 European Workshop on Fuzzy Decision Analysis for Management Planning an Optimization, Lotfi A. Zadeh with some participants and the organizing team. From left to right: Andreas Meiritz, Stefan Reddig, Rudolf Felix, NN, Lotfi A. Zadeh, Mechtild Watermann, Oliver Wagner, Rainer Albersmann, Stephanie Holz, NN, Frank Weber, Hans Hellendoorn.

In the subsequent years a number of customer-driven real world applications in almost one hundred factories, locations and business processes all over the world have been designed based on the fuzzy analysis of decision conflicts. There are applications in the automotive industry, in banking, in public transportation, in the maintenance of electrical grids, in the distribution of fashion products etc. The customers' requirements changed the focus of the applications, moving them toward the field of optimization as it turned out that optimization could be seen as a system of loops of many decisions on conflicting goals made one after another. Of course, in most cases I knew intuitively that a solution did exist long before the solution really worked. But here, when evaluating commercial and technical risks, one could participate in the advantage of fuzzy logic, in the sense that in many cases using it makes bridging the gap between the intuitive understanding and a final formal solution much easier – something I was not given at the beginning of my secondary school while studying propositional logic, but something I got later thanks to fuzzy logic.

Fig. 26.4. Dortmund, Germany, 1996: EFDAN ´96 European Workshop on Fuzzy Decision Analysis for Management Planning an Optimization, Lotfi A. Zadeh with the organizing team of the exhibition of ICD Informatik Centrum Dortmund and F/L/S Fuzzy Logik Systeme GmbH. From Left to right sitting: Martin Wehner, NN. From left to right standing: Rudolf Felix, Mechtild Watermann, NN, Lotfi A. Zadeh, Ralf Zimmermann, Jörg Kühlen, Frank Weber.

Fig. 26.5. Dortmund, Germany, 1996: EFDAN ′96 European Workshop on Fuzzy Decision Analysis for Management Planning an Optimization, Lotfi A. Zadeh and Rudolf Felix discussing the program of the workshop.

27

Fuzzy Relations and Cognitive Representations

Christian Freksa

Fuzzy set theory and fuzzy logics were able to demonstrate impressive achievements in control theory and in technical applications already in the 1970s; but Lotfi Zadeh's great concern was - and still is - to demonstrate the power of his radically different approach to representing human concepts as a representational foundation in Artificial Intelligence. With his approach of introducing graded compatibility values to describe relations between concepts and real-world entities, formal systems can characterize states of affairs in terms of a manageable number of concepts - much like humans who describe the world by concepts that are qualified through linguistic hedges, prosodic emphasis or attenuation, and many other subtle ways of describing situations in a differentiated way to capture their essential significance in a concise manner.

Such mechanisms enable humans to summarize complex events in a meaningful way. Without the ability to drastically eliminate details of events, people would be incapable of dealing with the complexity of the world. With this insight, Lotfi Zadeh described in the 1970s [7] his personal grand challenge for Artificial Intelligence: the ability to automatically summarize the content of a paper or a book as capable humans can do it. Zadeh realized that this would be extremely difficult to achieve by the dominating approaches in AI, at the time; the Fuzzy Set approach, in contrast, has a built-in approach to generalize from specific instances and to ignore insignificant details.

27.1 Conceptual Framework and Cognitive Requirements

The principles of a Fuzzy Set provide a rich framework within which we can discuss semantic relations in arbitrary knowledge representation systems. The relations between discrete linguistic labels, their finite or infinite domains of support, and the infinitely-valued membership values provide an excellent basis for describing and discussing representation-theoretical issues. These involve the symbols of language, the mental concepts associated with these symbols, and the entities that the symbols refer to.

At the time when Lotfi Zadeh introduced his notion of a linguistic variable with graded compatibility values, the cognitive psychologist Eleanor Rosch published seminal papers on concept categories and on more or less prototypical representatives of concept categories [4], [4]. The anthropologist Brent Berlin and the linguist

R. Seising et al. (Eds.): *On Fuzziness: Volume 1*, STUDFUZZ 298, pp. 177–183.
DOI: 10.1007/978-3-642-35641-4_27

Paul Kay described systems of color naming in different cultures and their interrelationships [1]. These are just two examples of human conceptualization that provide strong support for approaches of graded applicability of concepts and against "black or white" categories in cognitive systems.

Nevertheless, the notion of a 'fuzzy concept' is prone to considerable confusion and has not been universally well received. I would like to offer a representation-theoretic explanation why the notion of a fuzzy concept appears counterintuitive even to people who fully agree with gradual applicability of concepts to specific states of affairs. I will do so by comparing demands of complex technical systems with those of cognitive systems and I will refer to Stephen Palmer's notion of a knowledge representation system [3].

In describing and controlling technical systems, we are dealing with high-dimensional closed worlds, in which the ranges of the control variables are known. Although we cannot precisely describe all relations between potential control values, we can refer to them on an individual basis. In cognitive systems, however, we are confronted with multi-dimensional open worlds containing too many feature dimensions to enumerate and with open variable ranges for most of them. In these systems, we are less concerned with describing control values than with selecting, characterizing, and relating salient features; furthermore, due to the open characteristics of cognitive systems, we cannot specify contexts of applicability of concepts, in most cases; these must be derived implicitly. As a consequence, in cognitive systems we have to relate and compare feature dimensions and feature values and focus on those which are salient in a specific domain. It is neither desirable nor feasible to subdue the concepts we use to describe situations by specifying ranges of applicability in their specific contexts of use. For example, to understand what someone means by an 'expensive shop', we do not need to know the types of goods it sells or the price ranges of these goods; it is sufficient to know that 'expensive' is a (linguistic) value on the higher end of a prize scale, that this value is in contrast to 'inexpensive' and 'medium-prized', and that these values would be ordered *inexpensive* < *medium-prized* < *expensive*. Thus, the meaning is not derived from the hypothetical feature values in the specific situations in which they are used. Rather, concepts derive an important part of their meaning from their relation to other concepts.

In describing the world around us, we are concerned with the actual features and their values, rather than hypothetical feature values as in technical systems in which we modify the state of affairs. As a consequence, for cognitive systems, we require structures that support the concise representation of characteristic patterns describing given situations in comparison to contrasting situations.

Even if we are convinced that the relation between a symbolic cognitive notion and a set of real-world or hypothetical entities is best characterized by a fuzzy relation, there is a question with regards to the nature and the representation of the fuzziness For cognitive science this is an important issue, as the representations and processes need not only produce certain effects (this may be sufficient for technical applications), but they also serve to understand the details of the effects and their underlying mechanisms.

More specifically, we need to decide whether (1) fuzzy relations are or should be explicitly encoded as parts of mental concepts or whether (2) fuzzy relations will be caused by implicit effects of the perceptual and representational mapping characteristics between real-world entities and discrete mental concepts. For both alternatives we can conceive of representational and algorithmic implementations, some of which may appear cognitively more plausible than others. The choice of alternatives has important implications for the notion of a 'fuzzy concept' in cognition.

(1) implies that the mental concept itself is fuzzy. E.g., when I invoke the notion *tall*, I will invoke an entire set of more or less applicable potential instances (concept grounding); these potential instances form a part of the meaning of the concept; as some are more and some are less applicable, we obtain a fuzzy set of applicability. This is a quantitative interpretation of concepts as in classical fuzzy set theory; i.e., context of applicability and the degree of applicability of a concept in that context are part of the meaning of the concept.

(2) permits the abstract mental concept *tall* to be crisp. I.e. when I invoke the notion *tall*, I relate it to connatural notions like *short*, *medium-sized*, and *not tall* without extending these notions to real or potential instances; the meaning of the concepts is not derived by grounding them in the physical world but by relating them to comparable and contrasting concepts in the conceptual domain [2]; [6]. This is a qualitative relational interpretation of concepts. The meaning does not depend on a specific quantitative context, i.e., the relation between *short*, *medium-sized*, and *not tall* (e.g. an ordering relation) is invariant wrt. the specific reference set (e.g. tallness of professional basket ball players vs. first graders).

Fuzziness is not an issue on the abstract concept level, as the significance of concepts is in the relation to other concepts rather than in the direct relation to instances; in this view, concepts like *tall* do not have a quantifiable meaning on the concept level. Fuzziness may become an issue once we apply the relations between concepts to relations between instances; but on the level of relations between concepts, fuzziness is no longer an omnipresent property that cognitive systems have to deal with all the time. For example, if our real-world domain consists of three entities ordered by tallness, say 5, 6, and 7 feet tall, respectively, we will be able to apply the ordering relation of the corresponding concepts directly and characterize them as *short*, *medium-sized*, and *tall*, respectively, with no fuzziness entailed. A special charm of this relational view is the implicit context-adaptivity of the concepts involved.

At the end, the debate on whether or not we want to talk about 'fuzzy concepts' or about fuzzy relations between concepts and entities may boil down to the philosophical question which parts of a representational system we decide to call a 'concept': are concepts platonic crisp entities in the mind that allow us to think and dream about real and imagined objects and structures abstractly, or do concepts include relevant aspects of these objects and structures as well as the corresponding compatibility functions more concretely. I personally prefer to consider concepts in a qualitative and platonic way; the fuzziness gets introduced when I attempt to apply my network of concepts to real-world situations. I will explain my reasons in the following.

27.2 How Did I Arrive at the Field? – My Personal Background

As a high school student I developed a strong interest in cybernetics; this led me to enroll in physics, mathematics, and informatics at the Technical University of Munich. Concurrently with my undergraduate studies I had the great opportunity to join an interdisciplinary team of sleep researchers at the Max Planck Institute for Psychiatry in Munich as a computer programmer. In 1975, I was admitted as a Ph.D. student to the EECS Department at UC Berkeley where my first contact was Dr. Lawrence W. Stark, professor of physiological optics and engineering.

When I introduced myself to Dr. Stark's research group by presenting my work on real-time EEG sleep stage classification I discussed the issue of classifying boundary cases and how we dealt with them in our Munich team. Dr. Stark immediately advised me to talk to Professor Zadeh, the inventor of the fascinating Fuzzy Set Theory; he added that Zadeh had been confronted with much antagonism from colleagues who did not understand the value and importance of his contribution.

I talked to Professor Zadeh and became a regular participant of his weekly research seminar and his informal gatherings at the 3 C's café. He, in turn, became my Ph.D. advisor. In the year that I was admitted to Berkeley, the Sloan Foundation-funded Berkeley cognitive science program started with a regular highly interdisciplinary seminar series. Eleanor Rosch and Stephen Palmer from psychology, George Lakoff and Chuck Fillmore from linguistics, John Searle and Hubert Dreyfus from philosophy, Paul Kay from anthropology, and Lotfi Zadeh from computer science, among other brilliant Berkeley scientists were regular participants. This was the most outstanding and exciting academic environment I could imagine to foster my interests in artificial intelligence and cognitive science. Brilliant and eloquent minds from different fields publicly interacted as self-confident human beings.

My own research as a graduate student in Artificial Intelligence was heavily influenced by the cognitive science seminar, by courses in cognitive psychology that I took from Stephen Palmer and Eleanor Rosch, by Lotfi Zadeh's Seminar on Expert Systems, by an exciting computer science colloquium series, and by special panel discussions of the Bay Area Circle on Artificial Intelligence organized by Lotfi Zadeh. I became particularly fascinated by the insight that high-level cognition seems to work rather reliably on the foundation of severely underspecified knowledge. For example, people are pretty bad at geometrically reconstructing spatial environments, even the ones they feel most familiar with; nevertheless people rarely get lost in these environments.

Whereas the basic issue of imprecise, incomplete, and fuzzy knowledge appear closely related in the engineering and cognition domains, I developed a strong opinion that natural cognitive systems use knowledge in a rather different way than what our engineering approaches aim at. The fuzzy set approach aims at precisiating knowledge to resolve uncertainty, and to make meaning more precise in order to control a continuous space of options; in cognition, a large body of problems consists of identifying existing situations which form discrete islands in the large space of theoretically possible configurations. Continuous-valued fuzzy membership values are of theoretical importance to describe on the meta-level how concepts and potential instances are related; but for actually matching concepts and objects on the object level, weaker approaches should suffice.

The basic approach I pursued was to characterize labels of 'fuzzy concepts' not by grounding them in terms of feature values, but in terms of relations to other labels that characterize the same or similar feature dimensions. For example, we use the labels *short*, *medium-sized*, and *tall* in arbitrary contexts in the same ordering relation *short* < *medium-sized* < *tall*; from a cognitive point of view, it is important that we can use feature dimensions (here: *size*) to establish categories in a universally comprehensible way; in most non-technical situations (in which we do not use labels in a strictly defined way), it is not necessary to consider boundaries between concepts - and we can still get the main idea across; conceptually neighboring labels will be correctly resolved through the reference context. In other words: although the labels engage in a fuzzy relation to entities in a space of theoretical feature values, we can use them abstractly to form categories without having to decide precisely which entities belong into which category: it is important to conceptually distinguish between a mountain and a valley and it is helpful to assume that the mountain starts where the valley ends; but we do not have to decide exactly where the valley ends and where the mountain starts: the boundary region simply is not of interest for these two concepts. The important point is that when we categorize entities along a given dimension and categorize concept labels along the same dimension, we will obtain coherent patterns which can be easily matched when we take into account neighborhood relations. This will be the case, if the network of abstract concepts structurally matches the network of corresponding features.

Fuzzy set theory and knowledge representation theory serve as excellent frameworks to characterize the cognitive agenda of knowledge representation and reasoning. The agenda involves abstract mental concepts and specific real or imagined entities in a space of theoretically possible feature values.

27.3 Issues and Lessons to Be Learnt

Lotfi Zadeh advised his students against doing their doctoral theses in the area of fuzzy set theory when this area was faced with hostility from other areas; he wanted to protect his students against reduced professional opportunities.

In my view, Professor Zadeh's most important contributions to science are not the now classical fuzzy set theory and fuzzy logic; it is the epistemological framework that permits relating human concepts and knowledge to a large variety of theories and formalisms and to a large variety of application domains. Zadeh's approach of generalizing from existing theories opened avenues for asking new questions regarding the epistemological status of notions related to uncertainty such as compatibility, similarity, fuzziness, vagueness, probability, possibility, and for discussing and comparing these notions.

The personal lesson I learnt from the confrontation with these notions was that we should look much more closely at specific problem domains and the precise questions we want to answer. We need to investigate in which ways we can represent the epistemological features of interest and to what extent we can create conditions in which we do not need to represent them on the problem solving level. By creating a knowledge processing environment which will ensure a proper treatment of specific aspects of knowledge we may be able to avoid to make certain features explicit.

We may safely abstain from explicitly representing certain epistemological features if we employ *intrinsic* representations of crucial aspects in the sense of Palmer ([3]). By doing so, we can guarantee that certain properties 'automatically' hold due to structural properties of the representation employed. A vivid example for an approach in which we may safely neglect the explicit representation of important knowledge is the spatial domain: an architect who represents the spatial layout of a building by a geometric floor plan on a 2-dimensional sheet of paper does not have to make geometric knowledge explicit (e.g. 'The sum of the angles in a planar triangle equals 180°'). The structural properties of the representation medium implicitly will guarantee that the rule holds; or, expressed negatively: it will be impossible to represent a triangle in this medium which will violate the rule of the sum of angles in a triangle. Similarly, we will be able to find representations of concepts and suitable processes which will automatically yield fuzzy relations without the need of explicitly characterizing them. The perceptual domain appears to be a suitable domain to explore intrinsic representations of fuzziness. Lotfi Zadeh provided a suitable theory to make explicit what is going on in such representations on the epistemological level.

Fig. 27.1. Spontaneous gathering of friends and admirers with Lotfi Zadeh on the occasion of the congress "Wissensbasierte Systeme" (Knowledge-based Systems) in Munich, Germany, 29 October 1985. Photographer: anonymous.

27.4 Epilogue

After receiving my Ph.D. on representing fuzzy knowledge by means of discrete relations from Berkeley, I returned to Europe full of excitement about continuing my research in cognitive knowledge representation, a field that European researchers had not yet seriously approached. It took 15 more years before we were able to establish computer science-based cognitive science in Germany. Lotfi Zadeh recognized my early frustration regarding lack of support for my research initiatives after leaving Berkeley. In 1981 he wrote to me: *My advice to you is to accept as a fact of life that you're in a conservative environment. I have no doubt, though, that eventually your ideas will prevail and receive the recognition they deserve. In short, don't give up your efforts no matter what.*

Thirty years later I can say, this was an excellent advice. As always.

Acknowledgement. I acknowledge generous support for the Spatial Cognition Research Center SFB/TR 8 at the Universities of Bremen and Freiburg by the German Research Foundation. I thank Thora Tenbrink for critical comments on this paper.

References

1. Berlin, B., Kay, P.: Basic Color Terms: Their Universality and Evolution. University of Califonia Press, Berkeley (1969)
2. Freksa, C.: Communication about visual patterns by means of fuzzy characterizations. In: XXIInd International Congress of Psychology, Leipzig (1980)
3. Palmer, S.E.: Fundamental aspects of cognitive representation. In: Rosch, E., Lloyd, B.B. (eds.) Cognition and Categorization, pp. 259–303. Lawrence Erlbaum, Hillsdale (1978)
4. Rosch, E.: On the internal structure of perceptual and semantic categories. In: Moore, T.E. (ed.) Cognitive Development and the Acquisition of Language, Academic Press, New York (1973)
5. Rosch, E.: Principles of categorization. In: Rosch, E., Lloyd, B.B. (eds.) Cognition and Categorization, pp. 27–48. Lawrence Erlbaum, Hillsdale (1978)
6. Tenbrink, T., Freksa, C.: Contrast sets in spatial and temporal language. Cognitive Processing 10, 322–324 (2009)
7. Zadeh, L.A.: Personal communication (1970s) (citation missing)

28

A Beginner's View on Fuzzy Logic

Itziar García-Honrado

28.1 First Touch with Fuzzy Logic

The first time I heard the term *fuzzy logic* goes together with the term *fuzzy set* [14]. It was in the last year of my degree of Mathematics in the subject of Statistics, and my partners and me have to complete an opinion pull answering through fuzzy sets. I heard a brief explication of what a fuzzy set is and I wrongly related it with a distribution function.

Immediately after my degree I entered the European Centre for Soft Computing to work on my PhD Dissertation under the advice of Professor Enric Trillas. During this period I met several times Professor Lofti Zadeh, and I was also acquainted with textbooks in Fuzzy Logic like [2] and [3], as well as with Trillas' work on which I finally did my Dissertation.

Nowadays, I know that a fuzzy set is not related at all with probability. In fact, if we consider that a fuzzy set ($p_1 \in [0,1]^X$) is a probability distribution, we realize we can not compare these fuzzy sets under the typical pointwise ordering ($p_1 \leq p_2$ if and only if $p_1(x) \leq p_2(x)$, $\forall x \in X$), since two fuzzy sets representing two probability distributions are identical or not comparable [12]. It is enough to follow the chain, $p_1 \leq p_2 \rightarrow p_1(a) \leq p_2(a)$, and $p_1(a') \leq p_2(a') \rightarrow 1 - p_1(a) \leq 1 - p_2(a)$ and $p_1(a) \leq p_2(a) \rightarrow p_1(a) = p_2(a)$, that is, functions p_1 and p_2 coincide.

Even more, probability is used to represent *uncertainty* and fuzzy sets *imprecision*. Both terms model unknown aspects, but under my view they model two kinds of lack of knowledge. Uncertainty is the one that goes together with an event before its realization and it is based on some background knowledge on the event, usually information on the results from the experience of previous realizations, and after the realization of the experience the uncertainty disappears. In the case of uncertainty, probability is devoted to model those situations. On the other hand, imprecision, under my view, is not only a matter of modelling experiences, or physical phenomena, it is mainly related to the human linguistic way of describing real situations. For instance, when a word is stated in a concrete language all people knowing that language understands its meaning. Following that idea a model for the meaning of predicates can be shown by using fuzzy sets.

Fuzzy logic is different from probability [16]. In fact, for defining a probability it is necessary a boolean structure and this is not the case of fuzzy logic: predicates act on non necessarily structured universes of discourse. So, fuzzy logic is not the same, and under my view it is not a kind of extension of the model of probability

R. Seising et al. (Eds.): *On Fuzziness: Volume 1*, STUDFUZZ 298, pp. 185–191.
DOI: 10.1007/978-3-642-35641-4_28

for representing unknown events. Currently, no definitive theory of probabilities for 'imprecise events' does exists [12].

Notwithstanding, there could be a relation between probabilities and fuzzy logic, but, which one? Models combining Fuzzy Logic and Probability. Although there are some models of probability of fuzzy sets [15], [17], [12], considering fuzzy events or fuzzy representation of the events, a lot of studies in that field are yet necessary.

To conclude this section, I would like to remark the necessity of distinguishing uncertainty from imprecision in order to built different models depending on the phenomena.

28.2 Fuzzy Sets for Representing Predicates in the Language

Currently, the evolution of fuzzy logic arrives into the field of Computing With Words [18]. Imprecision is in the language, and a possible way of representing gradable predicates is through fuzzy sets capable of building a model collecting their imprecision [7].

In fact, Wittgenstein considered that "The meaning of a predicate is its use in language" [13], and in order to built a model for a meaning of a predicate, following that definition, its use is analyzed, and is collected by a fuzzy set as it is shown in this section.

When stating a predicate (P) in a universe of discourse (X), it is established a relation, that allows to compare the elements in the universe, supposing that the relation is transitive and reflexive (mathematically it is a preorder), $\leq_P$. If the predicate is $P = tall$, we can say that "the element x in the universe of discourse *is less tall* than the element y", and represent that fact by $x \leq_P y$. So, the relation $\leq_P$ is known as the primary use or meaning of the predicate in the universe of discourse.

Once a structure $(X, \leq_P)$ is established, an important question lies in how to measure up to which extent x is P, allocating a value in the unit interval through a fuzzy set defined on the universe of discourse into the unit interval, $\mu_P : X \to [0, 1]$, and verifying that $\mu_P(x) \leq \mu_P(y)$ if $x \leq_P y$. In the particular case that $\mu_P(x) \leq \mu_P(y)$ if and only if $x \leq_P y$, it is built an accurate representation, and it is said that μ_P perfectly collects the meaning of P. Not always it is possible to have that sufficient and necessary condition.

It is usual to have non-comparable elements under a predicate in X, but the fuzzy set forces them to be compared because they allocate a degree between 0 and 1 to measure up to which extent x is P, and the unit interval has a total order. In order to avoid this problem, it could be enlarged the definition of fuzzy sets into an L-set, $\mu : X \to L$, where $(L, \leq)$ is any preorder. It is proven in [1], that there always exist a preorder $(L, \leq)$, perfectly collecting the meaning of each P, and it is known as the natural preorder associated to the predicate.

It is important to remark the importance of a careful design of fuzzy sets [10], or the L-set collecting the meaning of the predicate, it should be taken into account the population, the context, the purpose, the use,... So, depending on those variables different models of predicates could be built, that is, different fuzzy sets or L-sets, maybe analyzing the similarities keeping the idea of language games of Wittgenstein, as it is done in [11].

Here we show two possible representations of the predicate *tall* in two different contexts, a basketball team (μ_{A_1}) and a school (μ_{A_2}).

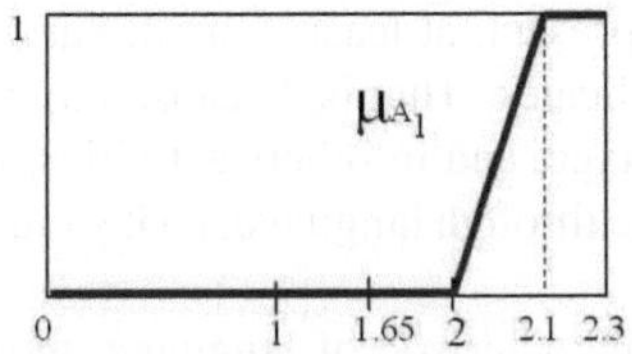

Fig. 28.1. Tall in a basketball team.

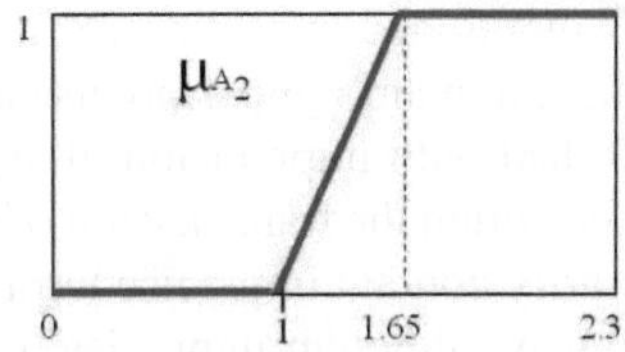

Fig. 28.2. Tall in a school.

It is easy to check in a simple example how a fuzzy set representing a predicate could vary depending on the population (Spanish people, a set of pygmy), the context (schools, adult population), the purpose (exaggerating what is considered tall),...

As well as a fuzzy set could represent a predicate, it also represents the collective generated by the predicate. For instance, in a big horse's farm, the predicate 'short' allows to talk of the collective of short horses in the farm.

Therefore, fuzzy sets could be seen as ways of representing imprecise concepts, which is, under my view, the main idea of that Logic. Additionally, some light is shed on the problem of what can characterize a function $\mu : X \to [0,1]$ to actually represent a fuzzy set.

28.3 Ideas Related with Fuzzy Logic

Fuzzy Logic collects, as a degenerated case, classical logic. In the jump between classical to Fuzzy Logic, many classical principles supposed "always" true, fail, because the flexibility of Fuzzy Logic breaks the rigid structure of Boolean algebras. This fact could seem to question the validity of Fuzzy Logic.

For instance, the flexibility of Fuzzy Logic allows elements to verify both a predicate and its negation, as it happens in real situations ("He is neither tall, nor not tall"). That idea represents the falsation of Non-contradiction (NC) principle in its classical form. Notwithstanding, going back into the Aristotelian Principle of NC, it is obtained a new interpretation under which Fuzzy Logic is not in contradiction with this classical principle [6]. Indeed, all principles that were considered "always" valid could depend on its interpretation.

In the field of Fuzzy Logic, the possibility of obtaining new interpretations grows because Fuzzy Logic is not defined in strong structures, and depending of the chosen structure for each model many properties could change, but this is not the case in classical Logic. This is, the positive side of Fuzzy Logic, but also it is necessary much more careful design when building models.

28.4 Models Following Those Ideas

Under my view, Fuzzy Logic allows us to build models closer to human activities, as it is mentioned in section 28.2. It allows us to represent, at least, a small part of language, non ambiguous precise and imprecise predicates. That is, because human beings deal with imprecision, it appears in the language, and in others activities, in fact, very often the transmission of knowledge is done through language, so it should necessarily contain imprecise terms.

Anyway, although imprecision is an important characteristic of language, there are other variables that are not included in the field of Fuzzy Logic, at least in the field as it is understood nowadays. But, in any case, Fuzzy Logic opens a door that allows us thinking in possible mathematical models more flexible and related with the real world.

For instance, the meaning of the particle *and* in the language can not be modeled by the common representation of intersection in Fuzzy Logic, since in that case it is mandatory to be commutative and when times appears, the particle *and* in language is not commutative ("He arrived and he left, or He left and he arrived").

Therefore, the idea of collecting imprecision through Fuzzy Logic could allow us to build models translating, in some particular cases (predicates or collectives) the way of speaking. And maybe, in the future, the ways of thinking through Commonsense Reasoning (or Conjectural [4]) Models using fuzzy sets [9], [8]. These models contain the deductive models of reasoning, but they also include new ways of obtaining conclusions from a piece of information, for instance, non-monotonic reasoning, when if there is new information, the number of conclusions decrease. It is important to notice that those models contain deduction since again it appears the idea of fuzzy logic with respect to the classical one, in the sense that it contains classical one but breaking the boundaries and its inflexibility.

With fuzzy sets a small part of the language can be represented, but this part is indeed larger than the one representable with classical sets.

28.5 The Importance of Fuzzy Logic

After only few years working in the field of Fuzzy Logic, I consider very important its development towards an experimental science. Since sciences usually require construction of models of the real world, and we are immersed in it and describe it by means of imprecise terms, indeed, only artificial models of reality are precise. Inside the concept of the real world, there are people which are also rounded of imprecision, nobody could define himself, his feeling, ways of reasoning, speaking,... into crisp terms. The main characteristics of human beings are variability and originality, and without imprecision they will disappear.

So, instead of hiding this logic, I think it would be crucial to show its potential. And, as a previous stage, to show that Maths, although with a strong logic structure, allows to model imprecision. As Zadeh uses to say 'Fuzzy Logic is not fuzzy'.

28.6 The Future

Following with the ideas shown in these lines and related with fuzzy logic, I propose to further develop models of Language and models of Commonsense Reasoning.

Regarding the topic of language, fuzzy logic allows to represent predicates, also connectives, conditional sentences [5], but it seems that fuzzy logic is not enough to represent any piece of language. Collecting the meaning of language into mathematical expressions is a very long path, but in the future fuzzy logic can allow new contributions of some interest for this purpose, such as, analyzing other particular words different to predicates or collectives, analyzing the whole meaning of a sentence not based on a union of words, etc.

Relating with Commonsense Reasoning models, there are not many results when these models deal with fuzzy information. Therefore, as in the way of modelling human ways of reasoning imprecision has to appear, this is why there is the necessity of improving these models in structures where fuzzy sets could exist, the strongest structure of fuzzy set is a De Morgan Algebra where the classical form of the principle of Non-Contradiction is not verified. So, it is necessary to model reasonings in structures with fewer number of laws.

To conclude, I want to express the duty of showing the importance of Fuzzy Logic to approach reality through mathematical models. This idea could be interesting to be transmitted in different stages of education, not only in the graduate and postgraduate studies. Nowadays, I have the opportunity of dealing with students that want to be teachers of primary school, and I think they should know the existence of this logic, in order to enlarge their vision of Mathematics, and not just teach mathematics trying to built reasonings with their students closer to reasonings coming from classical logic and deduction, but also taking into account, for example, real systems described in Natural Language. Approaching maths in ways closer to the current world.

Acknowledgement. This work has been supported by the Foundation for the Advancement of Soft Computing (ECSC) (Asturias, Spain), and by the Spanish Department of Science and Innovation (MICINN) under project TIN2011-29827-C02-01.

Fig. 28.3. Ana Belén Ramos, Ángela Blanco, Lotfi A. Zadeh and Itziar García-Honrado at November 2008 in Avilés (Spain) at the ceremony of Cajastur International Prize on Soft Computing

References

1. García-Honrado, I., Trillas, E.: An essay on the linguistic roots of fuzzy sets. Information Sciences (181), 4061–4074 (2011)
2. Klir, G.J., Yuan, B.: Fuzzy sets and fuzzy logic. Prentice Hall PTR, Upper Saddle River (1995)
3. Nguyen, H.T., Walker, E.A.: A First Course in Fuzzy Logic. Chapman & Hall/CRC (2000)
4. Popper, K.R.: Conjectures and Refutations. Routledge & Kegan Paul, London (1963)
5. Trillas, E.: Lógica borrosa y narrativa: Un párrafo de Vila-Matas. In: Proceedings ESTYLF 2004, pp. 15–20 (2004) (in Spanish)
6. Trillas, E.: Non Contradiction, Excluded Middle, and Fuzzy Sets. In: Di Gesù, V., Pal, S.K., Petrosino, A. (eds.) WILF 2009. LNCS (LNAI), vol. 5571, pp. 1–11. Springer, Heidelberg (2009)
7. Trillas, E.: On a Model for the Meaning of Predicates – A Naïve Approach to the Genesis of Fuzzy Sets. In: Seising, R. (ed.) Views on Fuzzy Sets and Systems from Different Perspectives. STUDFUZZ, vol. 243, pp. 175–205. Springer, Heidelberg (2009)
8. García-Honrado, I., Trillas, E.: On an Attempt to Formalize Guessing. In: Seising, R., Sanz González, V. (eds.) Soft Computing in Humanities and Social Sciences. STUDFUZZ, vol. 273, pp. 243–264. Springer, Heidelberg (2012)

9. Trillas, E., García-Honrado, I., Pradera, A.: Consequences and conjectures in preordered sets. Information Sciences 180(19), 3573–3588 (2010)
10. Trillas, E., Guadarrama, S.: Fuzzy representations need a careful design. International Journal of General Systems 39(3), 329–346 (2010)
11. Trillas, E., Moraga, C., Sobrino, A.: On 'family resemblances' with fuzzy sets. In: Proceedings IFSA-EUSFLAT, pp. 306–311 (2009)
12. Trillas, E., Nakama, T., García-Honrado, I.: Fuzzy Probabilities: Tentative Discussions on the Mathematical Concepts. In: Hüllermeier, E., Kruse, R., Hoffmann, F. (eds.) IPMU 2010. LNCS, vol. 6178, pp. 139–148. Springer, Heidelberg (2010)
13. Wittgenstein, L.: Philosophical Investigations, 3rd edn. Basil Blackwell, Oxford (1981)
14. Zadeh, L.A.: Fuzzy Sets. Information and Control 8, 338–353 (1965)
15. Zadeh, L.A.: Probability measure of fuzzy events. Journal of Mathematical Analysis and Applications 23, 421–427 (1968)
16. Zadeh, L.A.: Fuzzy sets versus probability. Proceedings of the IEEE 68(3), 421–421 (1980)
17. Zadeh, L.A.: Fuzzy probabilities. Information Processing & Management 20(3), 363–372 (1984)
18. Zadeh, L.A.: Fuzzy logic = computing with words. IEEE Transactions on Fuzzy Systems 4, 103–111 (1996)

29

Reciprocal and Linguistic Preferences

José Luis García-Lapresta

29.1 Introduction

In this note I first explain how I was interested by Fuzzy Set Theory. Among the huge variety of issues within the Fuzzy Set Theory, I briefly focus on the problem of dealing with the intensities of preference that human beings usually feel when they compare different alternatives. Clearly, Fuzzy Set Theory is an appropriate framework for modeling degrees of preference, both in the numerical and linguistic approaches.

29.2 How I Discovered Fuzzy Set Theory

The first conference where I presented a contribution was the *VII Congrés Català de Lògica* (7th Catalan Conference of Logic) which was held in Barcelona, April 15-16, 1988. I participated in that conference with the encouragement of my advisor, Josep Maria Font, and my talk was about Algebraic Logic, the topic of my thesis. I had the opportunity of meeting some Catalan mathematicians whose field of research was Fuzzy Logic (Claudi Alsina, Francesc Esteva, Lluís Godo, Joan Jacas, Enric Trillas, etc.). I realized that this group was very enthusiastic with Fuzzy Logic, and they behaved like a family, of which Lotfi Zadeh was the father. After that conference, I had the opportunity of attending some meetings and talks about Fuzzy Set Theory.

Once I defended my PhD Dissertation in 1991, I was interested in classic Preference Modeling and Social Choice. Reading some working papers of the *Institut des Mathématiques Economiques* (Dijon, France) by Claude Ponsard, and Richard Barret, Prasanta K. Pattanaik and Maurice Salles, I understood the importance of the fuzzy approach for modeling human behavior in decision-making, and then a part of my research focused on fuzzy preference modeling[1] and the aggregation of graded preferences.[2]

[1] Initially, I was very motivated by the papers of Nurmi [25] and Tanino [32].

[2] In `http://www.eco.uva.es/lapresta/index_ing.htm` such research is detailed.

R. Seising et al. (Eds.): *On Fuzziness: Volume 1*, STUDFUZZ 298, pp. 193–197.
DOI: 10.1007/978-3-642-35641-4_29

29.3 Graded Preferences

In the fields of Economics, decision making and related areas, the usual assumption about individual preferences is that, for every pair of alternatives, each agent can declare if one alternative is preferred to another; otherwise, the agent should declare indifference between these alternatives. Consequently, individuals cannot show intensities of preference between alternatives. Additionally, it is usually assumed that both preference and indifference relations are transitive, i.e., the weak preference (or preference-indifference) relation is a *weak order* (or *complete preorder*).

Within the fuzzy approach, there is a huge literature on how to model valued and graded preferences (see Zadeh [33], Orlovski [26], Fodor and Roubens [12], De Baets and Fodor [8], Bodenhofer *et al.* [4] and Dubois [9], among others).

We now focus on two types of graded preferences: reciprocal and linguistic preferences.

29.3.1 Reciprocal Preferences

Reciprocal preferences are an interesting tool for modeling numerical intensities of preference. Given two alternatives, say x and y, a reciprocal preference relation R assigns a number $R(x,y)$ between 0 and 1 for the intensity of preference between x and y in the following way: if both alternatives are indifferent, then $R(x,y) = 0.5$; but if one alternative is preferred to another, then it is possible to declare an intensity in a bipolar scale[3]. If x is preferred to y, then $0.5 < R(x,y) \leq 1$; and if y is preferred to x, then $0 \leq R(x,y) < 0.5$. It is assumed that $R(x,y) + R(y,x) = 1$, so if an individual prefers x to y with an intensity $R(x,y) > 0.5$, then the intensity of preference between y and x is $R(y,x) = 1 - R(x,y) < 0.5$. This assumption generalizes the asymmetry condition of ordinary preference relations, whenever $R(x,y) \in \{0, 0.5, 1\}$: $R(x,y) = 1 \Leftrightarrow R(y,x) = 0$.

In the setting of reciprocal preferences, different transitivity properties have been considered[4], some of them under the general condition $R(x,y) \geq g(R(x,z), R(z,y))$ for every alternative z such that $R(x,z) > 0.5$ and $R(z,y) > 0.5$, where g is an appropriate monotonic function[5].

In spite of the clear advantages of considering reciprocal preferences instead of ordinary preferences, usually human beings are more comfortable when the may express their preferences in a linguistic rather than a numerical fashion. It is more natural to say "I strongly prefer x to y" than "I prefer x to y with an intensity of $\sqrt{\pi}/2$". The problem is how to manage words instead of numbers.

---

[3] It is important to note that reciprocal preference relations are $[0,1]$-valued preference relations, but they are not fuzzy in strict sense.

[4] See Bezdek *et al.* [3], Nurmi [25], Tanino [32], Fodor and Roubens [12], Dasgupta and Deb [5], De Baets and Fodor [8], Switalski [29–31], De Baets and De Meyer [6], De Baets *et al.* [7], García-Lapresta and Montero [15] and Rademaker and De Baets [28], among others.

[5] See García-Lapresta and Meneses [14] for references, and for a real case analysis about the fulfillment of six transitivity conditions in the framework of reciprocal preferences.

29.3.2 Linguistic Preferences

Usually linguistic terms are managed by means of a semantics given by trapezoidal or triangular fuzzy numbers. In this approach, it is possible to aggregate linguistic information by means of the fuzzy number arithmetics (see Zadeh [34–36] and Herrera and Herrera-Viedma [18, 19], among others).

According to Herrera *et al.* [20]: *Two main different approaches are used to aggregate and compare linguistic values: the first acts by direct computation on labels [...]; and the second uses the associated membership functions. Most of the available techniques belong to the latter. However, the final results of these are fuzzy sets which do not correspond to any label in the original term set. To obtain a label, a "linguistic approximation" is needed.*

In Herrera and Martínez [21], linguistic terms are associated with integer numbers in order to allow different aggregation procedures. The aggregated value is associated with a 2-tuple defined by a linguistic term and a number. This procedure is completed with a linear order on the set of 2-tuples that permits rank order the outcomes generated by the aggregation process. Although the process manages linguistic information, it is mathematically equivalent (but not behaviorally) to work with numerical values, once each linguistic term is associated with an integer number.

However, there are also some proposals for aggregating linguistic information, within finite ordinal scales, under a pure ordinal approach where semantics have no place. In this way, it is interesting to mention the contributions of Ovchinnikov [27], Mas *et al.* [23, 24], Fodor [11], Marichal and Mesiar [22] and Grabisch [16, 17].

Within the ordinal approach, Balinski and Laraki [1, 2] have introduced a new voting system called *Majority Judgment* where voters assess candidates through linguistic terms within a finite scale. The authors manage linguistic information in a pure ordinal fashion (the median of individual linguistic assessments) in order to assign a collective linguistic assessment to each candidate, and they provide a tie-breaking procedure for ranking candidates. In order to avoid some drawbacks of *Majority Judgment* when this method is applied to small committees, García-Lapresta and Martínez-Panero [13] and Falcó and García-Lapresta [10] have modified the original voting system, by using OWA operators and the 2-tuple fuzzy linguistic representation [21], and distances between linguistic terms, respectively.

Acknowledgement. I am indebted with Bernard De Baets, Enrique Herrera-Viedma, Bonifacio Llamazares, Luis Martínez and Javier Montero. The financial support of the Spanish Ministerio de Ciencia e Innovación (projects ECO2009-07332, ECO2008-03204-E/ECON), and ERDF are gratefully acknowledged.

References

1. Balinski, M., Laraki, R.: A Theory of Measuring, Electing and Ranking. Proceedings of the National Academy of Sciences of the United States of America 104, 8720–8725 (2007)

2. Balinski, M., Laraki, R.: Majority Judgment. Measuring, Ranking, and Electing. The MIT Press, Cambridge (2011)
3. Bezdek, J., Spillman, B., Spillman, R.: A Fuzzy Relation Space for Group Decision Theory. Fuzzy Sets and Systems 1, 255–268 (1978)
4. Bodenhofer, U., De Baets, B., Fodor, J.C.: A Compendium of Fuzzy Weak Orders: Representations and Constructions. Fuzzy Sets and Systems 158, 811–829 (2007)
5. Dasgupta, M., Deb, R.: Transitivity and Fuzzy Preferences. Social Choice and Welfare 13, 305–318 (1996)
6. De Baets, B., De Meyer, H.: Transitivity Frameworks for Reciprocal Relations: Cycle-transitivity Versus FG-transitivity. Fuzzy Sets and Systems 152, 249–270 (2005)
7. De Baets, B., De Meyer, H., De Schuymer, B., Jenei, S.: Cyclic Evaluation of Transitivity of Reciprocal Relations. Social Choice and Welfare 26, 217–238 (2006)
8. De Baets, B., Fodor, J.C.: Twenty Years of Fuzzy Preference Structures. Rivista di Matematica per le Scienze Economiche e Sociali 20, 45–66 (1997)
9. Dubois, D.: The Role of Fuzzy Sets in Decision Sciences: Old Techniques and New Directions. Fuzzy Sets and Systems 184, 3–28 (2011)
10. Falcó, E., García-Lapresta, J.L.: A Distance-based Extension of the Majority Judgement Voting System. In: Acta Universitatis Matthiae Belii. Mathematics, vol. 8, pp. 17–27 (2011)
11. Fodor, J.: Smooth Associative Operations on Finite Ordinal Scales. IEEE Transaction on Fuzzy Systems 8, 791–795 (2000)
12. Fodor, J., Roubens, M.: Fuzzy Preference Modelling and Multicriteria Decision Support. Kluwer Academic Publishers, Dordrecht (1994)
13. García-Lapresta, J.L., Martínez-Panero, M.: Linguistic-Based Voting Through Centered OWA Operators. Fuzzy Optimization and Decision Making 8, 381–393 (2009)
14. García-Lapresta, J.L., Menesess, L.C.: Individual Valued Preferences and Their Agregation: Analysis of Consistency in a Real Case. Fuzzy Sets and Systems 151, 269–284 (2005)
15. García-Lapresta, J.L., Montero, J.: Consistency in Preference Modelling. In: Bouchon-Meunier, B., Coletti, G., Yager, R.R. (eds.) Modern Information Processing. From Theory to Applications, pp. 87–97. Elsevier, Amsterdam (2006)
16. Grabisch, M.: Representation of Preferences Over a Finite Scale by a Mean Operator. Mathematical Social Sciences 52, 131–151 (2006)
17. Grabisch, M.: How to Score Alternatives When Criteria are Scored on an Ordinal Scale. Journal of Multi-Criteria Decision Analysis 15, 31–44 (2008)
18. Herrera, F., Herrera-Viedma, E.: Linguistic Decision Analysis: Steps for Solving Decision Problems Under Linguistic Information. Fuzzy Sets and Systems 115, 67–82 (2000)
19. Herrera, F., Herrera-Viedma, E.: Choice Functions and Mechanisms for Linguistic Preference Relation. European Journal of Operational Research 120, 144–161 (2000)
20. Herrera, F., Herrera-Viedma, E., Verdegay, J.L.: A Linguistic Decision Process in Group Decision Making. Group Decision and Negotiation 5, 165–176 (1996)
21. Herrera, F., Martínez, L.: A 2-tuple Fuzzy Linguistic Representation Model for Computing with Words. IEEE Transactions on Fuzzy Systems 8, 746–752 (2000)
22. Marichal, J.-L., Mesiar, R.: Aggregation on Finite Ordinal Scales by Scale Independent Functions. Order 21, 155–180 (2004)
23. Mas, M., Mayor, G., Torrens, J.: T-operators and Uninorms on a Finite Totally Ordered Set. International Journal of Intelligent Systems 14, 909–922 (1999)
24. Mas, M., Monserrat, M., Torrens, J.: On Bisymmetric Operators on a Finite Chain. IEEE Transactions on Fuzzy Systems 11, 647–651 (2003)

25. Nurmi, H.: Approaches to Collective Decision Making with Fuzzy Preference Relations. Fuzzy Sets and Systems 6, 249–259 (1981)
26. Orlovski, S.: Decision-making With a Fuzzy Preference Relation. Fuzzy Sets and Systems 1, 155–168 (1978)
27. Ovchinnikov, S.: Means on Ordered Sets. Mathematical Social Sciences 32, 39–56 (1996)
28. Rademaker, M., De Baets, B.: Aggregation of Monotone Reciprocal Relations with Application to Group Decision Making. Fuzzy Sets and Systems 184, 29–51 (2011)
29. Switalski, Z.: Rationality of Fuzzy Reciprocal Preference Relations. Fuzzy Sets and Systems 107, 187–190 (1999)
30. Switalski, Z.: Transitivity of Fuzzy Preference Relations – an Empirical Study. Fuzzy Sets and Systems 118, 503–508 (2001)
31. Switalski, Z.: General Transitivity Conditions for Fuzzy Reciprocal Reference Matrices. Fuzzy Sets and Systems 137, 85–100 (2003)
32. Tanino, T.: Fuzzy Preference Orderings in Group Decision Making. Fuzzy Sets and Systems 12, 117–131 (1984)
33. Zadeh, L.A.: Similarity Relations and Fuzzy Orderings. Information Sciences 3, 177–200 (1971)
34. Zadeh, L.A.: The Concept of a Linguistic Variable and its Applications to Approximate Reasoning. Information Sciences Part I: 8, 199–249, Part II: 8, 301–357, Part III: 9, 43–80 (1975)
35. Zadeh, L.A.: Fuzzy logic = Computing With Words. IEEE Transactions on Fuzzy Systems 4, 103–111 (1996)
36. Zadeh, L.A.: From Computing With Numbers to Computing With Words – From Manipulation of Measurements to Manipulation of Perceptions. In: Wang, P.P. (ed.) Computing with Words, pp. 35–68. John Wiley & Sons, New York (2001)

30

On a Meeting Point between Fuzzy Sets and Statistics

María Ángeles Gil

30.1 Approaching Fuzzy Set Theory

The first time that I heard about fuzzy sets was my brother, Pedro Gil, who talked about. To get the position of Full Professor at the University of Oviedo he had to pass different habilitation exams, among them one in which he had to prepare presentations on challenging topics. Professor Sixto Ríos, who was my brother's mentor and the scientific ancestor of a large number of current statisticians/operational researchers in Spain, decided that one of these topics would be Fuzzy Sets (based on Zadeh's seminal paper [6]).

When Pedro Gil joined the University of Oviedo as Full Professor in 1976, he created a research team to which I belonged from the very beginning. Since his main research expertise concerned Statistical Information Theory, group members started working on this field for the PhD and delayed for some years our approach to Fuzzy Set Theory. In 1983, after getting the PhD and the tenure, some of us consider to initiate such an approach. For this purpose, the friendly attitude of many Spanish researchers (namely, Enric Trillas, Llorenç Valverde, Miguel Delgado, Amparo Vila, Curro Verdegay, Francisco Azorín, among others) was determining. We could be lost in a new world, but they support us a lot so we could benefit from their expertise.

The aim of our first research developments involving fuzzy sets was that of performing statistics on random experiments/variables on the basis of fuzzy information from them (i.e., we considered what is usually referred to as the 'epistemic' view of the fuzzy information/data associated with a random experiment, see Dubois and Prade [1]). To deal with such a general problem we made use of the concept of *fuzzy information system* in Okuda *et al.*'s sense [3], which consisted of a Ruspini fuzzy partition of the sample space often along with Zadeh's probability of fuzzy events [7].

This first approach to Statistics with fuzzy data was mainly carried out in collaboration with our colleagues Norberto Corral (parametric estimation problem), María Teresa López (comparison of experiments) and María Rosa Casals (testing statistical hypotheses). Our interest on the topic was increasing quickly mainly thanks to the support of the above mentioned Spanish researchers who encouraged us to disseminate the results in international journals and conferences being familiar with it.

R. Seising et al. (Eds.): *On Fuzziness: Volume 1*, STUDFUZZ 298, pp. 199–204.
DOI: 10.1007/978-3-642-35641-4_30

In this way, in July 1985 we had the chance of participating in the First IFSA World Congress (held in Palma de Mallorca) and meeting many reputed researchers, realizing they were existing nice people beyond the books and papers we know from them. And, of course, Professor Zadeh was there. And we asked Elie Sanchez to mediate for having a picture with him (the one in Figure 30.1).

At that stage we ignored how easy is to have a picture with Zadeh. He never complains about... and probably he decides to take his own.

Fig. 30.1. On the occasion of the First IFSA Conference (Mallorca, July 1985), M.R. Casals, M.A. Gil, J. Bezdek, M.L. McAllister, L. A. Zadeh, T. Riera, M. Delgado, E. Sanchez, J.L. Verdegay, J. Bolaños and J. Montero.

We continued working on the same direction for a few more years and we met again Professor Zadeh in 1988, on the occasion of the IPMU'1988 held in Urbino. At that time, I had previously contacted him to ask for the possibility of spending a stay of around two years working under his guidance and he had agreed with.

Once more I should state that in response to my request, Zadeh wrote that he had hosted before several Spanish scientists (I remember he mentioned Ramón López de Mántaras, Enric Trillas, Llorenç Valverde, Teresa Riera, and others) so, for sure, they were the best reference for me.

30.2 Approaching UCBerkeley-$\mathscr{C}al$ and New Research Direction

By the middle of August 1988 I started my stay in the University of California at Berkeley (EECS Department-Computer Science Division). Professor Zadeh had several PhD students (I can remember Soumitra Dutta, Pratap Khedkar, Chuen-Tsai Sun, Lung Albert Chen, Chuen-Chin Lee, Yung-Yaw Chen and Jyh-Shing Jang), and many visiting scholars and also visiting senior researchers (among them, Enrique Ruspini).

Since he couldn't find appropriate desk space for so many people he took care of all the visitors by looking for very good hosts, in my case Professors David Blackwell (Statistics Department) and Alice M. Agogino (Mechanical Engineering Department). It was a real privilege to work with them, to attend their group meetings, and exchanging discussions with their group members. It is impossible to summarize the influence of this stay in so many respects. To realize about the mutual admiration between Blackwell (a highly reputed Bayesian) and Zadeh, as well as the warm respect of Agogino and Ruspini to Zadeh, was something that should be lived (instead of described), and I had the chance to.

Fig. 30.2. Lotfi and Fay Zadeh along with David and Anne Blackwell, Alice Agogino (with little Arianne), Dale Gieringer and María A. Gil (in a chinese restaurant in Berkeley, July 1989)

The stay in Berkeley was unforgettable. A deep honor and pleasure. And it was crucial to join a new research direction: the development of statistics with fuzzy data by means of *random fuzzy sets* (or *fuzzy random variables* in Puri and Ralescu's sense [4]). Random fuzzy sets determine a well-formalized model within the probabilistic context. They are a special case of what Fréchet called random elements [2]) and an extension of the so-called random sets.

The aim of this second direction was that of performing statistics on random attributes being intrinsically fuzzy-valued (i.e., we considered what is referred to as the 'ontic' view of the fuzzy information/data associated with a random experiment, see Dubois and Prade [1]). It should be remarked that at present interesting approaches have been developed in connection with the epistemic view, but we have no longer investigated on the fuzzy information systems approach.

We used random fuzzy sets for the first time to model the utility function in a single-stage statistical decision problem in which we consider utilities/losses of the consequences were essentially imprecise. For the last two decades we have dealt with random fuzzy sets, first to study several probabilistic aspects (mainly in collaboration with Miguel López-Díaz, Ana Colubi and Luis J. Rodríguez-Muñiz) and to analyze fuzzy data from a statistical perspective (in collaboration with Hortensia López-García, María Asunción Lubiano and Manuel Montenegro, and more recently with all the members of the SMIRE research group.[1]

Besides the many scientific benefits of my stay in Berkeley, where I counted with the permanent support, guidance and advice of Zadeh, I learned many human lessons that I have tried to apply as often as possible. Furthermore, whenever I have met Zadeh, he has always provided me with interesting ideas and suggestions for new challenging problems to examine. He is a master as a scientist and as a human being, and I will feel indebted with him for life.

30.3 Zadeh, the University of Oviedo and Asturias

After staying in Berkeley for around two years, I met Professor Zadeh in several conferences (so, new advices and wise comments from him along with some new pictures).

And there were some key meetings held in Oviedo. In July 1994, the University of Oviedo agreed, at the request of its Faculty of Sciences (Maths and Physics), to award Professor Zadeh with a Honoris Causa Doctorate "... in recognition of the extraordinary work developed as the creator of the Fuzzy Logic, Mathematics and Technology, as well as of the human and intellectual magistery on research groups in the University of Oviedo." For this reason, we were honored in December 1995 with Zadeh's first visit to Asturias (the region Oviedo belongs to). He received the award in a solemn ceremony. Many of his disciples joined him for the event (readers can see some of them in Figure 30.3).

In September 2004, Professor Zadeh was invited to open the 2nd International Conference on Soft Methods in Probability and Statistics held in Oviedo. Enric

[1] See `http://bellman.ciencias.uniovi.es/SMIRE` for details and [5] for references.

Trillas was the person in charge of introducing him and chairing Zadeh's plenary speech. The conference was partially supported by the main Savings Bank in Asturias, CajAstur. Zadeh and Trillas had the opportunity to talk with some CajAstur representatives (who were highly concerned with the advancement of the R+D and the innovation in Asturias) about an idea they had conceived: the creation of a research institution on Soft Computing in Europe, Spain being a quite suitable choice for that purpose.

Fig. 30.3. Zadeh in his Honoris Causa Doctorate by the University of Oviedo (December 1995) with several of his disciples attending the ceremony

Around one year later, in December 2005, CajAstur, the Ministry of Industry of Spain and the Principality of Asturias promoted the Foundation for the Advancement of Soft Computing in Mieres (Asturias) and launched a scientific research centre, the European Centre for Soft Computing (`http://www.softcomputing.es/`). Since the creation of this institution Professor Zadeh has been the Chair of the Scientific Committee of the ECSC, and since 2010 he becomes its Honorary Chair. For the years he has acted as Chair, he has often visited Asturias and this opportunity has been a gift for all of us.

To conclude, I should state we don't consider ourselves as researchers but rather users of Fuzzy Set Theory. In the SMIRE research group we all are statisticians and enjoy developing statistics. We consider fuzzy set concept means a realistic, easy-to-use, easy-to-interpret and expressive model for many imprecise experimental data. And the use of fuzzy sets for this purpose is allowing us to exploit much more existing statistical information than the use of other traditional scales to model imprecise data.

Acknowledgement. I wish to thank deeply Professor Zadeh for all I have learned from him. I would like extending my acknowledgement to all those helping me in knowing Zadeh and Fuzzy Set Theory.

References

1. Dubois, D., Prade, H.: Gradualness, Uncertainty and Bipolarity: Making Sense of Fuzzy Sets. Fuzzy Sets Syst. and Systems 192, 3–24 (2012)
2. Fréchet, M.: Les éléments aléatoires de nature quelconque dans un espace distancié. Annales de I'Institut Henri Poincaré X(IV), 215–310 (1948)
3. Okuda, T., Tanaka, H., Asi, K.: A formulation of Fuzzy Decision Problems with Fuzzy Information Using Probability Measures of Fuzzy Events. Information Control 38, 135–147 (1978)
4. Puri, M.L., Ralescu, D.A.: Fuzzy random variables. Journal of Mathematical Analysis and Applications 114, 409–422 (1986)
5. SMIRE Research Group publications, `http://bellman.ciencias.uniovi.es/SMIRE/Publications.html`
6. Zadeh, L.A.: Fuzzy sets. Information and Control 8, 338–353 (1965)
7. Zadeh, L.A.: Probability measures of fuzzy events. Journal of Mathematical Analysis and Applications 23, 421–427 (1968)

31

Lotfi A. Zadeh and Economic Uncertainty

Jaime Gil Aluja

One particular day whose date I cannot specify, which would have been in the interval [1968-1970], I received a call from Professor Kaufmann advising me the following: *"I have mailed you a work from Professor Lotfi A. Zadeh which I believe is worth studying"*. A few days later I received a copy of the article, "Fuzzy Sets", a treasure which I keep very safe to this day, and the reading and studying of this would totally change the direction of our modest works, at the same time giving a new sense to my teaching and research tasks carried out up to that time.

We later had the great pleasure of meeting Professor Lotfi A. Zadeh in person and this meant receiving a lesson and at the same time a stimulus to work on a first objective: that our city should be the centre around which our studies, research and teaching of the *Economic and Management Systems* within the sphere of uncertainty, supported on *Fuzzy Sets*, would revolve. In the meantime, Arnold Kaufmann was putting the finishing touches to what would be the first book known to us on *Fuzzy Logic* written by a single author: *"Introduction à la théorie des sous-ensembles flous"*, Published by Masson in 1973. This would be followed by a further three volumes under the same title (1973-1978), translated into Spanish, English and Russian. A few years later, in 1980, Professor Enric Trillas had his work: "Conjuntos borrosos" [Fuzzy Sets] published by Vicens.

From successive contacts with Lotfi A. Zadeh and with his teaching and training we learnt that our future decision should take in us three directions: in teaching, in research and in the organisation of groups interested in this new conception of economic and business studies. On this hopeful horizon, we had the great fortune of receiving the enthusiastic cooperation of young professors from the Universities of Barcelona, Rovira i Virgili and Gerona, the latter two headed by the greatly missed Professor Carlos Cassú and by Professor Joan Carlos Ferrer in Gerona and Professors Antonio Terceño and Gloria Barberà in Reus-Tarragona.

Professor Lotfi A. Zadeh encouraged us during the first steps of our teaching activity and under his inspiration we organised seminars at universities and other institutions. The first was given by Professor Kaufmann at the University of Barcelona, Fundación Abad Oliba, at the headquarters of the U.B. in Reus (which would later become the Faculty of Economy and Business of the Rovira i Virgili University). At later seminars the teaching was shared by Professor Kaufmann and myself and at the same time the teaching was extended territorially to other communities: Andalucía, Galicia, the Basque Country, Extremadura, Valencia, Castilla-León. Little by little

R. Seising et al. (Eds.): *On Fuzziness: Volume 1*, STUDFUZZ 298, pp. 205–209.
DOI: 10.1007/978-3-642-35641-4_31

research groups were formed, the presence of which at national and international congresses became ever more notorious.

When the decade of the 1980s came to an end we could give Professor Zadeh some excellent news: Fuzzy Sets were to be incorporated into the teaching plans of the University of Barcelona, Faculty of Economic and Business Sciences, in the following subjects: "Operational Research (methods and models in uncertainty)"; "Operational Techniques for Management in Uncertainty"; "Investment in Uncertainty"; "Financial Management II (financial analysis in uncertainty)" and "Business Creativity".

The teaching of these subjects later gave rise to the drawing up of Doctoral Theses which have always warranted the highest consideration by the respective tribunals that judged them. We will mention as an example: "Financial Management in Uncertainty. From singular expertise to the R+expertons" (Ana Mª Gil Lafuente 1992); "Commercial management: the taking of decisions in a sphere of uncertainty" (Jordi Bachs Ferrer 1993); "Determination of the uncertainty that is inherent to commercial operations with Latin America based on the theory of fuzzy sub-sets" (Ricardo Onses 1994); "Instruments for the Analysis of Financial Operations with uncertain data" (Antonio Terceño 1995); "Numerical and non-numerical marketing in uncertainty" (Jaime Gil Lafuente 1996); "Expectations of agricultural businessmen on the price of raw materials as a basis for a model of optimization by means of the technique of Fuzzy Sets in programming" (Vicente Sanjosé Mitjans 1997). All of these have been read and defended at the University of Barcelona with the highest qualification.

The last thesis presented in this field at the University of Barcelona, was titled: "Modelos para el análisis de atributos contemplados por los clientes en una estrategia de marketing relacional" (Carolina Luis, December 15th 2011) and was directed by Professor Ana Mª Gil Lafuente, obtaining the maximum classification of First Class Honours "Cum Laude".

The quality, dedication and enthusiasm of the research professors who have taken over the baton allow us to visualise a splendid future for this "new" conception of economic and management studies, based on the inspired idea of "Fuzzy Sets" of Lotfi A. Zadeh.

Teaching can nearly always be found on the threshold of new research.

The initial works of Lotfi A. Zadeh and also those which appeared successively allow us to delve into the very roots of the structure of economic thinking. Indeed, the bases on which the idea of the Fuzzy Set of Lotfi A. Zadeh sit could constitute the starting point for the development of new logical operators which in turn would allow the development of important elements for the treatment of economic and management problems.

The seed planted by Lotfi A. Zadeh swiftly gives fruit in the economic field. Proof of this are the more than 200 works published in scientific magazines and proceedings of congresses and conferences during the first decade in which the new investigations took place.

It is also worth emphasising that, following the steps of Lotfi A. Zadeh, the first book on management was published in Spanish in 1986 under the title of "Introducción a la teoria de los subconjuntos borrosos a la gestión de la empresa" [Introduction to the theory of fuzzy sub-sets to business management], with the signature of A. Kaufmann and J. Gil-Aluja. Following this book and by the same authors there were 7 other books, until the passing away of Professor Kaufmann. In the latter years of his life two brilliant young professors had been incorporated into our working team: Antonio Terceño from the Rovira i Virgili University with the writing of the book "Matemática para la economía y gestión de empresas", (1994) [Mathematics for economy and business management] and Ana Mª Gil-Lafuente with the publication of the work "La creatividad en la gestión de empresas" (1994) [Creativity in business management] with later translations into several languages.

The works of Lotfi A. Zadeh also allow immersion into other fields of social activity. One of these is the economic-financial management of sport. In this field of study, Professor Jaime Gil-Lafuente published a book in the fuzzy field: "Algoritmos para la excelencia. Claves para el éxito en la gestión deportiva" (2002), [Algorithms for excellence. Keys to success in sports management] which has signified for the author the fact of being considered as one of the most important specialists in the world on the economy of sport (see University of Strasbourg).

With the work published as a book in 1986, in which the Fuzzy Sets of Lotfi A. Zadeh were systematically incorporated to the analysis and treatment of management problems, a new research activity began from which new theories have been born and the generalisation of others already existing has taken place. In this regard we should mention the theory of forgotten effects, the theory of affinities and the theory of expertons.

Geographic dispersion, but also ideological in an academic sense, of people and groups who became incorporated into the new Fuzzy lines of teaching and economic research required the creation of organisational structures, if what was desired was to successfully channel and give opportunities to researchers who desired the commencement of a promising scientific trajectory. Professor Lotfi A. Zadeh encouraged and inspired this project which materialized after multiple negotiations.

Considering the possibilities of the time and always according with Lotfi A. Zadeh, the decision was taken to establish the City of Reus as the centre and headquarters of our new activities, where the Faculty of Economic and Business Sciences of the Rovira i Virgili University was, and still is, located. The legal coverage of this scientific organisation would be provided by two institutions: an association which took the name of SIGEF (International Society for Fuzzy Management and Economy) which would organise a yearly meeting under the format of a congress and would publish a review, and a foundation (FEGI) (Foundation for the Study of Management in Uncertainty), among whose tasks included subsidising the incidental needs of SIGEF and the periodical granting of an award in order to honour researchers within the sphere of Fuzzy Economics and Management. In April 1994, the association and foundation were created and in the same city where they have their headquarters, Reus, on the 16-18 of November 1994 the 1st International Congress on Fuzzy Management and Economy took place. This congress has been followed yearly without interruption with congresses in different European and American cities.

Likewise, in 1994 The Fuzzy Economic Review was created to include scientific works, which at quarterly intervals today carries on the task of making known all the works that are considered of quality from among those who use elements of Fuzzy Logic to provide solutions to the problems that concern those responsible for economy and management at all times, both in the micro and the macro-economic world.

The FEGI foundation on the other hand, continues its task of encouraging and feeding economic research supported by the Fuzzy Sets of Lotfi A. Zadeh.

In 1994, on the occasion of the sudden death of Professor Arnold Kaufmann the FEGI foundation unanimously agreed to institute the Kaufmann Award in order to reward those scientists who stood out due to their research in the sphere of the study of economic or management systems with the use of Fuzzy Logic. The design of the medal was entrusted to the great sculptor Josep Mª Subirachs, author of the façade of the Passion of the temple of the Sagrada Familia by Gaudi. Subirachs accepted the task and designed and created the Kaufmann Medal, cast in gold. In the year 2004 we had the honour of awarding the Kaufmann Medal to Lotfi A. Zadeh during one of his visits to Catalunya. This year, 2012, on the occasion of the SIGEF Congress this prize will again be formally awarded to a personality to be designated once the Jury have gathered.

Fig. 31.1. Lotfi A. Zadeh with Professor Gil-Aluja at the presentation of the "Kaufmann Award", in December 2004

It is now nearly half a century since Lotfi A. Zadeh published his fundamental work "Fuzzy Sets". The message contained therein continues to be alive, and, what perhaps is most important, continues to be useful for awaking sleeping consciences, and to illuminate new paths towards a better knowledge of physical, biological and social phenomena. For those, like us, who have dedicated more than half a century of our lives in attempting to understand, explain and adequately treat economic and management realities, the work of Zadeh has meant an impulse, which we would hope to be permanent, in order to continue cooperating, through science, to attaining a better, a more just, freer and harmonious world.

32

What Is Fuzzy Logic – And Why It Matters to Us

The ALOPHIS Group: Roberto Giuntini, Francesco Paoli, Hector Freytes, Antonio Ledda, and Giuseppe Sergioli

32.1 The Aim

The aim of this short note is twofold: recounting how our research group became interested in fuzzy logic, and briefly discussing a definition of fuzzy logic suggested by Běhounek and Cintula (see [1]). Lest the anecdotal *incipit* should be dismissed (perhaps deservedly) with a blunt *So what?*, we remind that prospective contributors to this volume are required to mention how they arrived to the field of fuzzy logic and to present their views and expectations 'on fuzziness'. Both aims, therefore, seem to sit comfortably within the scopes of this book, especially in view of the fact that Lofti Zadeh has always been concerned with the problem of delimiting the boundaries of the subject he pioneered (see e.g. his [16]).

32.2 Why Do We Care?

In the 1980s, the logic scene in the Philosophy Department at the University of Florence, where the two oldest members of our group were trained in the trade, was dominated by two charismatic figures, Ettore Casari and Maria Luisa Dalla Chiara. Neither the former nor the latter is a fuzzy logician *stricto sensu* — nor, so far as we can remember, did they ever devote to fuzzy logic more than a passing reference in the undergraduate courses we attended. Both of them, however, had research interests that bordered on fuzzy logic, and by sharing their own views with us they contributed in a decisive way to turn us to this kind of investigation.

Ettore Casari was fascinated by the project of building a formal model for comparison in natural language, a project he fleshed out in several papers published from the early 1980s onwards (see e.g. his [5]). He wanted to account for such comparative sentences as 'c is at most as P as d is Q', where c,d are names and P,Q are predicates. If we accept that sentences may admit of different 'degrees of truth', the aforementioned sentence can be considered true when 'c is P' is at most as true as 'd is Q'. To formalise his idea, Casari used an implication connective which comes out true exactly when its antecedent is at most as true as its consequent. Although fuzzy logics share the same basic assumptions, for a number of reasons Casari was dissatisfied with such an approach: for example, the use of bounded algebras as systems of truth degrees in mainstream fuzzy logics prevents a proper treatment of comparative

R. Seising et al. (Eds.): *On Fuzziness: Volume 1*, STUDFUZZ 298, pp. 211–215.
DOI: 10.1007/978-3-642-35641-4_32

sentences of the form 'c is less P than d' when both c and d are clear-cut instances of P, yet it makes sense to say that d is more P than c is. The vicinity between comparative logic and fuzzy logic is further highlighted by the fact that the equivalent variety semantics of Casari's propositional comparative logic is the variety of *ℓ-pregroups*, a common abstraction of Abelian ℓ-groups and MV algebras. Not surprisingly, a crucial influence on the definitive form that comparative logic assumed by the end of the 1980s was played (according to Casari himself) by Daniele Mundici, then at the Mathematics Department of the University of Florence (to which he recently returned), a leading figure in the research on fuzzy logic in general, and on MV algebras in particular.

Marisa Dalla Chiara has advocated and actively participated in the development of the so-called *unsharp approach* to quantum theory since the seminal contribution by Ludwig (see [13]). In a nutshell: in standard (sharp) quantum logic *à la* Birkhoff-von Neumann, propositions ascribing properties are represented by projection operators (or, equivalently, by closed subspaces of a Hilbert space). In this approach, vagueness and truth degrees play no rôle: the possible values of a given physical magnitude are expressed by the eigenvalues of the corresponding self-adjoint operator, and projection operators have eigenvalues in $\{0,1\}$ — meaning that either the property at issue definitely holds or it definitely does not hold. In unsharp quantum theory and in unsharp quantum logic, however, a more general notion of property has been suggested. Projections are replaced by *effects*, whose eigenvalues may range throughout the whole real interval $[0,1]$. Unsharp quantum theory, therefore, accommodates 'vague' properties as well, which are not an all-or-nothing matter but may hold to a given degree. True to form, the mathematical structures that arise within this research stream are, more often than not, either closely related to fuzzy logical structures or even plain generalisations of such. GLP (See e.g. [9].) More recently, Marisa also championed another brand of quantum logic, called *quantum computational logic* (see for instance [7], Chapter 17.), which departs even more drastically from the standard Birkhoff-von Neumann approach. Meanings of sentences are no longer formalised through closed subspaces of a Hilbert space, but by means of *quantum information units* acting as quantum analogues of classical bits and registers: qubits, quregisters, and qumixes. Somewhat unexpectedly, however, fuzzy-like structures appear in this setting, too. And it is precisely this interplay that triggered most of the joint research work subsequently done by our group.

Over the last ten years or so, in fact, the Cagliari branch of the *équipe* led by Marisa has mainly focused on the algebraic models of quantum computational logics, as well as on the logics themselves but from the viewpoint of abstract algebraic logic. Most of the effort has gone into the investigation of *quasi-MV algebras* (see for instance [12]), generalisations of MV algebras connected with an irreversible disjunction connective arising in quantum computational logic, and their expansions by a genuinely quantum operator of square root of negation ($\sqrt{\prime}$ *quasi-MV algebras*: (see [10])). Since the 1-assertional logics of these varieties, or of some closely related quasivarieties (see [4] and [14].), are indeed weakenings or expansions, or expansions of weakenings, of infinite-valued Łukasiewicz logic, the belief that this domain and fuzzy logic are (to use a word cherished by quantum theorists) inextricably entangled is even more corroborated.

32.3 What Do We Think It Is(n't)?

The papers we devoted to quasi-MV algebras and their associated logics employ methods, concepts and tools that by any standard appear as closely related to the ones commonly adopted in present-day fuzzy logic. In spite of these evident similarities, these logics do not count as fuzzy according to the definition proposed by Běhounek and Cintula (see [1]). These authors confine themselves to what they call *weakly implicative logics*, i.e. (roughly speaking) propositional logics containing a connective $\rightarrow$ with properties that are reasonable for an implication (including modus ponens). In their opinion, a weakly implicative logic **L** is fuzzy iff it is strongly complete w.r.t. to the class of all *totally ordered* **L**-matrices, where the order is so defined as to have $x \leq y$ just in case $x \rightarrow y$ is a designated value of the matrix [1].

To its advantage, this suggestion has a fair amount of liberality. Their propounders resist the temptation to relegate fuzzy logic into the safe territory of the $[0,1]$ closed real unit interval, because any possible way to formally specify this idea would lead to unreasonable verdicts: if we require from a fuzzy logic that it be complete w.r.t. a $[0,1]$-based semantics, then many fuzzy predicate logics would not come out fuzzy (because they fail to be standard complete), while if we require that algebras on $[0,1]$ generate the corresponding variety, a prototypically fuzzy logic like product Łukasiewicz logic (see [11]) would not count as such. On the other hand, when it comes to classifying individual logics as fuzzy or not fuzzy, this criterion seems to tally in most cases with the usual practice in the community, as Běhounek and Cintula observe.

Nonetheless, it remains to be seen whether the choice of restricting the domain of application of this definition to weakly implicative logics is reasonable. Should anyone suggest, perhaps in accord with a more conservative viewpoint, that a logic *with weakening* **L** is fuzzy iff it is strongly complete w.r.t. to the class of all totally ordered **L**-matrices, Běhounek and Cintula would probably retort (and we would go with them) that such a delimitation is unjustified and even harmful, because it is precisely *outside* the class of logics with weakening that their own criterion makes the most interesting distinctions [2]. By parity of reasoning, one could rightly wonder if being able to decide whether some logic is fuzzy or not should really hinge on the presence of an implication satisfying modus ponens. Weakly implicative logics are

[1] This definition is mentioned in a slightly modified form in the more recent [2], written with Petr Hájek, but not as the proposed definition of fuzzy logic. Here, in fact, another definition is given, where Condition i) is replaced by a much more restrictive requirement: being an intuitionistic substructural logic, namely, having as algebraic counterpart a class of FL-algebras. Here we will discuss neither this stance, nor another position recently embraced by one of these authors in [6], where the plausibility of a sharp, formal definition of fuzzy logic is called into question.

[2] Some considerations along these lines are offered in [1], p. 608.

at least *protoalgebraic*, while some logics that are unanimously classified as fuzzy by the fuzzy logical community are not such, and therefore do not even fall within the scope of the criterion.

Consider the so-called 'infinite-valued Łukasiewicz logic that preserves degrees of truth', first introduced by Wójcicki [15] and deeply investigated by Josep Font and his collaborators.[3] This non-protoalgebraic logic uses the same valuations into the $[0,1]$ interval as the standard infinite-valued Łukasiewicz logic, but adopts a different consequence relation: whereas in Łukasiewicz logic a formula α follows from the set Γ just in case for all such valuations v, $v(\alpha) = 1$ whenever $v(\gamma) = 1$ for all γ in Γ, here a formula α follows from the set Γ just in case for all such valuations v, $v(\alpha)$ is greater or equal than the *minimum* of the set $\{v(\gamma) : \gamma \in \Gamma\}$. In other words, while valid inferences in standard Łukasiewicz logic preserve just *absolute truth* but allow degrees of truth to decrease from premisses to conclusion, here valid inferences always have a conclusion which is 'at least as true' as the 'falsest' premiss. If fuzzy logics are to be logics of truth degrees, it can be convincingly argued that not only this logic belongs to the class, but it also takes degrees of truth much more seriously than its absolute truth-preserving counterpart (see [8], p. 392). Furthermore, since in the times of Łukasiewicz it was commonplace to view logics as determined by a set of valid formulas, or theorems, rather than as consequence relations, it is not out of the question that Łukasiewicz himself, when thinking of his infinite-valued logic, had *this* logic in mind rather than the truth-preserving one that is nowadays associated with his name ([15, p. 279]).

The logics from [4] and [14] are in a similar situation. In general, they fail to be protoalgebraic and therefore their membership in the class of fuzzy logic cannot be determined by Běhounek's and Cintula's evaluation standard. Moreover, quantum computational logics are typically complete w.r.t. a class of *totally preordered* matrices, but this preordering may fail to be antisymmetric[4]. However, one could modify Běhounek's and Cintula's suggestion by relaxing in some way the precondition that the scope of the criterion is limited to weakly implicative logics, and, perhaps, by also loosening Condition ii) so as to let in logics that are complete w.r.t. totally *pre*ordered matrices. As regards both aspects, the framework suggested by Berman and Blok (see [3]) in their paper on algebras defined from ordered sets looks promising: one could simply require, for instance, that the ordering relation referred to in Condition ii) be equationally definable in the class of the algebra reducts of the matrix models of the logic at issue (but not necessarily through a condition of the type $x \to y \in D$, for some implication connective $\to$ and some equationally definable truth predicate D). In [4], a first attempt has been made to extend Berman's and Blok's framework to the case of equationally definable preorders. This is not the place, of course, to evaluate the merits of this proposal, nor to develop it further into an

[3] See [8] for a brief description and a philosophical assessment.

[4] There are important exceptions to this state of affairs. The quasivariety $\mathbb{C}$ of Cartesian $\sqrt{\prime}$ quasi-MV algebras is relatively 1-regular, and therefore its 1-assertional logic is regularly algebraisable with $\mathbb{C}$ as equivalent quasivariety semantics; $\mathbb{C}$, in turn, is generated by a single *lattice-ordered* algebra whose order is defined by the implication (generalised) connective of the logic.

alternative definition of fuzzy logic. What we wanted to point out is that Běhounek's and Cintula's criterion, though on the right track, is in need of some adjustment if it aims at a discrimination of many individual cases present in the logical landscape, of which quantum computational logics and logics preserving degrees of truth are interesting examples.

Acknowledgement. We thank Libor Běhounek and Petr Cintula for pointing out an inaccuracy in a first draft of this paper, and for their stimulating observations.

References

1. Běhounek, L., Cintula, P.: Fuzzy Logic as the Logic of Chains. Fuzzy Sets and Systems 157, 604–610 (2006)
2. Běhounek, L., Cintula, P., Hájek, P.: Introduction to Mathematical Fuzzy Logic. In: Hájek, P., Cintula, P., Noguera, C. (eds.) Handbook of Mathematical Fuzzy Logic. College Publications, London (forthcoming)
3. Berman, J., Blok, W.J.: Algebras Defined From Ordered Sets and the Varieties they Generate. Order 23(1), 65–88 (2006)
4. Bou, F., Paoli, F., Ledda, A., Spinks, M., Giuntini, R.: The Logic of Quasi-MV Algebras. Journal of Logic and Computation 20(2), 619–643 (2010)
5. Casari, E.: Conjoining and disjoining on different levels. In: Dalla Chiara, M.L., Kees, D. (eds.) Logic and Scientific Methods, pp. 261–288. Kluwer, Dordrecht (1997)
6. Cintula, P., Noguera, C.: Implicational (Semilinear) Logics I: A New Hierarchy. Archive for Mathematical Logic 49, 417–446 (2010)
7. Dalla Chiara, M.L., Giuntini, R., Greechie, R.: Reasoning in Quantum Theory. Kluwer, Dordrecht (2004)
8. Font, J.M.: Taking degrees of truth seriously. Studia Logica 91, 383–406 (2009)
9. Giuntini, R.: Quantum MV Algebras. Studia Logica 56, 393–417 (1996)
10. Giuntini, R., Ledda, A., Paoli, F.: Expanding Quasi-MV Algebras by a Quantum Operator. Studia Logica 87(1), 99–128 (2007)
11. Horčik, R., Petr Cintula, P.: Product Łukasiewicz Logic. Archive for Mathematical Logic 43(4), 477–503 (2004)
12. Ledda, A., Konig, M., Paoli, F., Giuntini, R.: MV Algebras and Quantum Computation. Studia Logica 82(2), 245–270 (2006)
13. Ludwig, G.: Foundations of Quantum Mechanics. Springer, Berlin (1983)
14. Paoli, F., Ledda, A., Spinks, M., Freytes, H., Giuntini, R.: Logics From Square Root Quasi-MV Algebras. International Journal of Theoretical Physics 50, 3882–3902 (2011)
15. Wójcicki, R.: Theory of Logical Calculi. Reidel, Dordrecht (1988)
16. Zadeh Lotfit, L.: Fuzzy Logic and Approximate Reasoning. Synthese 30, 407–425 (1975)

33

On Fuzziness

Fernando Gomide

When thinking about how to construct an artifact, we engineers usually believe that the precision should be the ultimate goal. Requirements, models, design, formulas, manufacturing and use of the artifacts should be cleanly exact. Soon we recognize the need to bear in mind the uncertain nature of the real world, an environment full of unknown, partially known, and imprecise things. Humanity is uncertain, responsive to confidence, creative, and subject to changeable truths. There is no precision in human existence. Precision cannot be the only thing offered to humans. To quote a well known statement (credited to Fernando Pessoa, a twenty century Portuguese writer): "To navigate is precise; to live is not precise" (Pessoa was referring to navigation and its necessary search for precision in crossing the seas, but alerting us to the fact that the same thing does not hold for human life, wherein this search for precision becomes undesirable). Actually, in the engineering world things are not much different: there is wisdom in uncertainty. In system modeling for instance, few engineers would disagree that a model is an approximation, and that reasoning with approximate models is approximate reasoning. Uncertainty helps to model the system complexity.

Consider the statements that might be required to describe a system. There may be millions of them, but they are likely to be linked in loops and paths of interdependence. In general, to some degree, the descriptive statements will be normative for that system in the sense that there will be a set of values for each variable beyond which the system becomes prohibitive or unfeasible. There is what we could call noninferior value attached to each variable, which is noninferior because it may favor system flexibility and survivability. Flexibility and survival should also be favored whenever changes occur, tending to keep variables values floating within the corresponding boundaries. Any extreme adjustment may cause one or more variables to attain extreme values, which always result in a stress to be alleviated. To achieve flexibility and survival, softer levels of tolerance for the bounds of a variable are a major requisite. Uncertainty helps flexibility and survivability.

Additionally, most of what we know about the world is descriptive, qualitative, difficult to quantify, and eventually never been stored beforehand. Such information is crucial to understand and to model complex systems. Yet often modelers limit themselves to hard variables, ones that can be measured directly and can be expressed as numerical values. They may find the rejection of soft variables as being more scientific than assuming values for the variables and relationships for which no

R. Seising et al. (Eds.): *On Fuzziness: Volume 1*, STUDFUZZ 298, pp. 217–222.
DOI: 10.1007/978-3-642-35641-4_33

numerical data are available. A typical issue raised by those who reject soft variables is "How can the accuracy of estimates about soft variables be verified and tested? How can statistical tests be performed with no numerical data?"

Naturally, all relationships and parameters in models, whether based on soft or hard variables, are imprecise to some degree. Many people may disagree as to the importance of different factors. Often, modelers perform model analysis to ponder how their conclusions might change if different reasonable assumptions were made. Analysis is not restricted to uncertainty in parameter values. It considers the variations of conclusions to alternative structural assumptions and choices of model boundary.

There is substantial agreement that exact, point prediction of the future is neither possible nor necessary. For instance, at present we are far from being able to predict social systems behavior except perhaps for carefully selected instances in the very short term. Effort spent on attempts at precise prediction is almost surely wasted, and results that claim to be such predictions are certainly misleading. On the other hand, much can be learned from models in the form of broad, qualitative, conditional understanding-and this kind of understanding is useful (and typically the only basis) for policy formulation. If the doctor tells us that we will have a heart attack if we do not stop eating fat food, this advice is helpful, even if it does not tell exactly when a heart attack will occur or how bad it will be.

33.1 Journey

I was born in Viçosa, a small city located in the state of Minas Gerais, Brazil. I grew up in Viçosa and moved to Belo Horizonte, the capital of Minas Gerais, five years later. I obtained my undergraduate degree in electrical engineering from the Polytechnic Institute of the Pontifical Catholic University of Minas Gerais, Belo Horizonte, during middle seventies. Combining precise mathematical concepts with the nature of engineering creation and design was a recurrent discussion item between classmates, professors and me.

I recall that we often argued how precise tools could be of help if components and parts that assemble a system are inherently imprecise. The professor of our fifth semester of mathematics was involved with the technical and philosophic issues of digital systems and system theory. He was aware of fuzzy sets and speculated that it could form a basis to develop approaches to treat imprecision and uncertainty in engineering and system theory. This was, I guess, the first time we heard about fuzzy sets, but we could not foresee our future involvement with it. At that time it was not clear how could fuzzy set theory give us an answer, but the idea remained latent. Later on, I was exposed to linear and nonlinear system theory, stochastic processes, system modeling, optimization and control during my master degree courses at the School of Electrical and Computer Engineering, University of Campinas, city of Campinas, São Paulo, Brazil. My focus was in optimization of large-scale static and discrete time dynamic systems. There I found professors and colleagues working in systems, theory of computation, computer and control technology, tools and optimization approaches. Again, fuzzy set theory was part of the agenda, but at a very superficial level. During this period, middle-late seventies, I was not aware of the professors of the Institute of Mathematics and Biology of the University of Campinas who were working actively in the area and cooperating closely with researchers and professors worldwide. A couple of years later, some of them would become good friends (and still are today), while others retired. One year after I obtained my master degree, I joined the graduate program in large scale and complex systems of the System Engineering Department, Case Western Reserve University, Cleveland, Ohio, USA. During this period my focus was on optimization of non-scalar-valued performance criteria, control theory, and applications. Again, imprecision and uncertainty were part of the agenda and I learned a bit more about stochastic processes and fuzzy sets, but decided to leave imprecision and uncertainty for later consideration. The deterministic approach we were involved with was complex enough, and refinements would make it too extensive to be embraced at that moment. Yet, the need to address the questions raised during early undergraduate years continued dormant, but still alive.

Fig. 33.1. Fernando Gomide during his undergraduate years, Belo Horizonte 1974

After graduation, early nineteen eighties, I started to work at the Center for Information Technology of the Minister of Science and Technology with colleagues that, like myself, have graduated recently in computer, systems and control engineering. Artificial and computational intelligence, particularly intelligent decision and information systems were major items on our agenda. During this period we found people and a fertile atmosphere to address fuzzy and intelligent systems theory. Applications in adaptive control, computer-aided control system analysis and design, scheduling, process and manufacturing control were developed. Linguistic fuzzy rule-based systems became central. They provided an effective and transparent way to handle process complexity and real world situations, especially the imprecise and uncertain nature of the engineering objects. Heuristics, operational knowledge, undesirable but eventually acceptable process variables ranges, conflicting objectives, smooth and safe state transitions, etc. could be handled and implemented using fuzzy rule-based and soft computing approaches. I returned as a full time professor to the School of Electrical and Computer Engineering of the University of Campinas four years later, December 1986. Since then I have been teaching and doing research in intelligent systems and applications. During the last decades the focus was on development and application of neural fuzzy network models and learning algorithms; genetic fuzzy systems; fuzzy modeling, optimization and control; fuzzy Petri nets; and evolving and granular fuzzy systems more recently. Rail transportation, electricity load, streamflow, economic and financial variables forecasting, crew scheduling, queuing and berth scheduling are amongst applications addressed lately.

33.2 International and National Cooperation

Broadly speaking, cooperation between Brazilian universities and European, North, Central, South American and Asian universities and research institutions has been lively since early nineties. Major milestones include the Brazil-Japan Symposium on Fuzzy Systems in 1994 and the 6th IFSA World Congress in 1995, held in the city of Campinas and São Paulo, respectively. Numerous special sessions and papers have been presented in major biannual national conferences such as the Brazilian Symposium on Intelligent Automation (SBAI) and Brazilian Congress on Automatics (CBA), both sponsored by the Brazilian Society of Automatics, the national member organization of IFSA. There are also membership relationships with international societies mainly with IEEE, EUSFLAT, and NAFIPS. The impact of the international and national cooperation activities has been enormous. Currently, many Brazilian universities include fuzzy set theory and soft computing courses at both, undergraduate and graduate levels, especially in engineering, computer science, and applied mathematics. Research crosses the boundaries of engineering, health sciences, pure and applied mathematics, computer science, economics, with applications in business, transportation, process control, decision support systems, petroleum, medicine, etc. Technical sessions, short courses, tutorials, plenary talks and round tables about fuzzy systems and soft computing became regular in the organization of the major national automation, control, applied mathematics, and computer science and engineering conferences sponsored by the Brazilian Society of Automatics (SBA), Brazilian Computer Society (SBC), Brazilian Society of Applied Mathematics (SBMAC), and the Brazilian Society for the Progress of Science (SBPC). Currently, a conference uniquely devoted to fuzzy sets theory and soft computing sponsored by

SBA, SBC, and SBMAC, with the support of IFSA, NAFIPS and EUSFLAT occurs biannually.

33.3 Back to Fuzziness

So what? Well, what are my current views and expectations on fuzziness? I believe that fuzzy set and system theory have indubitably brought a profound contribution to handle the inherent imprecision of many, if not most technical and human centered systems. Nowadays, companies and science and engineering schools worldwide do introduce fuzzy set theory and soft computing to their professionals and students earlier than three decades ago. But, in spite of the maturity of the field, there are various issues that deserve further thought, the semantic gap being one amongst the most relevant, I guess. Syntactically speaking, existing digital computers and their relatives are doing well, but semantically they still are very illiterate. Autonomous approaches to compute with words still pose considerable challenges. Similarly, granular computing is a promising idea to handle complex systems, but it seems to me that we do not have mechanisms to work and to compute with granules as atomic entities. We always end up with pointwise computing and this is somewhat contradictory with the main idea brought by soft computing as a way to achieve tractability, robustness and low cost. For instance, how could granular computing help to ease modeling and the computational tractability of large scale, complex decision making problems? Granular computing should provide an antidote to the

Fig. 33.2. A snapshot of the 6th IFSA World Congress, São Paulo 1995

curse of dimensionality, but effective approaches to cope with it are still lacking. The value of models and theories draws from the differences between them and the perceptions originated from mental models. When inconsistent results of a mental and a theoretical model are explored, and the essential causes of the differences are recognized, both the models and the theory can be enhanced. Fuzzy sets and systems are a key for science and engineering and will expand their potential further as soon as future computer technology simultaneously allows large scale hybrid processing and brainets.

34

Local Finiteness in T-Norm Based Bimonoides

Siegfried Gottwald

Abstract This paper offers a short discussion of the property of local finiteness for t-norm monoids and bimonoids. Such bimonoids are of interest in the context of weighted automata. The paper shows that, perhaps unexpectedly, the situation is more complex in the bimonoidal case than it is for monoids: there there are more possibilities for local finiteness.

34.1 Introduction

Triangular norms, *t-norms* for short, first showed their importance in the field of probabilistic metric spaces, cf. [13, 15]. These binary operations in the real unit interval there play a crucial role in the formulation of a probabilistic version of the triangle inequality.

Interestingly, also Lotfi Zadeh mentioned in his basic paper [17] on fuzzy sets besides his ordinary generalizations of the union and intersection operations to fuzzy sets, which are based upon the maximum and the minimum operation for the membership degrees, respectively, also an algebraic sum and an algebraic product of fuzzy sets: from a more general point of view these operations proved to be particular cases of a union operation and an intersection operation, based upon the conorm of the Łukasiewicz t-norm or the product t-norm, respectively.

It the beginning 1980s it became common use in the mathematical fuzzy community to consider t-norms as most suitable general candidates for connectives upon which generalized intersection operations for fuzzy sets should be based, see [1, 5, 14] or a bit later [12, 16]. Furthermore, this t-norm context allowed for a unified treatment of fuzzy set theory in the context of t-norm related operations of suitable many-valued logics, as done e.g. in [6–8]. Large classes of such logics since got axiomatized starting from the seminal work [11] of P. Hájek on the logic of all continuous t-norms. There is a large amount of further work done since in the field of mathematical fuzzy logics and fuzzy set theory, surveyed e.g. in [2, 10] and most recently in the handbook [3].

34.2 Preliminaries

An algebraic structure $\mathfrak{A}$ is *locally finite* iff each of its finite subsets G generates a finite subalgebra $\langle G\rangle_{\mathfrak{A}}$ only.

R. Seising et al. (Eds.): *On Fuzziness: Volume 1*, STUDFUZZ 298, pp. 223–228.
DOI: 10.1007/978-3-642-35641-4_34

In this paper we look at this property of local finiteness for t-norm based structures. A *t-norm* is a binary operation in the real unit interval which makes this interval into an ordered abelian semigroup which has the upper bound 1 of the order as unit element. And a *t-conorm* is a binary operation in the real unit interval which makes this interval into an ordered abelian semigroup which has the lower bound 0 of the order as unit element.

By $T_{\mathsf{L}}, T_{\mathsf{P}}, T_{\mathsf{G}}$ we denote the basic t-norms, i.e. the Łukasiewicz, the product, and the Gödel t-norm, respectively.

Each t-norm T has as its dual a t-conorm S_T determined by the equation

$$S_T(x,y) = 1 - T(1-x, 1-y)\,. \tag{34.1}$$

This duality determines a 1-1 relationship between t-norms and t-conorms.

We denote by $S_{\mathsf{L}}, S_{\mathsf{P}}, S_{\mathsf{G}}$ the conorms which correspond to $T_{\mathsf{L}}, T_{\mathsf{P}}, T_{\mathsf{G}}$, respectively.

34.3 The T-Norm Monoids and T-Conorm Monoids

For the t-norm monoids $\mathfrak{A} = ([0,1], T, 1)$ with the basic continuous t-norms $T_{\mathsf{L}}, T_{\mathsf{P}}, T_{\mathsf{G}}$ the local finiteness situation is rather simple. But this situation becomes more difficult in the cases of enriched structures.

Like the t-norm monoids $\mathfrak{A} = ([0,1], T, 1)$ one can consider their duals, i.e. the monoids $\mathfrak{A}^d = ([0,1], S_T, 0)$ which are determined by the corresponding t-conorms.

This, fortunately, does not offer a new situation in the present elementary context.

Proposition 1. *A t-conorm monoid* $([0,1], S_T, 0)$ *is locally finite iff its corresponding t-norm monoid* $([0,1], T, 1)$ *is.*

Proof: Let $\mathfrak{A} = ([0,1], T, 1)$ be a t-norm monoid and $G \subseteq [0,1]$. Because of (34.1) one gets that for each $a \in \langle G \rangle_{\mathfrak{A}}$ its dual $a^d = 1 - a$ is an element of $a \in \langle G^d \rangle_{\mathfrak{A}^d}$ for $G^d = \{a^d \mid a \in G\}$.

Proposition 2. *The t-norm monoid* $([0,1], T_{\mathsf{G}}, 1)$ *is locally finite.*

Proof: Obvious, because $T_{\mathsf{G}} = \min$.

Proposition 3. *The t-norm monoid* $([0,1], T_{\mathsf{P}}, 1)$ *is not locally finite.*

Proof: Any $a \in (0,1)$ generates an infinite submonoid $\langle a \rangle = \{a^n \mid n \in \mathbb{N}\}$ of $([0,1], T_{\mathsf{P}}, 1)$.

Proposition 4. *The t-norm monoid* $([0,1], T_{\mathsf{L}}, 1)$ *is locally finite.*

Proof: Instead of $([0,1], T_{\mathsf{L}}, 1)$ we consider the corresponding t-conorm monoid $([0,1], S_{\mathsf{L}}, 0)$. In $([0,1], S_{\mathsf{L}}, 0)$ each finite $G \subseteq [0,1]$ generates only a finite number of elements: 1, together with all the finitely many sums $k_1 a_1 + \cdots + k_n a_n$, $k_1, \ldots, k_n \in \mathbb{N}$, of S_{L}-multiples of $a_1, \ldots, a_n \in G$.

Via Proposition 1 these results immediately yield the

Corollary 5. *The Gödel-conorm monoid as well as the Łukasiewicz-conorm monoid are locally finite. The product-conorm monoid is not locally finite.*

These considerations can now easily be extended to t-norm monoids based upon arbitrary continuous t-norms. Such continuous t-norms can be uniquely represented as ordinal sums of Archimedean t-norms, i.e. t-norms T satisfying $T(a,a) < a$ for all $a \in [0,1]$. And each such Archimedean t-norm is an isomorphic copy of either the Łukasiewicz t-norm T_{L} or the product t-norm T_{P} as explained in [13] and also in [9].

Theorem 6. *A t-norm monoid $([0,1],T,1)$ with a continuous t-norm T is locally finite if and only if T does only have locally finite summands in its representation as ordinal sum of archimedean summands.*

The proof for this result shall be given in an extended version of this note.

This theorem immediately yields the following corollary.

Corollary 7. *A t-norm monoid $([0,1],T,1)$ with a continuous t-norm T is locally finite if and only if T does not have a product-norm isomorphic summand in its representation as ordinal sum of archimedean summands.*

Even more is now easily available.

Proposition 8. *If a continous t-norm T has a product-isomorphic summand in its ordinal sum representation then any extension of the t-norm monoid $([0,1],T,1)$ is not locally finite.*

It seems to be hard to omit the continuity assumption made up to now for the t-norms. A core problem is the lack of a sufficiently well developed structure theory for left continuous t-norms. Nevertheless one can successfully discuss particular examples. We offer only a very elementary one which has an obvious proof.

Proposition 9. *The t-norm monoid $([0,1],T_{\mathsf{nM}},1)$ based upon the nilpotent minimum*

$$T_{\mathsf{nM}}(x,y) = \begin{cases} \min\{x,y\}, & \textit{if } u+v>1 \\ 0 & \textit{otherwise} \end{cases}$$

is locally finite.

34.4 The T-Norm Bimonoids

Now we are interested in the t-norm based bimonoids $([0,1],T,S_T,1,0)$. In general, a bimonoid is an algebraic structure $\mathfrak{A} = (A,*_1,*_2,e_1,e_2)$ such that both $(A,*_1,e_1)$ and $(A,*_2,e_2)$ are monoids. For the following results we omit the–sometimes simple–proofs, again referring to an extended version of the paper.

Proposition 10. *The Gödel-bimonoid $([0,1],T_{\mathsf{G}},S_{\mathsf{G}},1,0)$ is locally finite.*

Proposition 11. *The product-bimonoid* $([0,1], T_{\mathsf{P}}, S_{\mathsf{P}}, 1, 0)$ *is not locally finite.*

Proposition 12. *The Łukasiewicz-bimonoid* $([0,1], T_{\mathsf{L}}, S_{\mathsf{L}}, 1, 0)$ *is not locally finite.*

Compared with Propositions 4 and 1 one recognizes that a crucial point for this last result is the simultaneous availability of the operations T_{L} and S_{L}.

Additionally, the failure of local finiteness really comes from the irrational numbers.

Proposition 13. *The rational Łukasiewicz-bimonoid* $([0,1] \cap \mathbb{Q}, T_{\mathsf{L}}, S_{\mathsf{L}}, 1, 0)$ *is locally finite.*

Interestingly, a result like Theorem 6 does not hold true for t-norm based bimonoids.

The following example, rather arbitrary but simple, shall illustrate this situation.

Example: The t-norm bimonoid $([0,1], T^*, S_{T^*}, 1, 0)$ with the continuous t-norm

$$T^* = \sum_{i \in \{1\}} ([\frac{1}{2}, 1], T_{\mathsf{L}}, \varphi^*) \tag{34.2}$$

and the order isomorphism $\varphi^* : [\frac{1}{2}, 1] \to [0,1]$ given by $\varphi^*(x) = 2x - 1$ is locally finite.

Our choice of T^* is such that T^* acts on the square $u_r = [\frac{1}{2}, 1] \times [\frac{1}{2}, 1]$ as (an isomorphic copy of) T_{L}, and acts as the min-operation over the remaining part of the unit square. Accordingly, the corresponding t-conorm S_{T^*} acts on the square $l_l = [0, \frac{1}{2}] \times [0, \frac{1}{2}]$ as (an isomorphic copy of) S_{L}, and acts as the max-operation over the remaining part of the unit square.

This means that it is impossible to have (the isomorphic copies of) T_{L} and S_{L} simultaneously available. So the bimonoid $([0,1], T^*, S_{T^*}, 1, 0)$ remains locally finite.

A more general result, however, is also available.

Theorem 14. *Suppose that T is a continuous t-norm with ordinal sum representation $T = \sum_{i \in I}([l_i, r_i], T_i, \varphi_i)$ without product-isomorphic summands. Assume furthermore that for each Łukasiewicz summand $([l_k, r_k], T_{\mathsf{L}}, \varphi_k)$ the interval $[1 - r_k, 1 - l_k]$ does not overlap with any domain interval $[l_i, r_i]$ for a Łukasiewicz summand $([l_i, r_i], T_{\mathsf{L}}, \varphi_i)$, $i \in I$. Then the t-norm bimonoid $([0,1], T, S_T, 1, 0)$ is locally finite.*

As in the case of t-norm monoids, also for bimonoids a certain extension to particular cases of left continuous t-norms is possible. A general approach is again lacking. Similar to Proposition 9 we mention only the following result.

Proposition 15. *The T_{nM}-bimonoid, based upon the nilpotent minimum T_{nM}, is locally finite.*

34.5 Concluding Remarks

Starting point for these investigations was the fact that the local finiteness of bimonoids is an interesting property for weighted automata, cf. e.g. [4], and hence the problem which t-norm based bimonoids have this property.

The results offer an interesting picture for the case of continuous t-norms and monoids as well as bimonoids based upon them. For t-norm based bimonoids there is, however, only a sufficient condition which characterizes a series of such bimonoids as locally finite. A complete picture is missing.

And missing is also a discussion of t-norms which are only left continuous. But this problem seems to be strongly linked to the lack of a sufficiently well developed structure theory for left continuous t-norms.

References

1. Alsina, C., Trillas, E., Valverde, L.: On Non-distributive Logical Connectives for Fuzzy Sets Theory. Busefal 3, 18–29 (1980)
2. Cintula, P., Hájek, P.: Triangular Norm Based Predicate Fuzzy Logics. Fuzzy Sets and Systems 161, 311–346 (2010)
3. Cintula, P., Hájek, P., Noguera, C. (eds.): Handbook of Mathematical Fuzzy Logic. Studies in Logic, vol. 37-38. College Publications, London (2011)
4. Droste, M., Stüber, T., Vogler, H.: Weighted Finite Automata Over Strong Bimonoids. Information Sciences 180, 156–166 (2010)
5. Dubois, D.: Triangular Norms for Fuzzy Sets. In: Klement, E.P. (ed.) Proceedings of the 2nd International Seminar Fuzzy Set Theory, pp. 39–68. Johannes Kepler University, Linz (1980)
6. Gottwald, S.: T-Normen und φ-Operatoren als Wahrheitswertfunktionen Mehrwertiger Junktoren. In: Wechsung, G. (ed.) Proceedings of the International Frege Conference 1984, Schwerin, Mathematical Research, vol. 20, pp. 121–128. Akademie-Verlag, Berlin (1984)
7. Gottwald, S.: Fuzzy Set Theory with t-norms and φ-operators. In: di Nola, A., Ventre, A.G.S. (eds.) The Mathematics of Fuzzy Systems. Interdisciplinary Systems Research, vol. 88, pp. 143–195. TÜV Rheinland, Köln (1986)
8. Gottwald, S.: Fuzzy Sets and Fuzzy Logic. Artificial Intelligence. Verlag Vieweg, Tecnea, Wiesbaden, Toulouse (1993)
9. Gottwald, S.: A Treatise on Many-Valued Logics. Studies in Logic and Computation, vol. 9. Research Studies Press, Baldock (2001)
10. Gottwald, S., Hájek, P.: T-norm Based Mathematical Fuzzy Logics. In: Klement, E.P., Mesiar, R. (eds.) Logical, Algebraic, Analytic, and Probabilistic Aspects of Triangular Norms, pp. 275–299. Elsevier, Dordrecht (2005)
11. Hájek, P.: Metamathematics of Fuzzy Logic. Trends in Logic, vol. 4. Kluwer Acad. Publ., Dordrecht (1998)
12. Klement, E.P.: Construction of Fuzzy σ-algebras Using Triangular Norms. Journal Mathematical Analysis Applications 85, 543–565 (1982)
13. Klement, E.P., Mesiar, R., Pap, E.: Triangular Norms. Kluwer Academic Publishers, Dordrecht (2000)

14. Prade, H.: Unions et intersections d'ensembles flous. Busefal 3, 58–62 (1980)
15. Schweizer, B., Sklar, A.: Probabilistic Metric Spaces. North-Holland Publ. Comp., Amsterdam (1983)
16. Weber, S.: A General Concept of Fuzzy Connectives, Negations and Implications Based on t-norms and t-conorms. Fuzzy Sets and Systems 11, 115–134 (1983)
17. Zadeh, L.A.: Fuzzy Sets. Information and Control 8, 338–353 (1965)

35

Around the BISC Roundtable

Sergio Guadarrama*

35.1 Introduction

The first time I meet Prof. Lotfi Zadeh was back on 2001 at the FUZZ-IEEE International Conference held at Melbourne University. I got introduced to him by my mentor and Ph.D supervisor Prof. Enric Trillas. Prof. Trillas and Prof. Zadeh have kept a close relation since the 70's, since they first meet at Barcelona at the first conference on Fuzzy Logic organized in Spain. Prof. Trillas has been one of the Fuzzy Pioneers and one of the persons who has done more to spread it, specially in Spain.

I had known about Prof. Zadeh from the papers I read about Fuzzy Sets and Fuzzy Logic and from the stories my mentor told me. So it was a very important experience for me (as for many other people who have read or heard about Prof. Zadeh) to meet him in person. He has very kind and charming personality, but he starts asking questions about you and your interests very soon.

Some of the first things I noticed from Prof. Zadeh was his curiosity, his strive for understanding complex problems, especially those ill-defined and involving imprecision, and his willingness to point them out to other researchers. Since he thinks that these kind of problems are more realistic and relevant than artificial or idealized problems, typically used to satisfy assumptions needed by the models.

35.2 The Beginnings of the BISC Roundtable

When Lotfi was visiting the Institute for Advanced Study, Princeton, NJ in 1956-57 he attended a discussion class with the eminent logician Stephen Kleene, and other 4 students. That class was not composed of regular lectures, it had no final exam and there was no homework. Instead, it consisted on avid discussions among the attendees. According to Lotfi "This discussion class was the best education he has received, where one great logician was discussing very few students". That strong memory stayed in his mind and served years later to come out with the idea of setting up a BISC roundtable discussion, where visiting scholars, postdocs and faculty could present their own ideas (without slides or detailed preparation) and foster a fruitful discussion.

* Affiliated researcher at the European Centre for Soft Computing (ECSC) and at International Computer Science Institute (ICSI).

R. Seising et al. (Eds.): *On Fuzziness: Volume 1*, STUDFUZZ 298, pp. 229–234.
DOI: 10.1007/978-3-642-35641-4_35

Fig. 35.1. BISC Roundtable discussion on March 13th 2012

The idea of roundtable is based on the lack of places to present and discuss openly current research problems or questions without using slides or blackboards, that is, without going into all the little details as it is usually done in seminars or conferences. The idea that a dialogue between the presenter (or discussion leader, in Lotfi's terms) and the participants (or discussants) of the roundtable can be established, a dialogue that will allow the participants to get a better understanding of what other researchers are working on, and the presenters to clarify their own ideas and get, most of times, interesting questions and comments (since the participants are not experts in the presenter's topic).

By the end of 2003 the concept of roundtable discussion was formed and implemented. Probably because that year Lotfi had some health problems that impeded him to fly. That was probably a pity for all the people around the world who had invited him, and to whom he has to kindly decline the invitation[1], but at the same time it was a great gift for all the BISC visitors at that time (Prof. Takagi, Prof. Alteking, Dr. De Cock, Dr. Diaz, ... and myself). We all got the chance to see and discuss with him every week, to attend the BISC roundtable discussions, and to participate in the organization of the FLINT-Conference. During the development of my Ph.D. disser-

[1] Something very rare on him, since he has travelled so many times around the world to deliver invited talks and spread the work of Fuzzy Logic and Soft Computing, and had probably accumulated more than 1 million miles

tation I was lucky enough to stay in Berkeley for one year between 2003 and 2004 and work under Lotfi supervision (in the way Lotfi supervise his students, by asking tough questions, by posing difficult problems and by encouraging you to purse your own ideas).

After the summer of 2004 Lotfi started to travel again, specially to Spain, where he fostered the creation of the European Centre for Soft Computing (ECSC). ECSC is a research center devoted to Soft Computing and established in 2005 in Mieres (Asturias, Spain) with the support of the CajAstur Saving Bank, the Mining Unions, the European Union, the Spanish Government and the Asturias Government.

During more than 45 years Berkeley (and 20 years since the creation of BISC) has served as a reference point for all the researchers interested in Fuzzy Logic and Soft Computing, a place to visit if you want to get a glimpse of the ideas of Prof. Zadeh. It has been a place where so many visitors along the years have been able to present and discuss their ideas and to meet other visitors and faculty.

35.3 The Continuation of the BISC Roundtable

Since the end of 2008 when Lotfi suffered a heart attack and needed surgery to recover, he has not travelled abroad any more and has greatly reduced his trips within the US. This has allowed him to recover well and to regain some of his previous endeavors, like the BISC roundtable, with renewed energy.

The BISC group has hosted dozens of visiting scholar and postdocs since it inception (some years it was hosting up to 10 visitors at the same time) but recent changes in the EECS Department regulations have made very difficult to get new visitors. So last year occurred to Lotfi that the roundtable could be open to ECSC students and other people interested in Soft Computing or Fuzzy Logic in the Bay Area. He is still amused why that idea didn't occur to him earlier, but as he said several "sometimes simple ideas take longer to appear than obvious ones".

Nowdays the BISC roundtable is open to Berkeley students (both undergraduates as well as graduates), Berkeley alumni, professors (from any department beyond EECS) and to anyone interested in the topics treated. This serves a great purpose to expose what researchers are doing (it is sad that these days many professors and researchers don't know what their colleagues are working on). For this purpose a specialized seminar will server of little help, since as the incompatible principle expressed by Lotfi says "In complex problems precision and relevance become incompatible", that is, when talking about a complex problem a very precise description will be irrelevant, and a relevant description should be imprecise.

On the first occasion I served as discussion leader at the BISC roundtable back in 2009, I felt somehow "naked" to talk without slides. Since I have became so used to present my research results and ideas using slides or at least using a board full of equations and diagrams, that my ability to express in words (in a plain and understandable way) what is the focus of my research, which is the problem I am tackling and how it connects with other problems, was greatly reduced.

One of my best teachers told me once "If you cannot explain a problem using plain words to your youngest cousin in a way that she can understand it, then you don't fully understand the problem".

Fig. 35.2. Dr. Sergio Guadarrama at the BISC RoundTable

Some of the most interesting roundtable discussions I have been able to attend have been leaded by: Brian Barsky, Trevor Darrell, Isabel Guyon, Bjoern Hartmann, Marti Hearst, Srini Narayan, Terry Regier, Sherri Rousch, Enric Trillas, and Lotfi Zadeh himself.

Some of the most interesting topics discussed at the roundtable have been: using inverse optics and computer graphics to correct eye distortions, challenges and open problems in computer vision, causality and deep learning, new ways to interact with computers using gestures and touch screens, human computer interaction in humanities, changing the focus from representations to actions, conceptual mappings across cultures, reasoning about reasoning, guessing and conjecturing, computing with words, actions and perceptions, and how to deal with imprecision and uncertainty at the same time using Z-numbers.

Fig. 35.3. Prof. Lotfi Zadeh at the BISC Roundtable

During his career Lotfi has taken a counter-traditional approach to try to solve complex and ill-defined problems. Instead of trying to formalize them in a precise way, which require making many assumptions and simplifications to be more attainable to classical models, he has tried to take advantage of the tolerance for imprecision to solve them using linguistic approximate solutions.

Table 35.1. List of BISC Roundtable Discussion Leaders

7-10-2008	David Wilkinson	1-26-2011	Sergio Guadarrama*
7-31-2008	Euntai Kim	2-3-2011	Niloofar Razaee
8-21-2008	Adolfo de Soto	2-9-2011	Anant Sahai
9-11-2008	Zhiheng Huang	2-16-2011	Jean Walrand
10-2-2008	Asli Celikyilmaz*	2-23-2011	Teemu Mutanen
10-15-2008	Bert de Coensel	3-2-2011	Gautam Dasgupta*
10-22-2008	Terry Regier*	3-9-2011	Lotfi Zadeh*
10-29-2008	Isabelle Guyon*	3-16-2011	Abdolreza Abhari
11-5-2008	Jerry Feldman*	3-30-2011	Isabelle Guyon*
11-19-2008	Teed Rockwell	4-6-2011	Gerald Friedland*
12-3-2008	Dan Klein	4-13-2011	George Leitmann
12-18-2008	Olga Vybornova	4-20-2011	Sergio Guadarrama*
1-7-2009	Gerald Friedland*	5-4-2011	Trevor Darrell
2-11-2009	Guangping Zeng	5-12-2011	Ashok Deshpande
3-3-2009	Fumio Mizoguchi	6-7-2011	Erol Gelene
3-11-2009	Martin Wainwright	6-15-2011	Malik Ghallab
3-18-2009	Sue Liu	6-29-2011	Lotfi Zadeh*
4-1-2009	Murat Arcak*	7-21-2011	Donald Wunsch
4-8-2009	Dewang Chen	9-22-2011	Gerald Friedland*
5-13-2009	Gerald Friedland*	9-28-2011	John Canny
5-15-2009	Sanjoy Mitter	10-5-2011	Sergio Guadarrama*
5-19-2009	Claire Tomlin	10-12-2011	Bjoern Hartmann
5-27-2009	Monika Ray	10-14-2011	Maneesh Agrawala
6-5-2009	Gwen Wilke*	10-19-2011	Brian Barsky
6-18-2009	Irina Perfilieva	10-26-2011	Daniel Morozoff
7-2-2009	Matteo Brunelli	11-2-2011	Asli Celikyilmaz*
7-8-2009	Tom Griffiths	11-2-2011	Dilek Hakkani-Tur
7-13-2009	Manuel de Buenaga	11-9-2011	Ken Goldberg
7-29-2009	Peter Abbeel	11-16-2011	Michael Franklin
8-12-2009	David Galvez Ruiz	11-30-2011	Isabelle Guyon*
9-1-2009	Raija Koivisto	12-7-2011	Murat Arcak*
9-11-2009	Sergio Guadarrama*	1-18-2012	Gautam Dasgupta*
9-17-2009	John DeNero	1-25-2012	Alireza Shabani
9-23-2009	Enric Trillas	2-1-2012	Sherri Roush
9-23-2009	Luis Magdalena	2-8-2012	Jerry Feldman*
9-30-2009	Sibel Yaman	2-16-2012	Ruzena Bajcsy
10-6-2009	Christer Carlsson	2-23-2012	Sayeef Salahuddin
10-7-2009	Michael Ellsworth	2-29-2012	Venkat Anatharam
10-21-2009	Vesa Niskanen	3-8-2012	Jonathan Shewchuk
10-28-2009	Gerald Friedland*	3-14-2012	Joseph Austerweil
11-2-2009	Olga Poleshchuk	3-23-2012	Terry Regier*
11-30-2009	Laszlo Koczy	4-4-2012	Ilan Adler
12-2-2009	Douglas Oard	4-19-2012	Srini Narayanan*
12-16-2009	Srini Narayanan*	5-3-2012	Marti Hearst
11-5-2010	Gwen Wilke*	5-15-2012	Edy Portman
1-19-2011	Chen-Ting Wu	5-21-2012	Ashok Deshpande

During the roundtable discussions Lofti always has many questions to rise, specially those regarding imprecision and uncertainty. For instance he usually points out that in many fields imprecision is disregarded as avoidable and uncertainty is reduced to probability measures, without deeper consideration. In most realistic problems researchers make several assumptions to avoid dealing with imprecision and to be able to simplify the underlying uncertainty.

It is amazing that at 91 years old, Lotfi is still working hard and proposing new ideas, the last one has been the concept of Z-numbers, where he proposed a new way to represent and compute with imprecision and uncertainty at the same time. Each Z-number is comprised of a tuple of approximate constraints imposed by imprecise and/or uncertain statements.

During the years the BISC roundtable has been changing and adapting to encourage researchers from all over the world and from many disciplines to present their questions and tentative answers around a table with friendly but inquisitive discussions. But it is clear to me that the BISC Roundtable has maintained loyal to its original purpose of serving as an open space to discuss ideas and complex problems.

Acknowledgement. The author would like to thank Prof. Lotfi Zadeh and Ixel Chavez for all their help and information provided to make this short paper possible, but even more for creating and maintaining the BISC Roundtable along the years.

36

Fuzzy Arithmetic for Uncertainty Analysis

Michael Hanss

36.1 Introduction

When the theory of fuzzy sets arose as a new mathematical concept in the field of information processing some 50 years ago, it rapidly advanced to becoming a well-established scientific discipline and a challenging object of both theoretical research and practical application. Since its introduction by Lotfi A. Zadeh [21], enormous progress has been made and numerous subdomains of fuzzy set theory have emerged, such as fuzzy logic and approximate reasoning, fuzzy pattern recognition and fuzzy modeling, expert systems and fuzzy control – and fuzzy arithmetic. Compared to most other fields, the topic of fuzzy arithmetic has received only little attention in the early years, and the scope of its practical application has barely exceeded the level of elementary academic examples. The reasons for this may be seen in the absence of a well-organized, systematic and consistent elaboration of the theory of fuzzy arithmetic, the lack of practical approaches to its effective implementation, and the apparent underestimation of its potential for the solution of real-world problems.

Over the intervening years, however, the significance of fuzzy arithmetic has changed drastically. Starting off with Zadeh's extension principle [22] and continued by the pioneer works of Dubois and Prade [1, 2], Kaufmann and Gupta [14], and others, fuzzy arithmetic has evolved into a well-established and powerful tool, characterized by a clearly defined axiomatic basis [11]. Today it represents a well-founded and systematic method for computing with fuzzy-valued quantities and it can be applied successfully to sophisticated industrial applications in the framework of uncertainty analysis for dynamical systems.

36.2 Uncertainties in Dynamical Systems

A common problem in the numerical simulation of dynamical systems is the fact that exact values for the parameters of the models can hardly be provided, and that instead, the parameters can exhibit a high level of uncertainty. In a rather coarse and general sense, uncertainty can be defined as 'everything that is not absolutely known' and it may manifest itself in quite a number of different forms. The various forms of appearance, however, can practically be split into two major categories [17]: aleatory uncertainties and epistemic uncertainties.

R. Seising et al. (Eds.): *On Fuzziness: Volume 1*, STUDFUZZ 298, pp. 235–240.
DOI: 10.1007/978-3-642-35641-4_36

Aleatory uncertainties result from natural variability or scatter in the physical properties of a system over time or space. They are random in nature and generally related to the uncertainty of the outcome of an event or experiment. Against this background, an efficient representation of aleatory uncertainties can be realized by the use of random numbers with their probability density functions derived from measurements and experimental data. As widely acknowledged in literature (e.g., [19]), the most effective and predominantly used methods for the quantification of the propagation of aleatory uncertainties through systems are based on probability theory as well as on Monte Carlo simulations for practical applications.

Epistemic uncertainties, on the other hand, arise from insufficiency or even complete absence of knowledge, resulting from vagueness in parameter definition, from subjectivity in numerical implementation, or from simplification and idealization as it usually occurs in the procedure of system modeling. Due to this significant and indisputably different character of epistemic uncertainties compared to aleatory uncertainties, probability theory may not be appropriate to effectively represent epistemic uncertainties [13]. Furthermore, practical data for a randomness-based quantification of these uncertainties are usually not available.

For these reasons, a promising alternative strategy consists in quantifying epistemic uncertainties by fuzzy numbers and propagating the uncertainty through the system, i.e. evaluating the model with fuzzy-valued parameters, by the use of fuzzy arithmetic (see Section 36.3). In the first instance, the representation of epistemic uncertainties by ordinary intervals seems to be the most practical and straightforward approach if only worst-case bounds and no further information about a possible distribution within the interval is available. Apart from the fact that the evaluation of models with interval-valued parameters by the use of classical interval arithmetic proves to be rather problematic because of the overestimation effect (see Section 36.3), the sharp boundedness of the intervals acts quite contrary to the predominant human perception of quantifying imprecision. The somehow blurred bounds of fuzzy numbers, instead, comply much better with this view. Moreover, uncertainty propagation on the basis of only one particular set of intervals for the uncertain parameters will automatically raise the question about how the results of the propagation will change (in a qualitative and quantitative way) with the amount of initial uncertainty, i.e. with the lengths of the intervals assumed. Against this background, fuzzy numbers are perfectly suited to answer this question, as they can be seen as a set of nested intervals ranging from a worst-case scenario in case of maximum uncertainty to a crisp nominal value in case of complete certainty.

Finally, it should be noted that although fuzzy numbers are primarily appropriate to represent uncertainties of epistemic type, they can, nevertheless, be used to quantify aleatory uncertainties in equal measure. Even though potential information about the probability distribution of the parameter values is not included in this case and only some worst-case statements are taken into account, the additional representation of aleatory uncertainties by fuzzy numbers allows for the interaction of aleatory and epistemic uncertainties to be studied. This may be of particular interest in cases where worst-case scenarios are requested, rather than statements about the probability of failure, as well as in cases where the effects of aleatory uncertainties, such as variability or scatter, are supposed to be predominant, but then turn out to be of less importance than epistemic uncertainties arising from subjectivity in implementation or from the simplification of models.

36.3 Fuzzy Arithmetical Concept

A special application of the theory of fuzzy sets, which is rather different from the well-established use of fuzzy set theory in the context of fuzzy control, is the numerical implementation of uncertain model parameters as fuzzy numbers. Fuzzy numbers are defined as convex fuzzy sets over the universal set of real numbers with their membership value equal to unity only for one single value, the so-called center value or nominal value. Practically, fuzzy numbers with linear membership functions, so-called triangular fuzzy numbers, are often used, for they represent a linear transition from a nominal value, in case of complete certainty, to a worst-case scenario, in case of maximum uncertainty. Any other shape of membership function, however, may be selected if appropriate to quantify the uncertainty of a specific model parameter. The calculation with fuzzy numbers is referred to as fuzzy arithmetic and proves to be a non-trivial problem, especially with regard to the evaluation of large and complex mathematical models with fuzzy-valued operands.

The problem of incorporating uncertainties into complex numerical models of dynamical systems, such as finite element models, has already been addressed in a number of publications, of which the vast majority is based on stochastic descriptions of the uncertainties (e.g., [3], [19]).

The alternative concept of using fuzzy descriptions of the uncertainties emerged more recently [16], and Rao and Sawyer [18] presented an approach for its incorporation into the finite element method. However, since that approach uses the conventional concept of standard fuzzy arithmetic, based on interval computation, it suffers considerably from the overestimation effect, also referred to as the dependency problem or conservatism [8, 11].

With the objective of reducing this effect while maintaining the computational effort to an acceptable level, Moens and Vandepitte [15] presented a fuzzy finite element approach which is based on the application of special optimization strategies of an approximative character. The achievements of this method have been exemplified in the context of the calculation of frequency response functions of undamped structures; however, its successful general applicability to arbitrary finite element problems and especially to the solution of complex real-world problems in both the frequency domain and the time domain still seems to pose a significant challenge.

As a successful practical implementation of fuzzy arithmetic, which allows the evaluation of any mathematical expression and arbitrary models of engineering systems with uncertain, fuzzy-valued parameters, the Transformation Method [8] can be used. The basic idea of the Transformation Method is to choose crisp parameter values from the interval range of the fuzzy-valued parameters, decomposed into α-cuts, and to combine these values in well-defined parameter combinations, for which the problem of interest is evaluated. This method is available in a general, a reduced and an extended form, with the most appropriate form to be selected depending on the type of model to be evaluated [8, 9, 11]. In addition to this 'simulation part' of the Transformation Method, there is also an 'analysis part', which can be used to quantify the influence of each fuzzy-valued model parameter on the overall fuzziness of the model output. For these purposes, so-called standardized mean gain factors and normalized degrees of influence have been introduced [8, 11], quantifying in an absolute and in a relative character, respectively, the effect of the uncertainty of each model parameter on the overall uncertainty of the model output.

Due to the fact that the Transformation Method is based on multiple crisp-valued computations, the method can be coupled to any software environment, such as finite element solvers, without time and cost expensive rewriting of the program code for the use of fuzzy-valued parameters. For this purpose, the in-house software package FAMOUS (Fuzzy Arithmetical Modeling Of Uncertainty System) has been developed, representing an efficient implementation of fuzzy arithmetic based on the Transformation Method. It has been applied successfully to various engineering problems (see e.g., [4, 12, 20]).

As an extension of the Transformation Method of fuzzy arithmetic, an inverse fuzzy arithmetical method has been proposed by Hanss [10] and Haag et al., to estimate the uncertain parameters of a simplified model on the basis of the output of an advanced model, or based on measurement data of a real system [5, 6]. Making use of this inverse fuzzy arithmetical approach in conjuction with measured output data, the fuzzy-valued parameters of the simplified model can be identified in such a way that the re-simulated output of the fuzzy-parameterized model conservatively covers

Fig. 36.1. The author, Michael Hanss, with Lotfi A. Zadeh on the occasion of the 22nd International Conference of the North American Fuzzy Information Processing Society – NAFIPS 2003, Chicago, IL, USA, July 2003.

the range of possible output values of the advanced model or the real system, respectively. Thus, a model with simplified structure, but fuzzy-valued parameters, can be used instead of a structurally complex model with crisp parameters. Based on this procedure, a novel criterion has been formulated to generally assess the appropriateness and the quality of models [7]. Unlike the conventional way of proceeding, which usually focuses on the L_2-norm of the output deviation only, the presented quality criterion also takes into account the uncertainty of the model parameters which are the source of the output deviation assuming a special model structure. By this method, the quality of models with different type or structure for the same real system can effectively be quantified and rated. Furthermore, the models can be optimized more comprehensively than this is possible with the rather narrow view of optimizing the output deviation only. For engineering applications, for example, the quantification of the model uncertainties, which cause the output deviation, enables the engineer to launch adequate measures in the actuator domain, rather than in the output domain, where this is impossible.

36.4 Conclusions

In recent years, fuzzy arithmetic has emerged as a powerful methodology to comprehensively model systems with uncertainties. It allows for the inclusion of uncertainties – in particular of those of epistemic type – from the very beginning of the modeling procedure and thus leads to a somehow more sincere and honest numerical simulation which reflects both the benefits and the limitations of the available information about the system model. Moreover, making use of an inverse fuzzy arithmetical approach in conjuction with a special model validity criterion, the quality of models can be assessed and rated with respect to their appropriateness.

Acknowledgement. It is the author's great pleasure to express his distinct appreciation and sincere thanks to Professor Lotfi A. Zadeh for his lifetime achievement, an invaluable work that provided decisive inspiration to a great number of scientists and accomplished the creation of a new scientific community.

References

1. Dubois, D., Prade, H.: Fuzzy Real Algebra: Some Results. Fuzzy Sets and Systems 2, 327–348 (1979)
2. Dubois, D., Prade, H.: Fuzzy Sets and Systems: Theory and Applications. Mathematics in Science and Engineering, vol. 144. Academic Press, New York (1980)
3. Elishakoff, I., Ren, Y.J.: The Bird's Eye View on Finite Element Method for Structures with Large Stochastic Variations. Computer Methods in Applied Mechanics and Engineering 168, 51–61 (1998)
4. Fleissner, F., Hanss, M., Haag, T., Eberhard, P.: Uncertainty Analysis for a Particle Model of Granular Chute Flow. Computer Modeling in Engineering & Sciences (CMES) – Special Issue on Contact Mechanics in the Engineering Sciences II 52(2), 181–196 (2009)

5. Haag, T., Reuß, P., Turrin, S., Hanss, M.: An Inverse Model Updating Procedure for Systems with Epistemic Uncertainties. In: Proceedings of the 2nd International Conference on Uncertainty in Structural Dynamics, Sheffield, UK, pp. 116–125 (2009)
6. Haag, T., Herrmann, J., Hanss, M.: Identification Procedure for Epistemic Uncertainties Using Inverse Fuzzy Arithmetic. Mechanical Systems and Signal Processing 24(7), 2021–2034 (2010)
7. Haag, T., Carvajal González, S., Hanss, M.: Model Validation and Selection Based on Inverse Fuzzy Arithmetic. Mechanical Systems and Signal Processing (2012)
8. Hanss, M.: The Transformation Method for the Simulation and Analysis of Systems with Uncertain Parameters. Fuzzy Sets and Systems 130(3), 277–289 (2002)
9. Hanss, M.: The Extended Transformation Method for the Simulation and Analysis of Fuzzy-parameterized Models. International Journal of Uncertainty, Fuzziness and Knowledge-Based Systems 11(6), 711–727 (2003)
10. Hanss, M.: An Approach to Inverse Fuzzy Arithmetic. In: Proc. of the 22nd International Conference of the North American Fuzzy Information Processing Society – NAFIPS 2003, Chicago, IL, USA, pp. 474–479 (2003)
11. Hanss, M.: Applied Fuzzy Arithmetic – An Introduction with Engineering Applications. Springer, Berlin (2005)
12. Hanss, M., Turrin, S.: A Fuzzy-based Approach to Comprehensive Modeling and Analysis of Systems with Epistemic Uncertainties. Structural Safety 32(6), 433–441 (2010)
13. Hemez, F.M., Booker, J.M., Langenbrunner, J.R.: Answering The Question of Sufficiency: How Much Uncertainty is Enough? In: Proc. of the 1st International Conference on Uncertainty in Structural Dynamics – USD 2007, Sheffield, UK, pp. 23–48 (2007)
14. Kaufmann, A., Gupta, M.M.: Introduction to Fuzzy Arithmetic. Van Nostrand Reinhold, New York (1991)
15. Moens, D., Vandepitte, D.: Fuzzy Finite Element Method for Frequency Response Function Analysis of Uncertain Structures. AIAA Journal 40, 126–136 (2002)
16. Moens, D., Hanss, M.: Non-probabilistic Finite Element Analysis for Uncertainty Treatment in Applied mechanics: Recent Advances. Finite Elements in Analysis and Design – Special Issue on Uncertainty in Structural Dynamics 47(1), 4–16 (2011)
17. Oberkampf, W.L., DeLand, S.M., Rutherford, B.M., Diegert, K.V., Alvin, K.F.: Error and Uncertainty in Modeling and Simulation. Reliability Engineering & System Safety 75(3), 333–357 (2002)
18. Rao, S.S., Sawyer, J.P.: Fuzzy Finite Element Approach for the Analysis of Imprecisely Defined Systems. AIAA Journal 33, 2364–2370 (1995)
19. Schuëller, G.I.: On the Treatment of Uncertainties in Structural Mechanics and Analysis. Computers and Structures 85(5-6), 235–243 (2007)
20. Turrin, S., Hanss, M., Selvadurai, A.P.S.: An Approach to Uncertainty Analysis of Rockfall Simulation. Computer Modeling in Engineering & Sciences (CMES) – Special Issue on Contact Mechanics in the Engineering Sciences III 52(3), 237–258 (2009)
21. Zadeh, L.A.: Fuzzy Sets. Information and Control 8, 338–353 (1965)
22. Zadeh, L.A.: The Concept of a Linguistic Variable and its Application to Approximate Reasoning. Information Sciences 8, 199–249 8, 301–357 9, 43–80(1975)

37

"Fuzzy Cloud": Sfumato versus Chiaroscuro in Music

Hanns-Werner Heister

37.1 Introduction

Interestingly L. Zadeh himself initially conceived Fuzzy Logic (including Fuzzy Set Theory and Soft Computing) in a transdisciplinary fashion: It was meant for application beyond the borders of his discipline. By now there have been numerous attempts and contributions from a great variety of scientific disciplines (Cf. for example [11].) All of this shows the many-fold and productive significance of Fuzzy Logic. Up to 22 November 2008 I myself had made contact with Fuzzy Logic every now and then without even being aware of it. Fuzzy Logic is applied in video cameras or refrigerators etc. Consciously and theoretically I was made aware of this only on the day of R. Seising's presentation.[1] Some aspects of what interests me, even fascinates me in Fuzzy Logic have been discussed in two publications of mine from 2009/2010 ([4], [6]). Some further aspects I will briefly sketch here.

37.2 'Tolerances': Sensual-Practical Realization versus Cerebral-Notational Conception

From the point of view of a scholar of the arts, the realistic reference of Fuzzy Logic to necessary blurs is crucial in the actual production of art. Within this context the prevalence of the workmanship and the engineering over the mathematical and formally logical, of the sensual-practical over the purely cerebral, is undeniable. The mathematics in the background are just as fascinating – but here I remain somehow skeptical regarding the absolute claim to sole representation of authority and unquestionable exactness; in the realm of the mind, to which mathematics principally belong, it might apply, at least in principle. But taking into account reality, be it nature or society, it lastly doesn't sum up to more than a seeming exactness. Projected onto

[1] [10] In earlier publications R. Seising gives a detailed account of the initial conception of Fuzzy Logic by Zadeh, see [9]. Cf. an early but already nearly endless list e.g. in [1], Content p. VII-XI and 362: Aid to creativity / Analysis of scientific literature / Applied Operations Research / Artificial Intelligence and Robotics Image / Biological and Medical Sciences / Control / Damage assessment of structures / Economics and Geography / Linguistics / Processing and Speech Recognition / Psychology / Semiotics / Sociology. (The list, a little bit unsystematically ordered and uneven, and long, but nevertheless perhaps and gappy, is given here in alphabetical order.)

R. Seising et al. (Eds.): *On Fuzziness: Volume 1*, STUDFUZZ 298, pp. 241–251.
DOI: 10.1007/978-3-642-35641-4_37

the principles of art and its procedures the mathematics resemble the chiaroscuro, the clear and distinct demarcation of light and darkness in painting, whereas Fuzzy Logic appears more like sfumato, the blending of contours and emphasis on transitions, or in a more general sense, they relate to each other like clock and cloud (see below). In music it is interesting to observe the quantification of the divergence, but above all qualitative aspects, i.e. why and on the basis of which factors these divergences from the notation come to be the prescriptions, the blueprint for the performance.

If this constitutes a practical form of fuzzyfication, then in the production, the creation of an art work in the scope of concept, idea, sketch and finalized work, the converse process is decisive: The development of concrete, distinct forms from the amorphous, the formless, the evolving of the chaos of the material into the cosmos of artwork. In the context of a 'theory of similarity' it's about degrees of relatedness in their quantification, and thereby at least indirectly also qualitative forms of relatedness between individual results of this producti on process. This, as well as the overall context of forms of existence of the art process up to the resonances in forms of memory and imagination, are in the arts one of the most important duties of Fuzzy Logic.

37.3 "Nuages", Musical Prose – Fuzzyfication

Clouds, symptoms and symbols of weather, wind, water and processes of heat circulation and more, have from the beginning been significant and fascinating phenomena of nature, even before the Neolithic revolution with the ensuing development of agriculture that depended on climate and weather. In regard to Fuzzy Logic the clouds' two-fold character as form and amorphous object is significant. The groundwork for the classification of clouds was established by Luke Howard as early as 1802, following the example set by Carl Linné's model for the classification of animals and plants. Today's settings are based on this; one example: figure 37.1[2]

The specification of vague forms in detail, the more precise classification of transitions and caesuras in the real ,structured continuum' is one of the classical tasks of Fuzzy Logic, which can contribute quite a bit to this.

In visual arts representations of clouds[3] in the genre of landscapes are indeed natural, especially in Eastern Asia and Europe (in the latter since the Renaissance). However, musically clouds are only really representable once a high standard of development regarding material, technology and language of music has been reached. Required are loose regulations of musical compositional technique and structure. They are generally to be understood as tendency towards musical prose, and enable a conscious reproduction of fuzziness. This level is given since the time of accomplished industrialization (in the centres) and the emerging monopolization, as in impressionism and symbolism as well as naturalism on the other hand - otherwise opposing trends. Exemplary are works such as Charles Baudelaire's prose poem L'étranger of 1869 with lines like "J'aime les nuages ... les nuages qui passent ...

[2] http://en.wikipedia.org/w/index.php?title=File%3ALowcloudsymbols.gif

[3] In detail cf. among others [12].

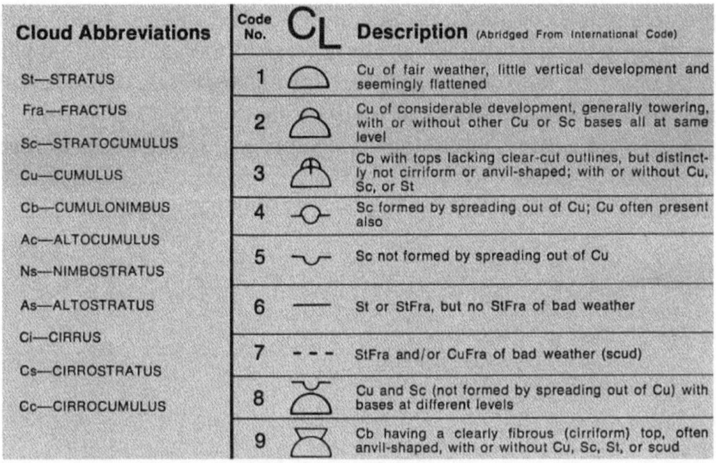

Fig. 37.1. Illustration 1

là-bas ... là-bas ... les merveilleux nuages!"[4], or Claude Debussy's movement Nuages (Clouds) from the Trois Nocturnes (1897/1899) – Nuages – Fêtes - Sirènes (last movement with women's choir; Debussy employed only wordless vocalizations, like György Ligeti later did in his Clocks and Clouds). The representation and imagination of ‚clouds' of all kinds reappeared increasingly in "sound composition" in the time around 1960, only now shifted from the impressionist and programmatic into the structural. Ligeti made these entities more concrete through programmatic titles, as in his orchestra piece Atmosphères, prominently employed in Stanley Kubrick's movie "2001: A Space Odyssey" in 1968.

"Fabric" or "texture" is an old metaphor for the multi-voiced structure of compositions. Ligeti re-installs it with the concrete and sensual, by recourse to synaesthetic conceptions and by applying orchestral sound spectra. As "texture", i.e. "Gewebe" in a more specific sense, he compiles his dense, interwoven, overlapping, interlooping carpets of tone colours. He uses "micropolyphony" with a sfumato-like blurring of the voices instead of the chiaroscuro with its distinct profile in classical polyphony.[5] However, almost simultaneously he counters this with a counterpoint in the form of poignantly delimited musical gestures and entities, explicitly in 1972/1973 in Clocks and Clouds.

37.4 Schönberg's "War-Clouds Diary": Anthropomorphic Projection as Reduction of Complexity – Defuzzyfication

The moving, flowing, and ever-changing character of clouds, again and again solidifying into forms with distinct contours, invites anthropomorphic projections:

[4] Charles Baudelaire: *L'étranger*, in *Le Spleen de Paris. Petits poèmes en prose*, I (1869). "I love the clouds . . . , the clouds that pass by . . . there ..., there. . . : the wonderful clouds!"

[5] Cf. among others [3] on Ligeti and [5] on the problem of polyphony.

All people imagine seeing faces, or animals, etc. The anthropomorphic projection thereby seemingly clarifies the actual blurredness of the amorphous appearance of these natural phenomena. Arnold Schönberg formulates nearly delusional cloud representations in his “War-clouds diary” of 1914 (figure 37.2[6]) For a while he notates (verbally and graphically) meteorological conditions. He attempts to read and interpret the “book of nature” - from a kind of universal perspective that establishes correspondences between everything, lastly in aesthetically mediated analogies.

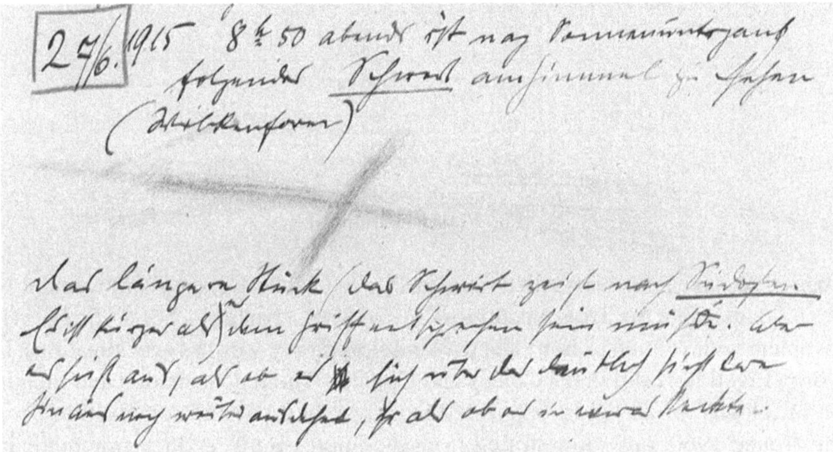

Fig. 37.2. Illustration 2: War-clouds diary 19/5/1915

> “27/6.1915 8 h 50 in the evening After sunset the following sword can be seen in the sky (form of a cloud) the longer part (the sword points South-West [)] is shorter than it should be in relation to the hilt. But it appears to extend beyond what is clearly visible, as if it were stuck in something.”

Schönberg amalgamates “Faith”, “God”, and “War”.[7]

> “War clouds diary; commenced on 24 September 1914 / Many people will have tried, like me, to read the events of the war in the skies today, now

[6] Illustration and transscription: http://www.usc.edu/libraries/archives/schoenberg /painting/exteriorhtms/noritter27.htm. 12.1.2012. Arnold Schoenberg: “War-Clouds Diary” *Journal of the Arnold Schoenberg Institute* 9/1 (June 1986): p. 55, plate no. 5 (color). Owner and Location: Arnold Schönberg Center, Vienna, Austria. Text there in German.

[7] With this anthropomorphic projection Schönberg alludes to magical-religious divination, such as the Roman auspice or also the omina from the hetascopy from ancient Greece and ancient Babylonia and many other instances. Gavin Pretor-Pinney, author of Wolkengucken and founder of the “Cloud Appreciation Society”, rightly comments, “Clouds are the Rorschach inkblots of the sky, so to say abstract forms, into which we project our desired images.” http://www.sueddeutsche.de/panorama/guck-in-die-luft-club-gemaelde-am-himmel-1.921434. 16.9.2011.

> that finally the faith in higher powers and in God returns. [...] I [...] hope to find correspondences, once more exact reports are available, since a number of the war events so far have only been foreseeable by the "atmosphere/mood" of the sky. Repeatedly the "golden glow", the "wind of victory", a "dark blue sky", "bloody clouds" (at sunset) have caught my attention, which have unfailingly preceded victorious German events. Similarly worsening weather conditions with storm and rain, deep black clouds preceded all negative turns of events in the Austrian-Russian war zone. [...] I must note that not always does a cloudy sky give me the impression of unfortunate events."

Schönberg believes the natural object to be a socially significant sign. In his irrationality he remains sufficiently realistic in order to relativize the trivial attribution and also the dissonance of the non-beautiful nature in regard to usual perceptions of good weather as possibly positive quality of the bellicose. Surely his unreflecting chauvinist assumption that the skies are on the side of the Austrian-German troops is naive. There is even a glimpse of skepticism or at least reservation when Schönberg hopes for empirical verification through lengthy empirical surveys. His master student Alban Berg attempts a rational explanation within the discourse of natural sciences in his letter to Schönberg on 8 October 1914:

> "Your suggestion to Webern to observe the clouds and the sky has also moved me strongly/touched me in a strong way. Already this summer - admittedly without connecting it with the war - I have noted incredible appearances... The consistently good weather of this summer... was striking to me too; only that I saw a more material [sic] connection to the war. For I believed, after often having had the opportunity to observe the strong effects of weather shooting ["Wetterschießen"[8]], that the millions of shots over a period of weeks – including almost the entire area of central Europe – across battle lines that stretch over several hundreds of kilometers might

[8] Apparently there is no English aequivalent for "Wetterschießen". Since the Renaissance this speciality was common mainly in the Alps. It was primarily directed against thunder storms and hail. The making of noise as archaic magic of defence was one part of it. Originally "Wetterschießen" was derived from magical-Christian conceptions, like assuming weather witches to be the root cause. Even though the mortar shells hardly even reached the clouds, this matter was rationalized in the context of the industrialization, as in the case of Alban Berg's comment. The flipside is the magical or technical conjuration of rain in times of drought, from the "rainmaker" rattle of the Aborigines up to the "cloud seeding" undertaken since the 1950ies. The more general "climate engineering" since the 1990ies combines both. The following militarist technocratic megalomania reminds of the aforementioned witch mania: "The misters of the American institute belive it to be principally possible, to dissect a hurricane or taifun into several storms of a harmless size. They have calculated that one would need to launch twenty atomic bombs every second over a períod of time to meet an approaching storm with equal force." (http://www.spiegel.de/spiegel/print/d-41124261.html, 2.2.2012) These considerations follow neither Aristotelian logic nor Fuzzy Logic, but obviously the logic of anti-humane insanity.

> have influenced the weather. Surely, what I now see in the sky, seems to have other, deeper reasons. Very particular was for example the night sky some days ago. The clouds were juxtaposed in regular rhombus shapes; about so (figure 37.3): …[9]

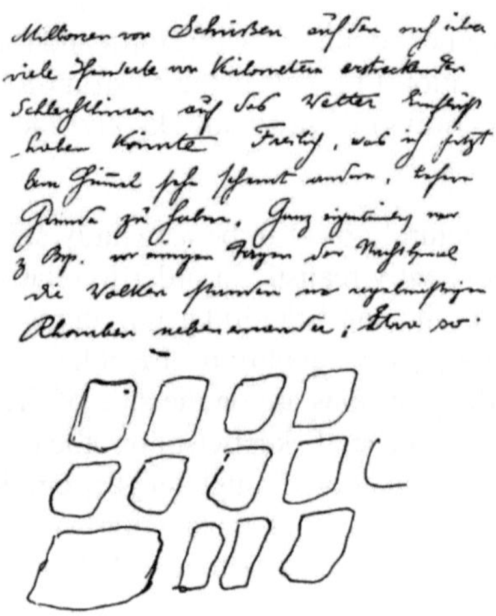

Fig. 37.3. Illustration 3

Alban Berg depicts the clouds in very regular shapes, almost in a constructivist manner. On top of this, the sum total of the cloud forms, if we exclude the vaguely indicated scribble to the right of the second row, as chance would have it, equals 12, which is a highly significant number not only for the Schönberg school and its twelve tone technique.

37.5 "Clocks and Clouds"

Without (explicit) knowledge of Fuzzy Logic Ligeti in 1972/1973 composed Clocks and Clouds in a rich orchestration with 12 female voices. (The also musically famous 12, as e.g. in equally tempered tuning[10] as in dodekaphony!)

> "The title of Ligeti's Clocks and Clouds refers to an essay by the Anglo-Austrian philosopher Karl Raimund Popper, "On Clocks and Clouds" [8]. Popper's essay describes two different kinds of processes that occur in nature, one that can be measured exactly ("clocks") and the other, made up of

[9] Letter and sketch quoted from [7], p. 135. Text there in German.

[10] One of the favorite objects of Fuzzy Logic-Studies in music.

indefinite occurrences that can only be described in a statistical approximation ("clouds")."[11]

Sir Karl R. Popper, foremost known for his fight against objective knowledge and dialectics, for once had a productive idea.[12] Even if he had understood the difference to be a contrary or even contradictory dichotomy - Ligeti in any case configured this polarity dialectically:[13]

> "'I liked Popper's title and it awakened in me musical associations of a kind of form in which rhythmically and harmonically precise shapes gradually change into diffuse sound textures and vice-versa, whereby then, the musical happening consists primarily of processes of the dissolution of the 'clocks' to 'clouds' and the condensation and materialization of 'clouds' to 'clocks'.'"[14]

Ligeti therefore designed the polarity as process:

> "These transformations are not clearly delineated occurrences of "now clocks" and "now clouds." Instead, through minutely shifting rhythmic patterns, Ligeti presents the listener with a malleable texture whereby the homogeneous character of the musical material allows for little distinction between a clearly defined, periodic ticking rhythm and the blurred dissolution into clouds."

---

[11] Steve Lacoste (Archivist for the Los Angeles Philharmonic Association) http://www.laphil.com/philpedia/piece-detail.cfm?id=2742 (9.1.2011). Also the following quotes (including the Ligeti quotes) are taken from the contribution of Lacoste.

[12] In consideration oft he fact that Zadeh had published his concept as early as 1965, a direct connection is quite possible.

[13] An astonishing inversion from a different culture, on the subject of Gustav Mahler's Lied von der Erde (1908/1909), movement VI, *Der Abschied* (*The Farewell*): "In Chinese poetry, clouds can carry a different symbolic meaning from that which we would expect in Western literature. While clouds are commonly seen as *transitory* in Western literature, they are often to be understood as a *permanent* phenomenon in Chinese texts. Despite the fact that clouds appear in *different shapes every day*, they are a natural object that will *always* show up in the sky. Furthermore, they can be seen by people from anywhere. [. . .] In Wang Wei's *Farewell*, the clouds [...] have the [. . .] two meanings: their friendship will *always* be there and it will accompany the departed friend *no matter where* he goes." [13] Sun 2009, Footnote 95, p. 79. The Chinese poet Wang Wei stressed this stabile forever: *Staying at a Teacher's Mountain Retreat, Awaiting a Friend in Vain*, Last stanza = Strophe "But our friendship lasts / *forever*, like the white *clouds* / in the sky." (Literal translation from the Chinese) "Und ewig, ewig sind die / weißen Wolken" (Hans Bethge, Translation from the French) "And the white clouds are forever, forever" (Translation from Hans Bethge) (Sun 2009, p. 79) Mahler omits the clouds seeming to him too transitory and substitutes them by "Allüberall und ewig / Blauen licht die Fernen!" - "Everywhere and forever the distance shines bright and blue!" (Derryck Cooke, Translation from the German) The translations in themselves are, by the way, examples of Fuzzy Logic.

[14] Ibidem. Also the following quotes (including the Ligeti quotes) are taken from the contribution of Lacoste.

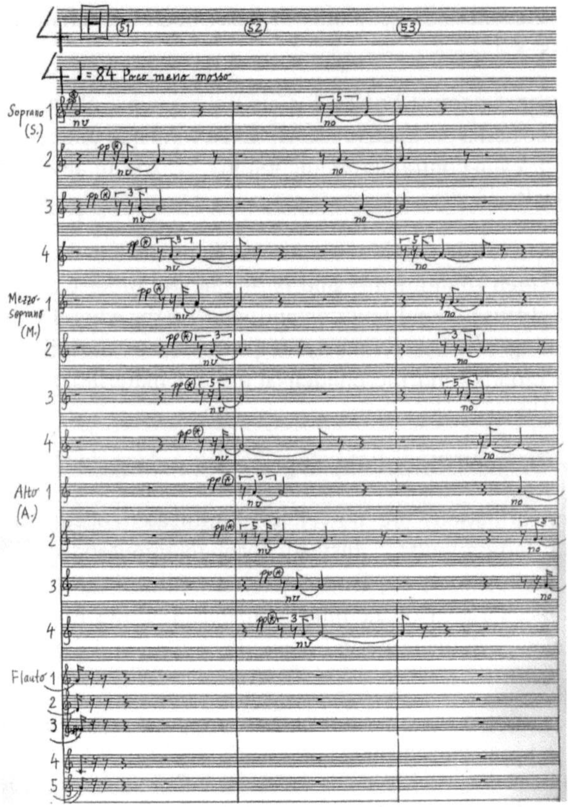

Fig. 37.4. Illustration 4: Ligeti, Clocks and Clouds, Excerpt women's choir + 5 flutes, T. 51-53 C: Schott International, Mainz

Ligeti carries this interference even a step further. Even in clearly outlined objects the contours can dissolve – soft composing instead of hard composing. Ligeti states in a 1978 interview: "I should like to refer to the soft, limp watches of Dali's painting (The Persistence of Memory, 1931), which had associative value in the composition of this piece..."

Likewise Ligeti employs microtonality, intervals smaller than half tones, as a partial moment of the dialectics of the clearly outlined and the amorphous.[15] The usage of complementary contrasts in the composition is reflected in the instrumentation.

> "The composer also creates "clocks" and "clouds" harmonically, moving from harmonies based on standard tuning to those based on non-traditional intervals, thereby creating harmonies that are once clouded, once discrete.

[15] Here the problem of actual deviations from a normative twelve-tone system of equal tempered tuning is intensified. In regard to this as framework cf. among others the studies of Teresa León and Vicente Liern, Mathematics and soft computing in music, in: [11], pp. 425-440; Josep Lluís Arcos: Music and Similarity Based Reasoning, in: [11], pp. 441-452.

> Because of their capacity to realize these subtle shifts, five flutes form the backbone of Clocks and Clouds' harmonic skeleton. The five clarinets and twelve-voiced women's chorus nearly match the flutes' flexibility in realizing the harmonies as well as combining with them in high-pitched tone color (along with harmonics in the cellos and double basses) in "fluid" textures. In contrast, the other instruments, especially the two harps, four bassoons, two trumpets, and strings produce precise rhythmic patterns."

Finally, Fuzzy Logic can also be consciously comprehended and even directly stimulate the production of music, in compliance with Lotfi Zadeh's inter- and multidisciplinary conception. In a conversation on 25 November 2011, a friend and colleague, Reinhard Flender, composer and musicologist, reported the commission of a double (string) quartet[16] (he had already composed two string quartets). A fundamental problem regarding organization of movements and sound is, among others, to differentiate between the orchestration, the choral multiplication of single voices, and the chamber music in the strict sense of consisting only of solo voices. After all, two string parts playing the same tone or melody will necessarily be minutely out of tune, since the pitches cannot by intonated in an absolutely pure fashion: Once again we are confronted with the opposition realization vs. conception/notation, including the - in this case unintentional – already mentioned microtonality. From about three instruments of the same kind onwards one can achieve an effect of coalescence, which conjures up a fuller sound from the "mistakes" or rather deviations – so again the transition or interference of the dimensions pitch and timbre. Since in the case of a double quartet apart from the violins (4) only two string instruments are available – 2 violas, 2 violoncellos – R. Flender intends to consciously make use of this fuzziness in his composition; how exactly he is not sure yet. I proposed to him occasionally to employ a massive octave unisono as demonstrative contrast to the filigree texture of the 8 voices, more particularly the "Viennese Unisono", a parallel movement in an interval of no less than 3 octaves.[17] This idea did not completely convince him, if I observed his reaction correctly. But my already mentioned contribution to the volume [6] edited by him obviously had inspired him, and in our conversation he had the idea to provisionally title his composition Clouds, more precisely[18] even Fuzzy Clouds.[19]

[16] As other double quartets or octets – of string, wind, wind and string, even violoncello – at the limits of chamber music.

[17] Mozart, who probably originally conceptualized this compositional technique, carries the main theme over even more than four octaves in the beginning of the I. movement of *Eine kleine Nachtmusik*.

[18] For the translation of the manuscript into English I thank my son Hilmar Heister.

[19] "Voraussagen sind schwierig, vor allem wenn sie die Zukunft betreffen." [Predictions are difficult, especially if they concern the future.] (This sentence – remember the Schoenberg-/Berg-problem of weather-/war prediction – is arbitrarily ascribed to Mark Twain, Kurt Tucholsky, Karl Valentin; or Niels Bohr – whom it suits best, also in regard to Heisenberg's uncertainty principle – "Unschärferelation".) Therefore I cannot be sure if Reinhard Flender will maintain the preliminary title in the final form of existence. It would be desirable as an extension of the already available musical tributes to Lotfi Zadeh.

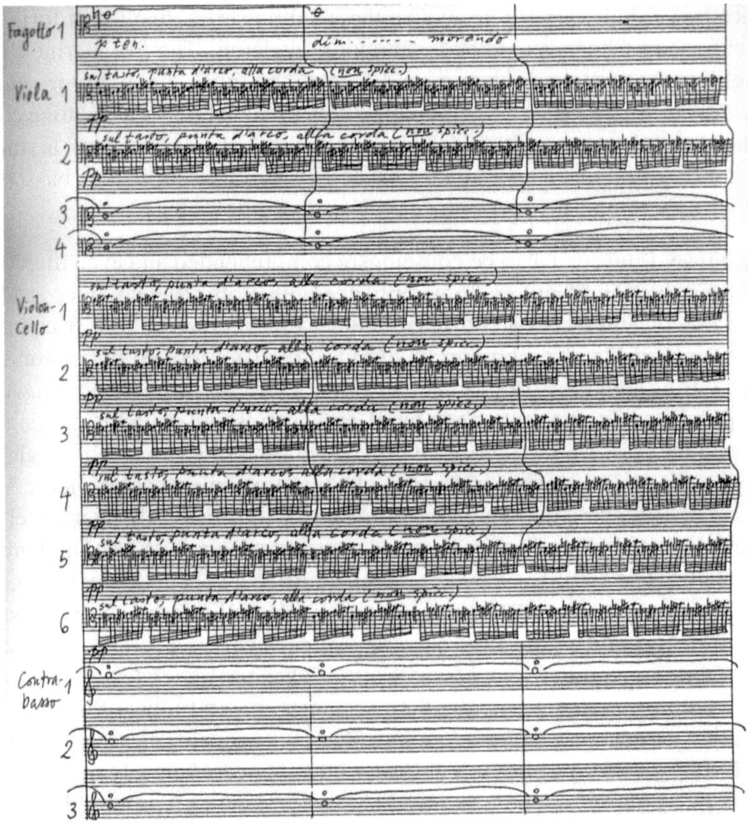

Fig. 37.5. Illustration 5: Ligeti, Clocks and Clouds, excerpt bassoon + strings, T. 51-53 C: Schott International, Mainz.

References

1. Dubois, D., Prade, H.: Fuzzy Sets and Systems: Theory and Applications. In: Bellman, R. (ed.) Mathematics in Science and Engeneering, vol. 144. Academic Press, New York (1980)
2. Heister, H.-W., Politik, K.: Religion: Gewalt bei Schönberg. In: Meyer, C. (ed.) Arnold Schönberg und sein Gott / and His God. Bericht zum Symposium, Wien, Juni 26-29, pp. 31–64 (2002, 2003); Journal of the Arnold Schönberg Center, Bd. 5
3. Heister, H.-W. (with Ines Gellrich, photographies): Un/Endlichkeit. Begegnungen mit György Ligeti 1989-2003. Modo-Verlag, Freiburg (2008)
4. Heister, H.-W.: Kybernetik, Fuzzy Logic und die Existenzformen der Musik. Ein vorläufiger Entwurf. In: Flender, R., Theede, M. (eds.): Innovation aus Tradition – Festschrift Hermann Rauhe zum 80. Geburtstag, Schott International, Mainz, pp. 93-115 (2010)

5. Heister, H.-W.: “Das ist Polyphonie.” Zur Natur in und von Mahlers Tonsatz. In: Heister, H.-W., Hochstein, W. (eds.) Wo die Schönen Trompeten Blasen - vokale and Instrumentale Semantik im Werk Gustav Mahlers (Musik und. Eine Schriftenreihe der Hochschule für Musik und Theater Hamburg. Neue Folge), vol. 11, pp. 205–228. Weidler Buchverlag, Berlin (2011)
6. Heister, H.-W.: Invariance and Variance of Motives: A Model of Musical Logic and/as Fuzzy Logic. In: Seising, R., Sanz, V. (eds.) Soft Computing in Humanities and Social Sciences. STUDFUZZ, vol. 273, pp. 423–450. Springer, Heidelberg (2012)
7. Nono-Schoenberg, N. (ed.): Arnold Schönberg. 1874-1951. Lebensgeschichte in Begegnungen. Verlag Ritter, Klagenfurt (1992)
8. Popper, K.R.: Of Clouds and Clocks. In: Popper, K.R. (ed.) Objective Knowledge: An Evolutionary Approach, pp. 2–66. Clarendon, Oxford (1972)
9. Seising, R.: Die Entstehung der Fuzzy Set Theorie und ihrer ersten Anwendungen – Ihre Entwicklung bis in die 70er Jahre des 20. Jahrhunderts, Stuttgart (2005); English as the Fuzzification of Systems: The Genesis of Fuzzy Set Theory and Its Initial Applications – Developments Up to the 1970s, Berlin etc (2007)
10. Seising, R.: Information - ein Begriff im Fluss. 60 Jahre 'Mathematische Theorie der Kommunikation, Vortrag zum Symposion 1948-2008: 60 Jahre Kybernetik - eine nach wie vor junge Wissenschaft?!, Berlin, November 21-22 (2008) (unpublished)
11. Seising, R., Sanz González, V. (eds.): Soft Computing in Humanities and Social Sciences. STUDFUZZ, vol. 273. Springer, Heidelberg (2012)
12. Spielmann, H., Westheider, O. (eds.): Wolkenbilder. Die Entdeckung des Himmels, Hamburg (2004)
13. Sun, S.-N.: Gustav Mahler's “Das Lied von der Erde”: An Intellectual Journey Across Cultures and Beyond Life and Death, Florida State University. College of Music. Master Thesis, 116 p (2009), http://etd.lib.fsu.edu/theses/available/etd-04122009-212814/unrestricted/SunSThesis.pdf (January 18, 2009)

38

Towards a Science of Creativity: Homage to Lotfi Zadeh

Cathy M. Helgason and Thomas H. Jobe

It was in 1997 that we first discovered fuzzy logic and met Lotfi Zadeh. Our interest in the implication of fuzzy logic for medicine led us to pursue an academic interest in Zadeh's contribution to science in its general implication. The most obvious point of relevance was that all of the so called "evidence based" clinical practice in medicine and experimentation level were intricately rooted in Aristotelian logic, probabilities and had in a significant organized fashion removed itself from fundamental discovery. Moreover, why had physician scientists accepted a method which allowed for the possibility of "chance" when someone's life or all the future of therapeutic ventures were at stake? The underlying philosophy to contend with was that which said that no physician or scientist could ever completely master the truth of how or why the clinical state was as it presented before or after therapeutic intervention. This knowledge had to be founded in a gamble. All this was far removed from mastering the principles of physiology, chemistry and biologic systems.

This situation was in great contrast to that respect for the human mind embraced by Dr Zadeh. Of course Lotfi was not in the field of medicine nor biological research and focused on automation. But, his goal was to have an automation that could approximate the workings of the human mind. This approach was entirely different from one where human ingenuity in problem solving and performance were dismissed. And while Zadeh pursued the Generalized theory of Uncertainty, fuzzy theory and computing with perceptions and words, his followers expanded on the use of fuzzy sets in logical, mathematical and practical settings.

Zadeh's focus on how the human mind functioned for the purpose of perception and problem solving always preceded mathematical formulation and it was his acceptance of the vastly greater level of uncertainty inherent in this goal which led to his contributions to science at the highest level. To make this point we consider uncertainty. On the one hand, there is the uncertainty of chance in which all possibilities have to be assumed to be known ahead of time and because of this given, the uncertainty lies in which of these possibilities are realized during an event, versus, on the other hand, the uncertainty of a creative intuition of the human mind, in which all the possibilities are not known ahead of time, and radically new possibilities come into existence. To accept the uncertainty of creative intuition there has to be a respect for the capability of the human mind to come up with something radically new and unheard of. Those who reject this level of uncertainty and prefer the greater certainty

R. Seising et al. (Eds.): *On Fuzziness: Volume 1*, STUDFUZZ 298, pp. 253–255.
DOI: 10.1007/978-3-642-35641-4_38

of chance turn away from the source of human creativity and its ability to shape the future. The gift of fuzzy theory given to all of us by Lotfi Zadeh was that it provided for the first time a way to represent this greater level of uncertainty, inherent in creative intuition, and thereby place the human mind in the position of being able to represent the continuum of experience through subsethood.

For medicine, constant innovation and new and better methods for diagnosis and treatment are necessary to improve the human condition. Clearly, a closed system of knowns in a universe of chance can never deliver discovery.

Finally, Max Plank's great contribution to the philosophy of science was the statement that consciousness itself always precedes its own contents. As he put it so eloquently in 1931 in The Observer; "I regard consciousness as fundamental. I regard matter as derivative from consciousness. We cannot get behind consciousness. Everything that we talk about, everything that we regard as existing, postulates consciousness." Fuzzy logic is ultimately , first and foremost, the logic of consciousness, but also contains within itself the logic of the contents of consciousness as well. This is Professor Zedah's great contribution : to capture both consciousness and the contents of consciousness in a single net.

Neils Bohr's coat of arms has the latin expression "Contraria sunt complementa" which embodies his discovery of complementarity in quantum mechanics. This insight, that opposites do not necessarity exclude one another but can add to each other was taken to a whole new level by Zadeh, in conjunction with his student Bart Kosko, when they formulated the fuzzy hypercube. This formulation of the complementarity of opposites goes the distance of making opposites orthogonal to one another as right angles formed by separate dimensions. Thus giving opposites complete freedom to interact as two independent dimensions. Such a framework laid the basis for the expansion of fuzzy logic from fuzzy sets, subject to crisp operations, to fuzzy sets subject to fuzzy operations, and to Kosko's formulation of fuzzy subsethood.

As Granik and Caulfield have demonstrated, the Schrödinger equation can be treated as a deterministic entity with a fuzzy character and the complementarity principle and wave-particle duality can be treated as a fuzzy deterministic micro object. The irony is that whereas Zadeh took inspiration in the development of fuzzy sets from the multivalued logic of quantum mechanics, the development of fuzzy logic has come full circle to embrace the quantum world and clarify for the first time that the variational principles, such as that of least action, represent Nature's way of defuzzifying Nature's essentially fuzzy presence. The Hamilton-Jacobi equation can be regarded as the defuzzification of the Schrödinger equation. Perhaps the greatest contribution of fuzzy logic to the quantum world, as developed by Granik and Caufield, is the elimination of the probabilistic interpretation of the wave function and the elimination of hidden variables in favor of a novel form of defuzzification within an infinite dimensional fuzzy hypercube. Thus a determinant degree of fuzziness substitutes for a probabilistic degree of crispness.

This full circle, comes about because, once fuzzy presence is substituted for crisp presence as ontologically prior, then randomness dissipates.

References

1. Dictionary of Scientific Quotations, 2000: Niels Bohr (1885-1962), and Max Plank (1858-1947)
2. Granik, A., John Caulfield, H.: Quantum Physics (quant-ph/0107054.vol.1) (July 9, 2001)
3. Helgason, C.M., Jobe, T.H.: The Fuzzy Hypercube and Causal Efficacy, Representation of Concomitant Mechanisms in Stroke. Neural Networks 11, 549–555 (1998)
4. Zadeh, L.A.: Generalized Theory of Uncertainty (GTU) – Principal Concepts and Ideas. Computational Statistics 51, 15–46 (2006)

39

Concept of Fuzzy Atmosfield and Its Visualization

Kaoru Hirota and Fangyan Dong

Abstract. A concept of Fuzzy Atmosfield (FA) is proposed to express the atmosphere generated by many machines/robots and humans communication. It is characterized by a 3D fuzzy cubic space with “friendly-hostile”, “lively-calm”, and “casual-formal” axes. Its visualization is also proposed by shape-color-size graphics. The FA aims to be a tool for expressing the atmosphere state of humans-machines communication. A part of basic demonstration example, i.e., enjoying home party by five eye robots and four human beings, is also introduced.

39.1 Introduction

Many interesting robots have been developed/demonstrated such as HONDA ASIMO and TOYOTA trumpet playing robot. In most cases, however, the demonstrations/operations are done basically by one robot and one human operator or at most a few to a few. In the near future, it is expected to realize such a society that many robots/machines and many humans communicate with each other by using internet connection or face to face communication.

To make a smooth communication in human-robot/machine or human-human interaction, understanding the emotion of others is important. In the case of many to many communication, however, it may become very important to pay attention to not only the emotion of each individual but also the atmosphere of the whole communication society. Many studies have been done on the emotion of each individual from view points of cognitive science or human-machine interface. But the atmosphere generated by the communication society/field by many individuals, i.e., machines, robots, and humans, has not been studied enough.

The authors’ group at Tokyo Institute of Technology has been studied on many robots and many humans communication through internet, where the atmosphere of the communication field/society by many (huge number of) individuals plays an important role for the smooth communication. The concept of Fuzzy Atmosfield, FA, is proposed to express the atmosphere in such humans-robots communication field/society. The “Atmosfield” is a new word from “atmosphere” and “field”, and is created by the authors’ group. It is characterized by a 3D fuzzy cubic space with

R. Seising et al. (Eds.): *On Fuzziness: Volume 1*, STUDFUZZ 298, pp. 257–263.
DOI: 10.1007/978-3-642-35641-4_39

"friendly-hostile", "lively-calm", and "casual-formal" axes by doing a cognitive science experiments and applying principle component analysis. The atmosphere in the communication field/society is expressed by a point in the 3D fuzzy cubic space and maybe varying/moving in the space time by time. To understand easily such movement of the atmosphere, a graphical representation method is also proposed, where "friendly-hostile" information is represented by "shape", "lively-calm" by "color", and "casual-formal" by "size". To illustrate the FA and its visualization method, a demonstration scenario "enjoying home party by five eye robots and four humans" is introduced/performed.

39.2 Concept of Fuzzy Atmosfield "FA"

The term Atmosfield (i.e. 'atmos' field, is composed of "atmosphere" and "field") is proposed to describe the atmosphere in the space surrounding us. The Atmosfield could not only reflect the atmosphere states during the interactive communication, but also affect the emotional states of individuals to some extent. Therefore, it is defined as one kind of psychological field, producing psychological feeling that can give influence to the process and results of humans' behavior. Due to the psychological characteristics of the Atmosfield, it could not be calculated as accurately as

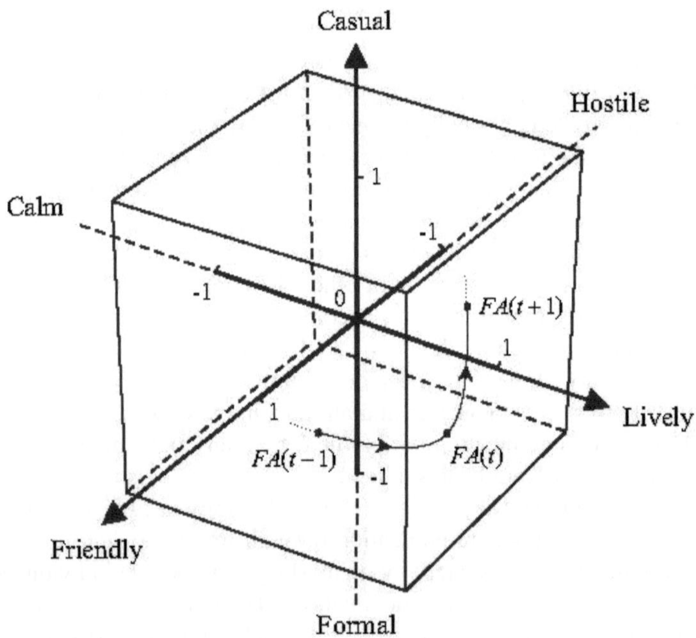

Fig. 39.1. Fuzzy Atmosfield, characterized by 3D fuzzy cubic space

some classical fields such as electrical field and magnetic field, and therefore, is defined as Fuzzy Atmosfield, which needs subjective comprehension to determine the attributes of the FA and fuzzy logic to deal with the reasoning from related factors to the FA.

The FA is characterized by a 3D fuzzy cubic space with three attributes (i.e. the three axes). Questionnaire surveys are carried out by ten peoples on twenty different occasions to determine the three axes of the FA, where some common atmosphere-related factors are enumerated as the candidates, e.g., friendly, lively, casual, harmonious, peaceful, noisy, warm, relaxed, and so on, whose ranges are defined as fuzzy domain, i.e., from -1 to 1. The results of questionnaires are analyzed by using principle component analysis and finally the most important 3 axes, i.e., "friendly-hostile", "lively-calm", and "casual-formal", are accepted to represent the FA. Finally, the FA is illustrated as shown in Fig. 39.1.

39.3 Visualization of Fuzzy Atmosfield

The FA aims to be a tool to express the real-time atmosphere during the communication with quantitative analysis, however, the dynamic states of atmosphere plotted in the 3D fuzzy cubic space (i.e. the points in 3D coordinate) are not easily observed and understood by humans.

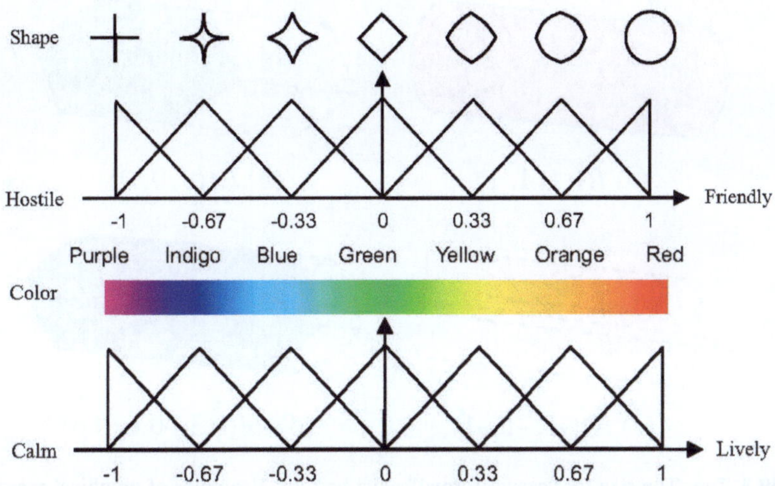

Fig. 39.2. Top: The shape for "friendly-hostile" axis; bottom: The color for "lively-calm" axis

Comparing to the coordinate values of atmosphere states, the graphics, which is one kind of visual presentations, is easier to use and understand, because of its functional, artistic, and easy-to-be-perceived characteristics. According to the elements of graphics, shape, color, and size are employed to describe the atmosphere states in "friendly-hostile" axis, "lively-calm" axis, and "casual-formal" axis, respectively, where the shape changes from circle to cross as the value of the "friendly-hostile" axis from 1 to -1; the color varies with the value of the "lively-calm" axis by using a color bar; and the size (i.e. thickness) describes the variation in the "casual-formal" axis. To associate "friendly-hostile" axis, "lively-calm" axis, and "casual-formal" axis, with shape, color, and size, respectively (cf. 39.2, and 39.3, top), a fuzzy domain of each axis is designed from 1 to -1, where fuzzy linguistic variables "highly", "medium", and "low" are adopted as the extent of each attribute. Several examples of the points in Fuzzy Atmosfield are shown in Fig. 39.3, bottom.

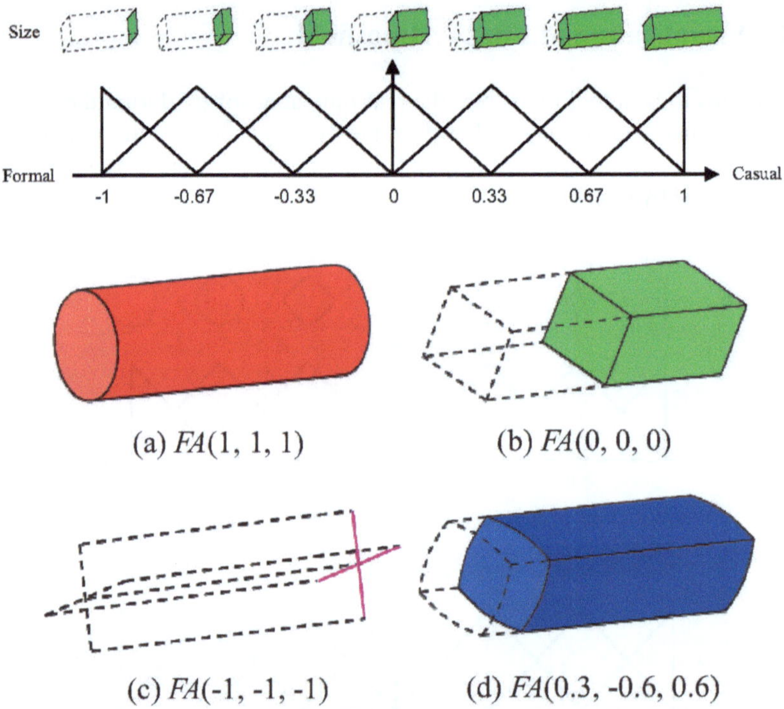

Fig. 39.3. Top: The size for "casual-formal" axis; bottom: Examples of graphical representation of FA

39.4 Enjoying Home Party by Five Eye Robots and Four Humans

A Mascot Robot System has been developed by the author's group as a part of the "Development Project for a Common Basis of Next-Generation Robots" (2005-2007) sponsored by NEDO (New Energy and industrial technology Development Organization) and "Development of Mascot Robot System with Casual Communication Ability" (2009-2012) sponsored by JSPS (Japan Society for the Promotion of Science). The main purpose of the projects is to perform casual communication between robots and humans. The system is implemented as multi-robots connected by RT middleware (RTM) on the internet. It consists of 5 robots, i.e., 4 fixed robots (placed on a TV, a darts game machine, an information terminal, and a mini-bar) and 1 mobile robot. Each of them includes an eye robot, a speech recognizer/synthesizer, a web camera, and a notebook PC.

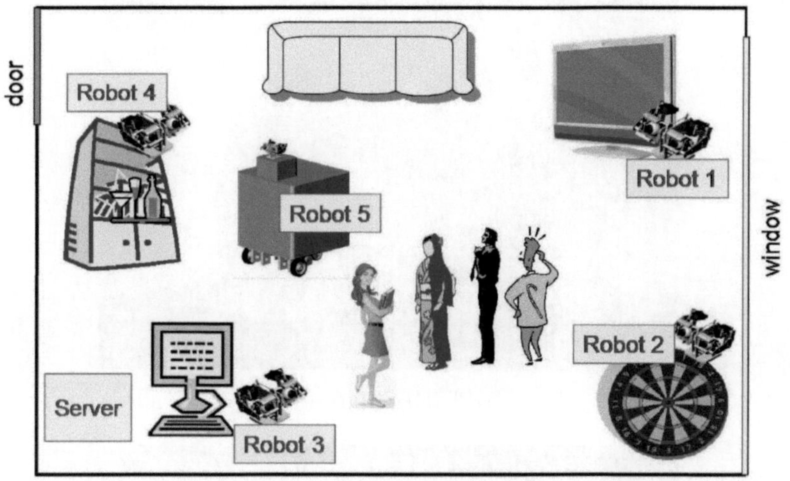

Fig. 39.4. Enjoying home party by 5 robots and 4 humans

These robots are connected together with a server through the internet by RTM, where fuzzy interruption technology makes it possible to perform smooth communications among plural robots. The Mascot Robot System's functioning is demonstrated in an ordinary living room, where casual communication between 5 robots and 4 human beings (1 host, 2 guests, and 1 walk-in) is conducted to enjoy home party scenario. Examples of demonstration scenes are shown in Fig. 39.5, bottom, where the fuzzy atmosphere is shown by the shape-color-size figure in the upside right.

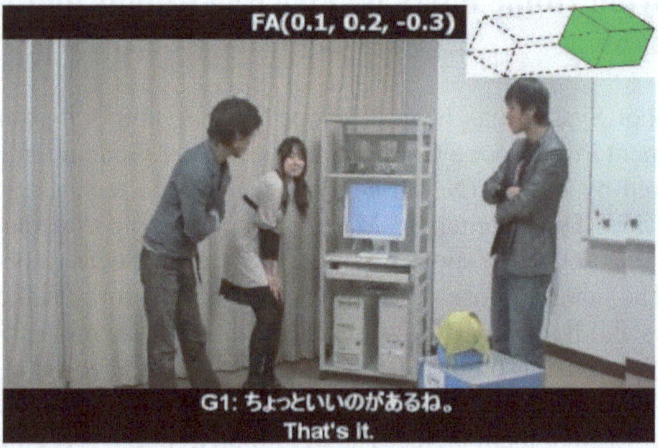

(a) *FA*(0.1, 0.2, -0.3)

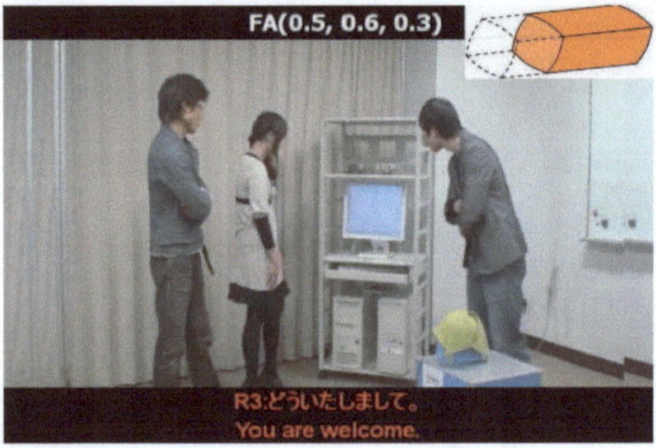

(b) *FA*(0.5, 0.6, 0.3)

(c) *FA*(0.3, 0.5, 0.3)

Fig. 39.5. Example of demonstration scenes with FA information

Fig. 39.6. Lotfi A. Zadeh, Fay Zadeh and Kaoru Hirota

39.5 Conclusions

A concept of Fuzzy Atmosfield is proposed paying attention to the atmosphere generated in the process of many to many interactive communication such as human-human interaction and human-robot interaction, and the state of the FA is described in a 3D fuzzy cubic space with "friendly-hostile", "lively-calm", and "casual-formal" axes. An application of the FA to multi-human-multi-robot interaction is implemented, where the Mascot Robot System is used to perform the experiment of home party with four humans and five eye robots in household environment. The FA is designed as a tool for users to understand the real-time atmosphere, where the state of the FA is visualized by graphical representation.

Acknowledgement. Authors are pleased to have an opportunity to write a small contribution to Professor Zadeh's book. He is a very kind leader in our fuzzy community. Lastly the first author would like to add a photo with Prof. and Ms Zadeh at Berkeley in 1990.

40

Computing with Words and Protoforms: Powerful and Far Reaching Ideas

Janusz Kacprzyk and Sławomir Zadrożny

Abstract. We show how Zadeh's computing with words and perceptions, the idea of an extraordinary power and far reaching impact, can lead to a new direction in the use of natural language in data mining, the linguistic data(base) summaries. We emphasize the relevance of Zadeh's protoform which may effectively and efficiently represent the user's intentions and interests, and show that various types of linguistic data summaries may be viewed as items in a hierarchy of protoforms of summaries.

40.1 Introduction

We wish to shortly present the essence and some applications of *computing with words* (CWW), and its inherent *protoforms*. These can be considered, in our opinion, to be the most influential and far reaching idea conceived by Zadeh, except for his "grand inventions" like fuzzy sets and possibility theories or foundations of the state space approach in systems modeling. To follow the spirit of this volume, our exposition will be concise and comprehensible.

Computing with words (and perceptions), or CWW, introduced by Zadeh in the mid-90s, and first comprehensively presented in Zadeh and Kacprzyk's books [17], may be viewed a new "technology" in the representation, processing and solving of various real life human centric problems. It makes it possible to use natural language, with its inherent imprecision, in an effective and efficient way.

Zadeh used the so-called PNL (precisiated natural language) in which statements about values, relations between variables, etc. are represented by constraints. Its statements, written "x isr R", may be different, and correspond to numeric values, intervals, possibility, verity and probability distributions, usuality qualification, rough sets representations, fuzzy relations, etc. For us, the usuality qualified statements have been be of special relevance. Basically, it says "x is usually R" that is meant as "in most cases, x is R". PNL may play various roles among which crucial are: description of perceptions, definition of sophisticated concepts, a language for perception based reasoning, etc. Notice that the usuality is an example of a modality in natural language. Clearly, this all is meant as a tool for the representation and processing of perceptions.

Another Zadeh's ingenious inception is the concept of a *protoform* [16]. In general, most perceptions are summaries, exemplified by "most Swedes are tall" which

R. Seising et al. (Eds.): *On Fuzziness: Volume 1*, STUDFUZZ 298, pp. 265–270.
DOI: 10.1007/978-3-642-35641-4_40

is clearly a summary of the Swedes with respect to height. It can be represented in Zadeh's notation as "most *A*s are *B*s". This can be employed for reasoning under various assumptions. One can go a step further, and define a protoform as an abstracted summary, like "*QA*s are *B*s", and now have a more general, deinstantiated form of our point of departure (most Swedes are tall), and also of "most *A*s are *B*s". Most of human reasoning is protoform based.

Basically, the essence of our work over the years was to show that the concept of PNL, and in particular of a protoform, viewed from the perspective of CWW, can be of use in attempts at a more effective and efficient use of vast information resources, notably through linguistic data(base) summaries which are very characteristic for human needs and comprehension abilities. In what follows we give an outline of our approach.

40.2 Linguistic Data Summaries via Fuzzy Logic with Linguistic Quantifiers

The linguistic summary is meant as a sentence [in a (quasi)natural language] that subsumes the very essence (from a certain point of view) of a set of data. Here this set is assumed to be numeric, large and not comprehensible in its original form by the human being. In Yager's approach (cf. Yager [12], Kacprzyk and Yager [3], and Kacprzyk, Yager and Zadrożny [4]), if $Y = \{y_1,\ldots,y_n\}$ is a set of records in a database, e.g., representing the set of workers, and $A = \{A_1,\ldots,A_m\}$ is a set of attributes characterizing the elements of Y, e.g., salary, age, etc., $A_j(y_i)$ denotes a value of A_j for object y_i, then a *linguistic summary* of a data set Y consists of: (1) a summarizer S, i.e. an attribute together with a linguistic green (fuzzy predicate) (e.g. "low salary" for attribute "salary"), (2) a quantity in agreement Q, i.e. a linguistic quantifier (e.g. most), and (3) truth (validity) T of the summary, $T \in [0,1]$; optionally, a qualifier R, i.e. another attribute together with a linguistic term (fuzzy predicate) may be added (e.g. "young" for "age").

The linguistic summaries, without and with a qualifier, may be exemplified by

$$T(\text{most of employees earn low salary}) = 0.7 \tag{40.1}$$
$$T(\text{most of young employees earn low salary}) = 0.85 \tag{40.2}$$

The core of a linguistic summary is a *linguistically quantified proposition* in the sense of Zadeh [15]; those corresponding to (40.1) and (40.2) may be written, respectively, as

$$Qy\text{'s are } S \tag{40.3}$$
$$QRy\text{'s are } S \tag{40.4}$$

The T, i.e., the truth value of (40.3) or (40.4), can be calculated by using either original Zadeh's calculus of linguistically quantified statements (cf. [15]), or other interpretations of linguistic quantifiers.

Formulas (40.3) and (40.4) may be seen as the most abstract protoforms, the highest in the hierarchy of protoforms, while (40.1) and (40.2) are examples of fully instantiated protoforms, "leaves" of their "hierarchy tree". Going down this hierarchy one has to instantiate particular components of (40.3) and (40.4), i.e., quantifier Q and fuzzy predicates S and R. The instantiation of the former one boils down to the selection of a quantifier. The instantiation of fuzzy predicates requires the choice of attributes together with linguistic terms and a structure they form when combined using logical connectives. Thus, in general, there is an infinite number of potential protoforms, though, due to a limited capability of the user only a reasonable number of summaries should be taken into account.

The concept of a protoform may provide a guiding paradigm for the design of a user interface supporting the mining of linguistic summaries. It may be assumed that the user specifies a protoform of linguistic summaries sought. Basically, the more abstract protoform the less should be assumed about summaries sought, i.e., the wider range of summaries is expected by the user. There are two limit cases, where:

- a totally abstract protoform is specified, i.e., (40.4),
- all elements of a protoform are totally specified as given linguistic terms.

In the former case the system has to construct all possible summaries for the context of a given database and show those with the highest validity T. In the second case, the whole summary is specified by the user and the system has only to verify its validity. The former case is usually more attractive for the user but more complex computationally. There is a number of intermediate cases that may be more practical. In Table 40.1 basic types of protoforms/linguistic summaries are shown, corresponding to protoforms of a more and more abstract form.

Table 40.1. Classification of protoforms/linguistic summaries

Type	Protoform	Given	Sought
0	QRy's are S	Everything	validity T
1	Qy's are S	S	Q
2	QRy's are S	S and R	Q
3	Qy's are S	Q and structure of S	linguistic terms in S
4	QRy's are S	Q, R and structure of S	linguistic terms in S
5	QRy's are S	Nothing	S, R and Q

Basically, each of fuzzy predicates S and R may be defined by listing its atomic fuzzy predicates (i.e., pairs of "attribute/linguistic term") and structure, i.e., how these atomic predicates are combined. In Table 40.1 S (or R) corresponds to the full description of both the atomic fuzzy predicates as well as the structure. For example: "*Q young* employees earn a *high salary*" is a protoform of Type 2, while "Most employees earn a *"?" salary*" is a protoform of Type 3. In the first case the system has to select a linguistic quantifier for which the proposition is true (valid) to a high degree. In the second case, the linguistic quantifier and (only) the structure of summarizer S are given and the system has to choose a linguistic term to replace the question mark ("?") yielding a highly valid proposition.

Thus, the use of protoforms makes it possible to devise a uniform procedure to handle a wide class of linguistic data summaries so that the system can be easily adaptable to a variety of situations, users' interests and preferences, scales of the project, etc.

An interesting extension of the concept of a linguistic summary to the linguistic summarization of time series data was shown in a series of works by Kacprzyk, Wilbik and Zadrożny [1, 2]. In this case the array of possible protoforms is much larger as it reflects various perspectives, intentions, etc. of the user. The protoforms used in those works may be exemplifed by: "Among all y's, Q are P", which may be instantiated as "Among all segments (of the time series) most are slowly increasing", and "Among all R segments, Q are P", which may be instantiated as "Among all short segments almost all are quickly decreasing", as well as more sophisticated protoforms, for instance temporal ones like: "E_T among all y's Q are P", which may be instantiated as "Recently, among all segments, most are slowly increasing", and "E_T among all Ry's Q are P", which may be instantiated as "Initially, among all short segments, most are quickly decreasing"; they both go beyond the classic Zadeh's protoforms.

It is easy to notice that the mining of linguistic summaries may be viewed to be closely related to *natural language generation* (NLG) and this path was suggested in Kacprzyk and Zadrożny [11]. This may be a promising direction as NLG is a well developed area and software is available.

40.3 Mining of Linguistic Data Summaries

In Kacprzyk and Zadrożny's [9] interactive approach, the mining of summaries proceeds via a user interface of a fuzzy querying add-on such as FQUERY for Access [5, 6, 10]. In such an add-on a dictionary of linguistic terms is maintained, such as "'young", "most" etc. This terms are then readily available as building blocks of a summary.

Thus, the derivation of a linguistic summary of Type 0 in Table 40.1 may proceed in an interactive (user-assisted) way as follows: (1) the user formulates a set of linguistic summaries of interest (relevance) using the fuzzy querying add-on, (2) the system retrieves records from the database and calculates the validity of each summary adopted, and (3) the most appropriate (highly valid) linguistic summaries are chosen.

Referring to Table 40.1, we can observe that the summaries of Type 1-4 may be produced by a simple extension of such a querying add-on as FQUERY for Access. On the other hand, the discovery of general Type 5 rules, which may be equated with the fuzzy IF-THEN rules, is difficult, and some simplifications about the structure of fuzzy predicates and/or quantifier are needed. Kacprzyk and Zadrożny [7, 8] proposed to distinguish a subclass of Type 5 summaries which may be interpreted as *fuzzy association rules* and mined using adapted versions of well-known algorithms, e.g., Apriori.

40.4 Concluding Remarks

We show how Zadeh's ingenious idea of computing with words and perceptions, based on his concept of a precisiated natural language (PNL), can lead to a new direction in the use of natural language in data mining, the linguistic data(base) summaries. We emphasize the relevance of Zadeh's protoform, and show that various types of linguistic data summaries may be viewed as items in the hierarchy of protoforms of linguistic summaries.

Acknowledgement. To Professor Lotfi Zadeh who – through his ingenious idea of computing with words and protoforms – has provided all of us with tools for an effective and efficient use of natural language in a vast array of systems modeling, data mining, knowledge discovery, ... tasks.

References

1. Kacprzyk, J., Wilbik, A., Zadrożny, S.: Linguistic Summarization of Time Series Using a Fuzzy Quantifier Driven Aggregation. Fuzzy Sets and Systems 159, 1485–1499 (2008)
2. Kacprzyk, J., Wilbik, A., Zadrożny, S.: An Approach to the Linguistic Summarization of Time Series Using a Fuzzy Quantifier Driven Aggregation. International Journal of Intelligent Systems 25, 411–439 (2010)
3. Kacprzyk, J., Yager, R.R.: Linguistic Summaries of Data Using Fuzzy Logic. International Journal of General Systems 30, 33–154 (2001)
4. Kacprzyk, J., Yager, R.R., Zadrożny, S.: A Fuzzy Logic Based Approach to Linguistic Summaries of Databases. International Journal of Applied Mathematics and Computer Science 10, 813–834 (2000)
5. Kacprzyk, J., Zadrożny, S.: FQUERY for Access: Fuzzy Querying for a Windows-based DBMS. In: Bosc, P., Kacprzyk, J. (eds.) Fuzziness in Database Management Systems, pp. 415–433. Springer, Heidelberg (1995)
6. Kacprzyk, J., Zadrożny, S.: The Paradigm of Computing with Words in Intelligent Database Querying. In: Zadeh, L.A., Kacprzyk, J. (eds.) Computing with Words in Information/Intelligent Systems. Part 2. Foundations, pp. 382–398. Springer, Heidelberg (1999)
7. Kacprzyk, J., Zadrożny, S.: On Combining Intelligent Querying and Data Mining Using Fuzzy Logic Concepts. In: Bordogna, G., Pasi, G. (eds.) Recent Research Issues on the Management of Fuzziness in Databases, pp. 67–81. Springer, Heidelberg (2000)
8. Kacprzyk, J., Zadrożny, S.: Computing with Words: Towards a New Generation of Linguistic Querying and Summarization of Databases. In: Sinčak, P., Vaščak, J. (eds.) Quo Vadis Computational Intelligence?, pp. 144–175. Springer, Heidelberg (2000)

9. Kacprzyk, J., Zadrożny, S.: Data Mining via Linguistic Summaries of Databases: An Interactive Approach. In: Ding, L. (ed.) A New Paradigm of Knowledge Engineering by Soft Computing, pp. 325–345. World Scientific, Singapore (2001)
10. Kacprzyk, J., Zadrożny, S.: Computing with Words in Intelligent Database Querying: Standalone and Internet-based Applications. Information Sciences 134, 71–109 (2001)
11. Kacprzyk, J., Zadrożny, S.: Computing with Words is an Implementable Paradigm: Fuzzy Queries, Linguistic Data Summaries, and Natural Language Generation. IEEE Transactions on Fuzzy Systems 18, 461–472 (2010)
12. Yager, R.R.: A New Approach to the Summarization of Data. Information Sciences 28, 69–86 (1982)
13. Yager, R.R., Kacprzyk, J. (eds.): The Ordered Weighted Averaging Operators: Theory and Applications. Kluwer, Boston (1997)
14. Yager, R.R., Kacprzyk, J., Beliakov, G. (eds.): Recent Developments in the Ordered Weighted Averaging Operators: Theory and Practice. Springer, Heidelberg (2011)
15. Zadeh, L.A.: A Computational Approach to Fuzzy Quantifiers in Natural Languages. Computers and Mathematics with Applications 9, 149–184 (1983)
16. Zadeh, L.A.: A Prototype-centered Approach to Adding Deduction Capabilities to Search Engines - The Concept of a Protoform, BISC Seminar. University of California, Berkeley (2002)
17. Zadeh, L.A., Kacprzyk, J. (eds.): Computing with Words in Information/Intelligent Systems, Part 1. Foundations. Part 2. Applications. Springer, Heidelberg (1999)

41

The Evolution of the Evolving Neuro-Fuzzy Systems: From Expert Systems to Spiking-, Neurogenetic-, and Quantum Inspired

Nikola Kasabov

Abstract. This chapter follows the development of a class of intelligent information systems called evolving neuro-fuzzy systems (ENFS). ENFS combine the adaptive/evolving learning ability of neural networks and the approximate reasoning and linguistically meaningful explanation features of fuzzy rules. The review includes fuzzy expert systems, fuzzy neuronal networks, evolving connectionist systems, spiking neural networks, neurogenetic systems, and quantum inspired systems, all discussed from the point of few of fuzzy rule interpretation as new knowledge acquired during their adaptive/evolving learning. This review is based on the author's personal (evolving) research, integrating principles from neural networks, fuzzy systems and nature.

41.1 Early Work on the Integration of Neural Networks and Fuzzy Systems for Knowledge Engineering: Neuro-Fuzzy Expert Systems

The seminal work by Lotfi Zadeh on fuzzy sets, fuzzy rules and intelligent systems [36–38] opened the field for the creation of new types of expert systems that combined the learning ability of neural networks, at a lower level of information processing, and the reasoning and explanation ability of fuzzy rule-based systems, at the higher level. An exemplar system is shown in Figure 41.1, where at a lower level a neural network (NN) module predicts the level of a stock index and a fuzzy reasoning module combines the predicted values with some macro-economic variables, using the following types of fuzzy rules [18]:

IF <the predicted by the NN module stock is high>
AND <the economic situation is good> (41.1)
THEN <buy stock>

These fuzzy expert systems continued the development of the hybrid NN-rule-based expert systems that used crisp propositional and fuzzy rules [13, 15, 17]. They represented a major topic at some conferences (Figure 41.2).

R. Seising et al. (Eds.): *On Fuzziness: Volume 1*, STUDFUZZ 298, pp. 271–280.
DOI: 10.1007/978-3-642-35641-4_41

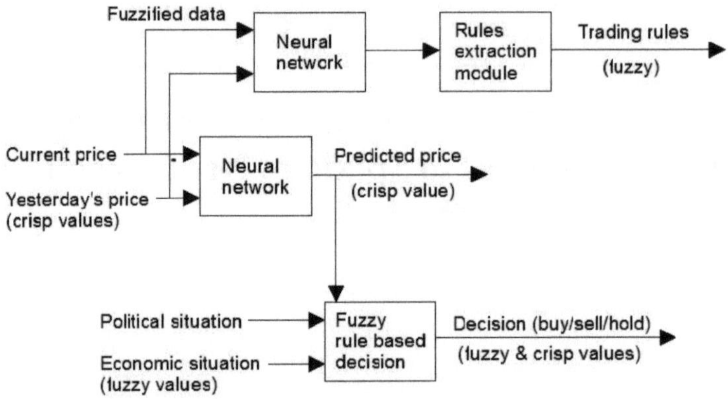

Fig. 41.1. A hybrid NN-fuzzy rule-based expert system for financial decision support (from [18])

Fig. 41.2. At the 1995 ANNES conference in New Zealand: Lotfi Zadeh with T. Yamakawa, Mrs T. Yamakawa, D. Mehandjiiska-Stavreva and N. Kasabov

41.2 Fuzzy Neurons and Fuzzy Neural Networks: Evolving Connectionist Systems

The low-level integration of fuzzy rules into a single neuron model and larger neural network structures, tightly coupling learning and fuzzy reasoning rules into connectionists structures, was initiated by Prof. Takeshi Yamakawa and other Japanese scientists and promoted at a series of IIZUKA conferences in Japan [35]. Many models of fuzzy neural networks were developed based on these principles [6, 18, 19].

The evolving neuro-fuzzy systems developed further these ideas, where instead of training a fixed connectionist structure, the structure and its functionality were evolving from incoming data, often in an on-line, one-pass learning mode. This is the case with the evolving connectionist systems (ECOS) [19–23, 31]. ECOS are modular connectionist based systems that evolve their structure and functionality in a continuous, self-organised, on-line, adaptive, interactive way from incoming information [20]. They can process both data and knowledge in a supervised and/or unsupervised way. ECOS learn local models from data through clustering of the data and associating a local output function for each cluster represented in a connectionist structure. They can learn incrementally single data items or chunks of data and also incrementally change their input features [22, 24]. Elements of ECOS have been proposed as part of the classical NN models, such as SOM, RBF, FuzzyARTMap, Growing neural gas, neuro-fuzzy systems, RAN (see [22, 24]). Other ECOS models, along with their applications, have been reported in [7, 24, 31, 32].

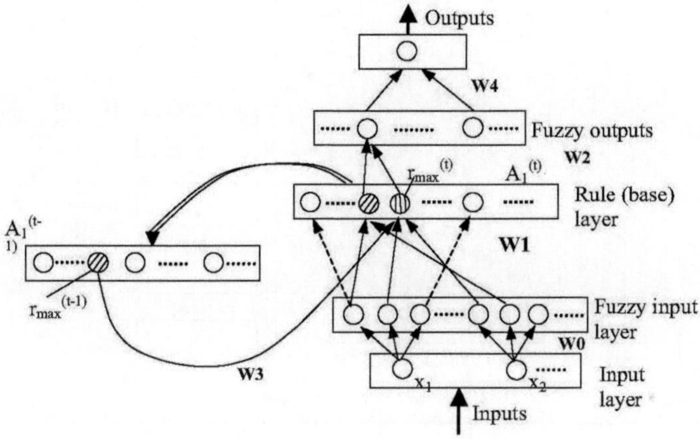

Fig. 41.3. An example of EFuNN model [21]

The principle of ECOS is for neurons to be allocated as centres of fuzzy data clusters and for the system to create local models in these clusters. Fuzzy clustering, as a mean to create local knowledge-based systems, was stimulated by the pioneering work of Bezdek, Yager and Filev [2–4, 34]. Here we will briefly illustrate the concepts of ECOS on two implementations: EFuNN [21] and DENFIS [23]. Examples of EFuNN and DENFIS are shown in Figure 41.3 and Figure 41.4 respectively. In ECOS clusters of data are created based on similarity between data samples either in the input space (this is the case in some of the ECoS models, e.g. the dynamic neuro-fuzzy inference system DENFIS), or in both the input and output space (this is the case e.g. in the EFuNN models). Samples that have a distance to an existing node (cluster center, rule node) less than a certain threshold are allocated to the same cluster. Samples that do not fit into existing clusters, form new clusters. Cluster centers are continuously adjusted according to new data samples, and new clusters are created incrementally. ECOS learn from data and automatically create or update a local fuzzy model/function, e.g.:

$$\mathit{IF} < \textit{data is in a fuzzy cluster } C_i > \mathit{THEN} < \textit{the model is } F_i >, \tag{41.2}$$

where F_i can be a fuzzy value, a linear or regression function (Figure 41.4) or a NN model [22–24].

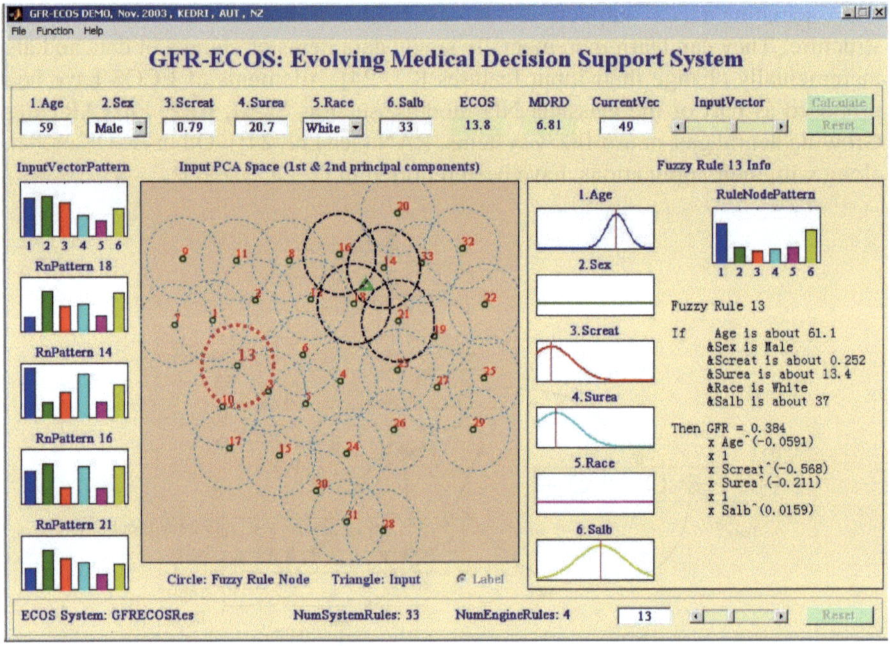

Fig. 41.4. An example of DENFIS model [24] for medical application

A special direction of ECOS was transductive reasoning and personalised modelling. Instead of building a set of local models (e.g. prototypes) to cover the whole problem space and then use these models to classify/predict any new input vector, in transductive modelling for every new input vector a new model is created based on selected nearest neighbour vectors from the available data. Such ECOS models are NFI and TWNFI [28]. In TWNFI for every new input vector the neighbourhood of closets data vectors is optimised using both the distance between the new vector and the neighbouring ones and the weighted importance of the input variables, so that the error of the model is minimised in the neighbourhood area [25].

While the classical ECOS use a simple McCulloch and Pitts model of a neuron, the further developed evolving spiking neural network (eSNN) architectures used a spiking neuron model using the same or similar ECOS principles and applications.

41.3 Evolving Spiking Neural Networks (eSNN) and Fuzzy Rule Extraction

A single biological neuron and the associated synapses is a complex information processing machine that involves short term information processing, long term information storage, and evolutionary information stored as genes in the nucleus of the neuron. A spiking neuron model assumes input information represented as trains of spikes over time. When sufficient input information is accumulated in the membrane of the neuron, the neuron's post synaptic potential exceeds a threshold and the neuron emits a spike at its axon (Figure 41.5). Some of the-state-of-the-art models of a spiking neuron include: early models by Hodgkin and Huxley [10], 1952; more recent models by Maas, Gerstner, Kistler, Izhikevich and others, e.g.: Spike Response Models (SRM); Integrate-and-Fire Model (IFM) (Figure 41.5); Izhikevich models; adaptive IFM; probabilistic IFM [11, 12].

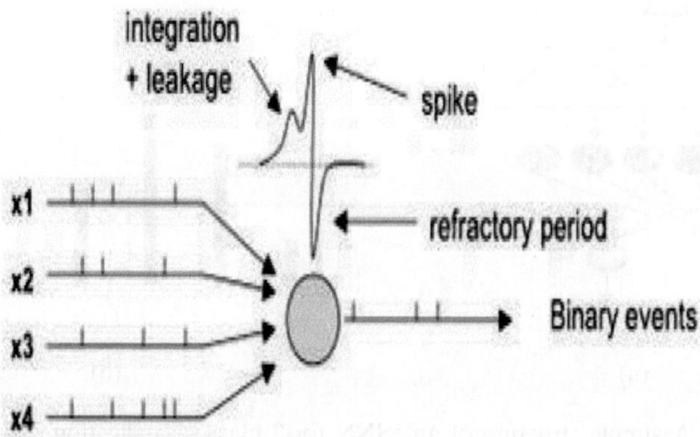

Fig. 41.5. The structure of the LIFM

Based on the ECOS principles, an evolving spiking neural network architecture (eSNN) was proposed in [24, 33] which was initially designed as a visual pattern recognition system. The first eSNNs were based on the Thorpe's neural model [29], in which the importance of early spikes (after the onset of a certain stimulus) is boosted, called rank-order coding and learning. Synaptic plasticity is employed by a fast supervised one-pass learning algorithm. Different eSNN models were developed, including: a reservoir-based eSNN for spatio- and spectro-temporal pattern recognition (Figure 41.6) [30]; eSNN an architecture that used both rank-order and time-based learning methods to account for spatio-temporal data [27]; specialised architectures for EEG modelling [24]; moving object recognition systems; etc.

Extracting fuzzy rules from an eSNN would make the eSNN not only efficient learning models, but also knowledge-based models. A method was proposed in [26] and illustrated in Figure 41.7). Based on the connection weights W between the

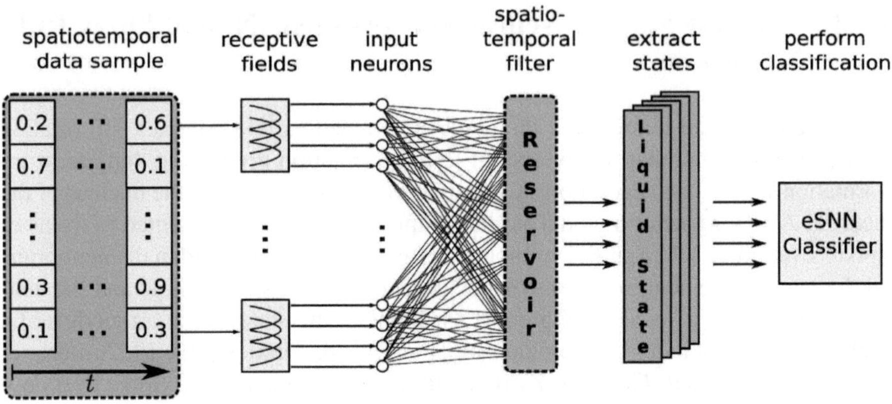

Fig. 41.6. A reservoir-based eSNN for spatio-temporal pattern classification

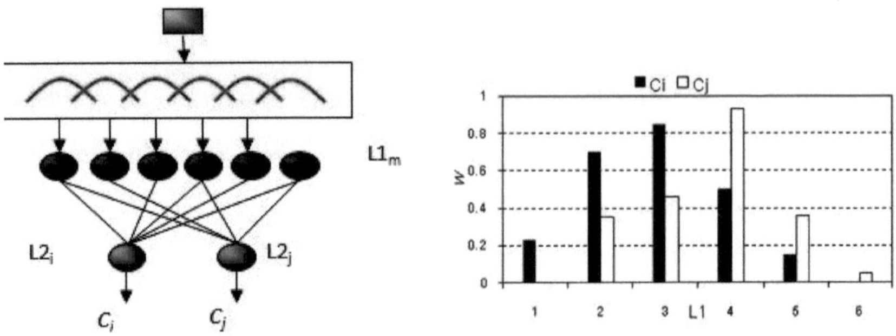

(a) (b)

Fig. 41.7. (a): A simple structure of an eSNN for 2-class classification based on one input variable using 6 receptive fields to convert the input values into spike trains; (b): The connection weights of the connections to class C_i and C_j output neurons respectively are interpreted as fuzzy rules

receptive field layer L1 and the class output neuron layer L2 the following fuzzy rules are extracted:

$$\begin{aligned} &IF\ (input\ variable\ v\ is\ SMALL)\ THEN\ class\ C_i;\\ &IF\ (v\ is\ LARGE)\ THEN\ class\ C_j. \end{aligned} \quad (41.3)$$

41.4 Computational Neuro-Genetic Models (CNGM) and Fuzzy Rules

A neurogenetic model of a neuron is proposed in [24] and studied in [1]. It utilises information about how some proteins and genes affect the spiking activities of a neuron such as fast excitation, fast inhibition, slow excitation, and slow inhibition.

An important part of the model is a dynamic gene/protein regulatory network (GRN) model of the dynamic interactions between genes/proteins over time that affect the spiking activity of the neuron - (Figure 41.8).

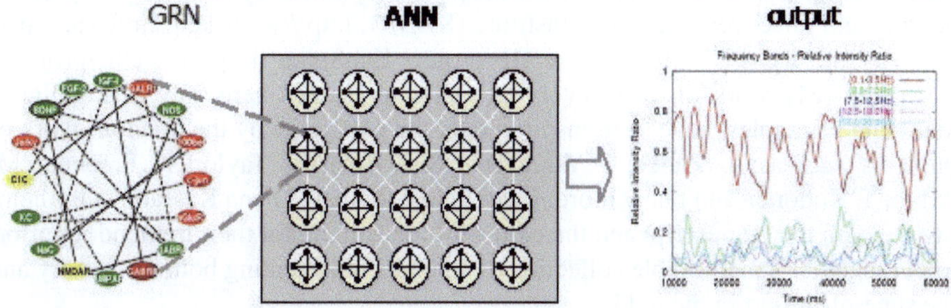

Fig. 41.8. A schematic diagram of a CNGM framework, consisting of a GRN as part of a eSNN [1].

New types of neuro-genetic fuzzy rules can be extracted from such CNGM in the form of:

IF < *GRN is represented by a function F* >
AND < *input is Small* > (41.4)
THEN < *Class C* >

41.5 Quantum Inspired SNN (QiSNN)

QiSNNs use the principle of superposition of states to represent and optimize features (input variables) and gene parameters of the SNN [24]. They are optimized through quantum inspired genetic algorithm [5] or QiPSO. Features or genes are represented as qubits in a superposition of 1 (selected), with a probability p_1, and 0 (not selected) with a probability p_0. When the model has to be calculated, the quantum bits 'collapse' in 1 or 0. Fuzzy rules in QiSNN would look like:

IF < *GRN is represented by a function F with a quantum probability p* >
AND < *input is Small with a quantum probability q* >
AND < *the model parameters are S with quantum probability s* > (41.5)
THEN < *Class C, with probability r* >

41.6 Conclusion

This chapter presented brief highlights of the development of neuro-fuzzy models for intelligent information systems. The main idea is to facilitate the discovery of new knowledge, along with the development of new connectionist models and systems integrating principles from neural networks, fuzzy systems, evolutionary computation and quantum computing.

Acknowledgement. The work on this chapter is supported by the Knowledge Engineering and Discovery Research Institute (KEDRI, http://www.kedri.info) and partially by the EU FP7 Marie Curie IIF project EvoSpike (http://ncs.ethz/projects/evospike/. My work on these evolving topics, presented here only as highlights, ideas and principles, have been inspired over the past years by the previous work of pioneers, such as: L. Zadeh, T. Yamakawa, S.-I. Amari, J. Taylor, W. Freeman, M. Arbib, T. Kohonen and many more. I would like to thank Diana Kassabova for helping me with the manuscript and the editors of the volume for their tremendous effort to put together a memorable collection of chapters representing both the history and the state-of-the-art in the field.

References

1. Benuskova, L., Kasabov, N.: Computational Neuro-genetic Modelling. Springer, New York (2007)
2. Bezdek, J.: Pattern Recognition with Fuzzy Objective Function Algorithms. Plenum Press, New York (1981)
3. Bezdek, J. (ed.): Analysis of Fuzzy Information, vol. 1, 2, 3. CRC Press, Boca Raton (1987)
4. Bezdek, J.: A Review of Probabilistic, Fuzzy, and Neural Models for Pattern Recognition. Journal of Intelligent and Fuzzy Systems 1, 1–25 (1993)
5. Defoin-Platel, M., Schliebs, S., Kasabov, N.: Quantum-inspired Evolutionary Algorithm: A multi-model EDA. IEEE Transactions on Evolutionary Computation 13(6), 1218–1232 (2009)
6. Furuhashi, T., Hasegawa, T., Shin-ichi, H., Uchikawa, Y.: An Adaptive Fuzzy Controller Using Fuzzy Neural Networks. In: Proceedings of Fifth IFSA World Congress, pp. 769–772 (1993)
7. Futschik, M.E., Kasabov, N.: Fuzzy Clustering in Gene Expression Data Analysis. In: Proceedings of the World Congress of Computational Intelligence WCCI 2002, Hawaii. IEEE Press (May 2002)
8. Gerstner, W.: Time Structure of the Activity of Neural Network Models. Physical Review E 51, 738–758 (1995)
9. Hebb, D.: The Organization of Behavior. John Wiley and Sons, New York (1949)
10. Hodgkin, A.L., Huxley, A.F.: A Quantitative Description of Membrane Current and its Application to Conduction and Excitation in Nerve. Journal of Physiology 117, 500–544 (1952)
11. Hopfield, J.J.: Pattern Recognition Computation Using Action Potential Timing for Stimulus Representation. Nature 376, 33–36 (1995)
12. Izhikevich, E.M.: Which Model to Use for Cortical Spiking Neurons? IEEE Transactions on Neural Networks 15(5), 1063–1070 (2004)
13. Kasabov, N.: Incorporating Neural Networks into Production Systems and a Practical Approach Towards the Realisation of Fuzzy Expert Systems. Computer Science and Informatics 21(2), 26–34 (1991)

14. Kasabov, N.: Hybrid Connectionist Production Systems. Journal of Systems Engineering 3(1), 15–21 (1993)
15. Kasabov, N., Shishkov, S.: A Connectionist Production System With Partial Match and its Use for Approximate Reasoning. Connection Science 5(3-4), 275–305 (1993)
16. Kasabov, N.: Connectionist Fuzzy Production Systems. In: Ralesu, A.L. (ed.) Fuzzy Logic in Artifical Intelligence. LNCS (LNAI), vol. 847, pp. 114–128. Springer, Heidelberg (1994)
17. Kasabov, N.: Hybrid Connectionist Fuzzy Production Systems – Towards Building Comprehensive AI. Intelligent Automation and Soft Computing 1(4), 351–360 (1995)
18. Kasabov, N.: Foundations of Neural Networks, Fuzzy Systems and Knowledge Engineering. MIT Press, Cambridge (1996)
19. Kasabov, N., Kim, J.S., Michaeland, W., Gray, A.: FuNN/2 – A Fuzzy Neural Network Architecture for Adaptive Learning and Knowledge Acquisition. Information Sciences - Applications 101(3-4), 155–175 (1997)
20. Kasabov, N.: Evolving Fuzzy Neural Networks – Algorithms, Applications and Biological Motivation. In: Yamakawa, T., Matsumoto, G. (eds.) Methodologies for the Conception, Design and Application of Soft Computing, pp. 271–274. World Scientific (1998)
21. Kasabov, N.: Evolving Fuzzy Neural Networks for on-line Supervised/unsupervised, Knowledge-based Learning. IEEE Transactions on Systems, Man, and Cybernetics 31(6), 902–918 (2001)
22. Kasabov, N.: Evolving Connectionist Systems: Methods and Applications in Bioinformatics, Brain Study and Intelligent Machines. Springer, Heidelberg (2002)
23. Kasabov, N., Song, Q.: DENFIS: Dynamic, Evolving Neural-fuzzy Inference Systems and its Application for Time-series Prediction. IEEE Transactions on Fuzzy Systems 10, 144–154 (2002)
24. Kasabov, N.: Evolving Connectionist Systems: The Knowledge Engineering Approach, 1st edn. Springer, London (2007) (2003)
25. Kasabov, N., Hu, Y.: Integrated Optimisation Method for Personalised Modelling and Case Study Applications. International Journal of Functional Informatics and Personalised Medicine 3(3), 236–256 (2010)
26. Soltic, S., Kasabov, N.: Knowledge Extraction From Evolving Spiking Neural Networks With Rank Order Population Coding. International Journal of Neural Systems 20(6), 437–445 (2010)
27. Song, S., Miller, K.D., Abbott, L.F.: Competitive Hebbian Learning Through Spike-timing-dependent Synaptic Plasticity. Nature Neuroscience 3, 919–926 (2000)
28. Song, Q., Kasabov, N.: TWNFI- A Transductive Neuro-fuzzy Inference System With Weighted Data Normalisation for Personalised Modelling. Neural Networks 19(10), 1591–1596 (2006)
29. Thorpe, S.J., Delorme, A., Van Rullen, R.: Spike-based Strategies for Rapid Processing. Neural Networks 14(6-7), 715–725 (2001)
30. Verstraeten, D., Schrauwen, B., D'Haene, M., Stroobandt, D.: An Experimental Unification of Reservoir Computing Methods. Neural Networks 20(3), 391–403 (2007)
31. Watts, M.J.: A Decade of Kasabov's Evolving Connectionist Systems: A Review. IEEE Transactions on Systems, Man and Cybernetics – Part C: Applications and Reviews 39(3), 253–269 (2009)
32. Widiputra, H., Pears, R., Kasabov, N.: Multiple Time-Series Prediction Through Multiple Time-Series Relationships Profiling and Clustered Recurring Trends. In: Huang, J.Z., Cao, L., Srivastava, J. (eds.) PAKDD 2011, Part II. LNCS, vol. 6635, pp. 161–172. Springer, Heidelberg (2011)

33. Wysoski, S.G., Benuskova, L., Kasabov, N.: Evolving Spiking Neural Networks for Audiovisual Information Processing. Neural Networks 23(7), 819–835 (2010)
34. Yager Ronald, R., Filev, D.: Generation of Fuzzy Rules by Mountain Clustering. Journal of Intelligent and Fuzzy Systems 2, 209–219 (1994)
35. Yamakawa, T., Uchino, E., Miki, T., Kusanagi, H.: A Neo Fuzzy Neuron and Its Application to System Identification and Prediction of the System Behaviour. In: Proceedings of the 2nd International Conference on Fuzzy Logic & Neural Networks, Iizuka 1992, Iizuka, Japan, pp. 477–483 (1992)
36. Zadeh, L.A.: Fuzzy Sets. Information and Control 8, 338–353 (1965)
37. Zadeh, L.A.: A Theory of Approximate Reasoning. In: Hayes, J.E.M., Donald, L.I.M. (eds.) Machine Intelligence (based on the International Machine Intelligence Workshop), vol. 9, pp. 149–194. Elsevier, New York (1979)
38. Zadeh, L.A.: Fuzzy Logic. IEEE Computer 21, 83–93 (1988)

"Jim, I Have a Question for You": My Travels with Lotfi Zadeh

Jim Keller

This will not be a technical article in any sense of that word. Lotfi Zadeh has always been a visionary and this article shares some of my interaction with a few of those visions and of Lotfi's influence on me and in fact many people in our community. I've always thought of myself as a newcomer to fuzzy sets. I never had a course in fuzzy set theory nor had any mentor during my formal education who told about the subject. However, after receiving the Fuzzy Systems Pioneer Award from the IEEE Computational Intelligence Society, I guess I now qualify as an old-timer. After I got my PhD in Math and joined an Electrical Engineering Department, I stumbled on fuzzy sets by reading a few (actually quite bad) engineering papers that claimed to be based on fuzzy sets. As naïve as I was, I thought, "oh boy, I bet I can prove a bunch of simple theorems about max and min" and become a player in the field. But it wasn't until I read a small 1984 monograph by Kurt Schmucker [1] on linguistic averaging and until I devoured Jim Bezdek's 1981 Pattern Recognition book [2] that I started to understand the potential for fuzzy systems in pattern recognition and image processing. Also, in my naïve approach to the subject, I saw none of the controversy surrounding fuzzy set theory. My first papers in fuzzy systems were to a small IEEE Workshop in 1985, and used linguistic averaging in multi-sensor and multi-temporal processing of imagery. After the talk I got seriously hammered in the hallway by a very passionate statistical pattern recognition advocate. Wow, that took me by surprise; so that night I got literally hammered and ended up meeting Jim Bezdek (in my stupor, all I could say was "I read your book"). Bez told me to blow that off and come to NAFIPS. It was after this experience that I went out and bought every fuzzy set text book and book of chapters I could find to see what I was getting myself into. I liked what I saw (seems so natural to me) and the rest is history.

But, did I actually think I might meet Professor Zadeh in person? That's the crux of this story. He is not only interested in meeting everyone who had an opinion on fuzzy sets, positive or negative, but he is genuinely concerned about them and their work. He has the uncanny ability not only to remember people, but to recall their research efforts. I remember after the 1987 NAFIPS Workshop at Purdue, my students came away inspired because Lotfi found time to chat one-on-one with them and gave them helpful suggestions as well as encouragement. He is truly interested in what each of us thinks and does. When necessary, he "circled the wagons" and took the brunt of the criticism leveled at the field and its practitioners. Lotfi Zadeh

R. Seising et al. (Eds.): *On Fuzziness: Volume 1*, STUDFUZZ 298, pp. 281–285.
DOI: 10.1007/978-3-642-35641-4_42

lived by his motto "If someone criticizes you, tell them you take it as a compliment." Until just recently, he was everywhere. So, there were many chances for many of us to chat with The Man. I soon realized that Lotfi was one of the most photographed people on the planet, but I didn't have a picture with him. I remedied that in 1990 (Figure 42.1) – who are those young people in that picture? Since then, I've joined the paparazzi and could easily fill 5, 10,.. pages with pictures. I'll sprinkle in a few representative favorites.

Fig. 42.1. My first picture with Lotfi: ISUMA 1990 at the University of Maryland

By now, you are probably wondering what the question in the title of this chapter has to do with anything. Over the years, I have gotten several phone calls from Professor Zadeh. I think the first time I was speechless; he must have thought he dialed the wrong number. After a brief initial exchange, some variation of that question usually surfaced. I will recount a few of these phone messages as my tribute to the personal influence Lotfi had on me. While the calls themselves were short and pleasant, most of the time there was a thought provoking message that I believe Lotfi wanted to challenge me personally to stretch my perceptions (is that a touch of arrogance? I hope not), and to cast the thought out onto the soft computing community. The first one I will mention is one I have a sneaking suspicion was a prank call for him, though I've never had the nerve to ask him.

Fig. 42.2. Lotfi, Abe Mamdani and two Spanish authors (Belin and Mercedes) at my 2003 FUZZ-IEEE in St. Louis

In early 2000, George W. Bush made a verbal attack on Al Gore in the US Presidential campaign, accusing Gore of using "fuzzy math", a pejorative term for that candidate. I was the editor of the IEEE Transactions on Fuzzy Systems at the time and Lotfi thought I would be the right spokesman to convince the Republican Party that fuzzy math was good, not bad. I remember he faxed me photocopies of the covers of several books on fuzzy arithmetic, etc. to use as evidence. For those of us who actually remember fax machines of that vintage can imagine the quality of faxed photocopies of colored covers of textbooks. Sorry, Lotfi, I couldn't bring myself to stride into that battle. I could just imagine the Secret Service showing up at my office. Hmmm, if had had jumped into action, could I have changed the course of that election? We'll never know now.

Zadeh coined the term "Recognition Technology" (RT) in 1998 [3]. In that paper and subsequent talks, he elaborated that these are current or future systems that have the potential to provide a "quantum jump in the capabilities of today's recognition systems". Lotfi claimed at that time that this can occur as a result of three converging developments:

(a)major advances in sensor technology;
(b)major advances in sensor data processing technology; and
(c)the use of soft computing techniques to infer a conclusion from observed data.

Fig. 42.3. One of my favorite pictures with Lotfi and Jim in Barcelona, 2005

How right he was. He called me to discuss the potential of a course sequence on RT that would truly benefit engineers and scientists. Lotfi anticipated a problems-based learning environment in Engineering. At MU, we have courses in each of the three pillars of RT and while we didn't create the course structure, we infused these ideas into our approach to landmine detection [4] and more recently into an interdisciplinary research effort on technology for eldercare. The RT framework provides the perfect structure to categorize and integrate the varied efforts of our large team. It's a great way to structure a talk on eldercare research.

During the new millennia, Lotfi called more than once with the question, "Jim, how do you define a cluster?" After stammering through the usual textbook definitions, he pushed on by asking if he produced a set of points, could I (Jim) say whether or not they formed a cluster. Of course, Lotfi was pressing to use this very common practical issue of my interest as the newest of those concepts that had no crisp definition. My initial reaction was that, while an interesting question, a cluster in the practical sense exists only in context of a group of clusters, found by some particular clustering algorithm, optimizing some particular criterion function. This was sort of an "I may not know how to define one, but I'll know it when I see it"-type response. As with many of Lotfi's comments, this question was posed to make us ponder common activities that turn out to be inherently fuzzy. In this case, even though there are many fuzzy set-based algorithms to perform clustering, the issue is deeper in that we are using fuzzy criteria in a mathematical optimization, but really don't have a suitable definition of the basic underlying structure. This question (and our feeble attempt at answering it [5]) gave rise to many interesting conversations that helped us ponder these simple sounding concepts and see the deeper issues. So, Lotfi, here's to you my friend. You have made a large difference in my life and my work. I hope this little tribute let's others see you as I have.

Fig. 42.4. Lotfi with my daughter, Amy, at the 2008 NAFIPS banquet in New York

References

1. Schmucker, K.J.: Fuzzy Sets, Natural Language Computations, and Risk Analysis. Computer Science Press, Rockville (1984)
2. Bezdek, J.: Pattern Recognition with Fuzzy Objective Function Algorithms. Plenum Press, New York (1981)
3. Zadeh, L.A.: Soft Computing, Fuzzy Logic and Recognition Technology. In: Proceedings of the IEEE International Conference Fuzzy Systems, Anchorage, AK, pp. 1678–1679 (May 1998)
4. Gader, P.D., Keller, J.M., Nelson, B.N.: Recognition Technology For the Detection of Buried Land Mines. IEEE Transactions on Fuzzy Systems 9(1), 31–43 (2001)
5. Keller James, M., Sledge, I.J.: A Cluster by any Other Name. In: Proceedings of the North American Fuzzy Information Processing Society Annual Meeting, San Diego, pp. 427–432 (June 2007)

43

My Journey into the World of Fuzziness

Etienne E. Kerre

In 1967 I graduated in mathematics from Ghent university in Belgium. I was very happy to immediately get a grant from the National Institute for Research in Industry and Agriculture (the predecessor of the current Flemish research sponsor IWT) in order to perform research on diffraction of electrons by crystals. In January 1970 I obtained the Ph D degree in mathematics from Ghent university on a dissertation entitled "A theoretical contribution to low energy electron diffraction (LEED) by crystals". Those days the so-called modern mathematics entered into the secondary schools and universities, i.e., mathematics based on Cantor's set theory. Despite the warm welcome by the famous David Hilbert: "No one will drive us from the paradise Cantor created for us", it took about 3 quarters of a century before set theory has been accepted as the language of contemporary science in general and mathematics in particular. Not earlier than 1963 I got a first course in "modern"mathematics including the basic concepts of set theory as well as the theory of matrices and determinants, unfortunately taught in a very confused way and hence not at all optimal to attract the attention to this beautiful world. I even experienced a colleague of this professor drawing a kind of a Venn diagram as a face with the union symbol as eyes and the intersection symbol as nose and writing next to the face it was equal to the empty set... Even in recent times I have met professors in mathematics that claimed without any hesitation that in mathematics nothing seriously happened after Riemann (1850)! It is difficult to convince such colleagues about the beauty and power of recent models such as fuzzy set theory! After my graduation I had to discover on my own the powerful expressiveness of Cantor's set theory by reading the books and papers written by J. Dieudonne, H. Cartan, M. Frechet a.o. about metric spaces, Banach spaces, Frechet derivatives, differential calculus, theory of groups, Boolean algebras... all impressive mathematics of structures laying bare the fundamental meaning of basic notions such as continuity, convergence, limits, derivatives and integration. During 3 to 4 years together with 2 colleagues I have put all my energy into the modernization of the basic courses on mathematical analysis for undergraduates in mathematics at Ghent university. This work resulted in 3 volumes of a systematic encyclopedia written in Dutch. At the same time I wrote some papers on Boolean algebras and the relationship between a mapping and its restrictions with respect to all kinds of continuity and limits.

Around 1974 I got a (bad smelling) Xerox copy of the seminal paper entitled "Fuzzy sets" by Lotfi Asker Zadeh.

R. Seising et al. (Eds.): *On Fuzziness: Volume 1*, STUDFUZZ 298, pp. 287–293.
DOI: 10.1007/978-3-642-35641-4_43

I immediately became very curious about the contents of this paper since its title reflected some kind of contradiction: how can a set (originally intuitively defined as: any collection into a whole X of definite , distinguishable objects x – called elements of X – of our intuition or thought) be fuzzy? After the first reading of this amazing paper I agreed with the restrictions mentioned in this paper with respect to the classical binary black-or-white logic and the mathematical apparatus built upon it. At the same time I became convinced about the enormous potential of this extension of classical set theory by using maximum and minimum to model the union respectively intersection of fuzzy sets. Indeed basic mathematical concepts based upon union and intersection of sets could immediately been fuzzified. So very soon straightforward fuzzifications of classical mathematical structures and concepts based on set theory appeared in the literature: fuzzy topological spaces, fuzzy groups, fuzzy vector spaces, fuzzy metric spaces, fuzzy geometries, fuzzy relations.... Most of those papers written during the seventies appeared in the *Journal of Mathematical Analysis and its Applications*. A typical example is C. L. Chang 1968 paper "Fuzzy topological spaces" where a fuzzy topology on a universe X has been defined as a subclass T of the class of all fuzzy sets on X that contains: (i) the universe X and the empty set, identified with the constant mapping from X to 1, respectively to 0; (ii) the Zadeh intersection (modeled by minimum) of every two elements of T and (iii) the Zadeh union (modeled by "maximum") of every arbitrary family of elements of T. Based on this extension basic topological notions such as closed set, interior, closure, neighborhood, compactness, normality, continuity... have been introduced in a straightforward way. My first steps into the world of fuzzy were undertaken in the domain of fuzzy topology, more specifically, the characterization of a fuzzy topology by means of a preassigned operation. It was easily shown that a fuzzy topology could be uniquely determined by means of the class of closed fuzzy sets, by means of an interior operator and by means of a closure operator. The remaining characterization in terms of neighborhood systems took much longer time and turned out to become one of my first fuzzy frustrations, because in some paper it was stated that indeed such a characterization holds; moreover the authors wrote next to its proof: straightforward! After several years we could find a counterexample as well as completely solve the characterization problem of a fuzzy topology in terms of the different neighborhood concepts of a point, a fuzzy point and a fuzzy singleton.

During those pioneering years I could also experience the power of a conference to disseminate knowledge and to act as a starting point for other researchers. Indeed every 4 years mathematicians organize their international congress (ICM), a huge event with several thousands of participants witnessing the awarding of the Field Medals, a kind of substitution for the missing Nobel prize for mathematics. At the ICM 1978 in Helsinki I presented my first fuzzy results in one of the 36 parallel sessions. I got 12 minutes to present my results for I guess about 20 people. Fortunately one of them was Prof. A. Mashhour from Mansoura university in Egypt, at that time the most famous Egyptian authority in classical topology. He was very pleased with my talk and he took my ideas back to Egypt and in this way the start was giving for the well-known flourishing Egyptian schools on fuzzy topology!

In the same spirit of direct fuzzification we contributed papers on fuzzy multivalued mappings, fuzzy relational calculus, fuzzy mathematical morphology and fuzzy information retrieval.

Fig. 43.1. Participants of the *Third Polish Symposium on Fuzzy Sets and Interval Analysis* in Poznan, Poland, September 20, 1989. One may recognise besides the author (second from left on the first row), Jerzy Albrycht, Maciej Wygralak and Tomasz Kubiak.

During the eighties a new period started in the development of the mathematics of fuzziness (I prefer this term to the term fuzzy mathematics in order to stress that there is nothing fuzzy about these mathematics. In the same spirit logic of fuzziness would be a better term than fuzzy logic). Due to the introduction of the concepts triangular norm and conorm an explosion in the possible fuzzifications of classical binary structures happened. So instead of the original max-min combination for the fuzzy union-intersection one could use an $S-T$ combination with S a triangular conorm and T a triangular norm, eventually dual to each other. I took part in this explosion process in the framework of an NSF project in 1990 at the university of Nebraska in Lincoln, USA. In the context of that project I was asked to teach during 90 hours the basics of fuzzy set theory to a selected group of undergraduate students and further guide them to perform original research. So in the morning I taught them the basics of the original max-min fuzzy set theory and in the afternoon I guided them in developing the bounded sum – Lukasiewicz intersection and the probabilistic sum – algebraic product fuzzy set theories and that worked perfect. I will never forget their efforts to find suitable counterexamples to show that some binary logical law did not hold anymore! As soon as a new fuzzification and hence a generalization of a binary domain was launched, authors started to study the deviations from the binary case, i.e., which properties still or no longer hold in the fuzzy world. For example the concept of fuzzy continuity of a mapping between two fuzzy topological spaces defined as the inverse image of each open fuzzy set in the codomain of the mapping being an open fuzzy set in its domain. With this definition it turned out that the chain rule for a composition of mappings still holds and hence many hereditary properties based upon this rule remained valid. On the other hand it turned out that the fuzzification of the topological notion of complete normality did not led to the fuzzy extension of the famous Tietze characterization theorem. As a result of this huge number of extensions of classical logic and set theory a lot of research has been performed in order to find that extension that kept most of the classical theorems. For example telling that two crisp subsets of some universe of discourse are disjoint, i.e., they have no element in common, is completely the same as telling that no element of one of them belongs to the other one. However this statement no longer holds in Zadeh's max-min fuzzy set theory and hence we get two different fuzzifications. Another example of the explosion in the alternatives of the fuzzy generalizations can be found in the huge number of ranking methods for fuzzy quantities and fuzzy numbers.

After 25 years fuzzy set theory became more and more mature and hence more strong and deep analysis started to be developed. Some examples that we developed starting from the nineties: the characterization of a fuzzy preference structure, the characterization of a fuzzy topology by means of a neighborhood system, a deep study of a fuzzy implication and its characterizations, an axiomatic system for a fuzzy relational database model, a comparative study of similarity measures, criteria and classification of defuzzification methods, an axiomatic system for ordering fuzzy quantities, a comparative study of fuzzy rough sets, the representation of intuitionistic triangular norms and conorms, the construction and classification of intuitionistic fuzzy implications, algebraic characterizations of fuzzy information relations, Smets-Magrez axioms for intuitionistic fuzzy residual implications, characterization of fuzzy temporal interval relations, triangle algebras as an axiomatization of interval-valued residuated lattices.

Fuzzy set theory has not been the only model to treat imprecise and uncertain information. During the past three decennia several new models have been introduced and developed: some models are extensions of Zadeh's fuzzy set theory and others started from a completely new idea. Chronologically ordered the following models have been introduced: L-fuzzy set theory by J. Goguen in 1967, flou set theory by Y. Gentilhomme in 1968, L-flou set theory by C. Negoita and D. Ralescu in 1975, type-2 fuzzy set theory by L. Zadeh in 1975, interval-valued fuzzy set theory by R. Sambuc in 1975, probabilistic set theory by K. Hirota in 1981, rough set theory by Z. Pawlak in 1982, intuitionistic fuzzy set theory by K. Atanassov in 1983, twofold fuzzy set theory by D. Dubois and H. Prade in 1987, grey set theory by J. Deng in 1989, fuzzy rough set theory by D. Dubois and H. Prade in 1990, rough fuzzy set theory by D. Dubois and H. Prade in 1990, theory of imprecise probabilities by P. Walley in 1991, soft set theory by K. Basu, R. Deb and P. Pattanaik in 1992, toll set theory by D. Dubois and H. Prade in 1993, vague set theory by W. Gau and D.

Buehrer in 1993 and bipolar fuzzy set theory by W-R. Zhang in 1994. Very soon I wondered if all these new theories could be completely independent of each other and already in 1987 I started research on exploring the interrelationships between them. In a series of papers most of them written in cooperation with my former Ph D student Glad Deschrijver we have proven many remarkable dependencies such as for example: vague set theory, grey set theory, interval-valued fuzzy set theory and intuitionistic fuzzy set theory are mutually equivalent and hence it makes no sense to continue developing all of them! Moreover these 4 models can be generalized or embedded into the type-2 fuzzy set theory.

Fig. 43.2. First *International Workshop on the Foundations and Applications of Possibility Theory* in Gent, Belgium in 1995: Lotfi and four of the author's Ph D students: Prof Gert de Cooman, Prof Bernard De Baets, Dr Elena Tsiporkova, late Prof Da Ruan

From the very beginning I became interested in the applications of fuzzy set theory and already in 1982 my paper "The use of fuzzy set theory in electrocardiological diagnostics" was published in an edited volume on approximate reasoning in decision analysis. In that paper the state of a cardiovascular system has been described by means of linguistic terms such as: the *QRS* duration is low, the *Q* amplitude is high, hence keeping the qualitative description by the cardiologist of the quantitative information obtained from the ECG analysis. It is worth to mention that in this paper I have introduced a new ranking method to order fuzzy quantities based on the famous extension principle in order to determine the optimal alternative or equivalently the final diagnosis. To my opinion the concept of a linguistic variable is the key concept for the many applications of fuzzy set theory, especially combined with its use in the so-called fuzzy rules to express knowledge in natural language, while the extension principle can be seen as the key concept to extend the classical mathematical toolkit to a colored or gradual world; both concepts are due to Lotfi Zadeh. Another early application (1985) concerned the use of fuzzy instructions to run a virtual ship between two continents that are dissipated by a small corridor with varying width; the steering of the ship has to be performed on using fuzzy instructions such as: come not too close to one of the coastlines, move to starboard, diminish speed. A complete algorithm was set up in order to reach these objectives and implemented in the good old FORTRAN!

We have contributed a lot to the development of fuzzy (relational) databases, mostly with my former Ph D student Guoqing Chen from Tsinghua university. Our research on the application of fuzzy set theory to image processing has led to five Ph D degrees: "Fuzzy morphological and fuzzy filtering techniques in image processing" in 2002 by Mike Nachtegael, "The use and the construction of similarity measures in image processing" in 2004 by Dietrich Van Der Weken, "Fuzzy and non-linear restorations and techniques for digital images" in 2007 by Stefan Schulte, "Color morphology with application to image magnification" in 2007 by Valerie De Witte and "Fuzzy techniques for noise removal in images sequences and interval-valued fuzzy mathematical morphology" in 2010 by Tom Melange. Currently two new students are working towards a Ph D degree on the application of fuzzy techniques to medical image processing in the framework of a big European Marie Curie project where in total 16 students are working in 8 different European universities on soft computing techniques to improve medical images and their retrieval.

I would like to end this journey with some thoughts and expectations about the future of fuzzy set theory. First of all the fuzzy community should agree about the basics of fuzzy set theory i.e., determine the essentials that should be known by newcomers in the field and present to them in affordable textbooks. I know there are already several good textbooks but I am not completely satisfied with any of them because most of them have been written from the personal background of the author(s). If one takes a textbook on Calculus then one can be sure to find concepts such as: limits, sequences, series, derivatives and integrals. So we have to define the analogue for fuzzy set theory: membership function, basic fuzzy set-theoretic operations, linguistic variables, extension principle, fuzzy arithmetic, ...Such a textbook could then be used for lecturing at universities and colleges. From the very beginning I realized the importance of offering courses on the basic issues for the dissemination of a new theory. Already in 1979 I started at Ghent university an optional graduate course on an introduction to the theory of fuzzy sets and its applications. From the academic year 1994-1995 on I have taught 4 courses at Ghent university on basic and advanced topics in fuzzy set theory to mathematicians, computer scientists, engineers and postgraduates in knowledge technology. Due to these courses I could attract clever students (29 in total) to work towards a Ph D degree on different topics in fuzzy set theory. Another point of attention concerns the growing number of journals publishing papers on soft computing. As said before in the first 13 years after the publication of Lotfi's seminal paper " Fuzzy sets", only a few journals were willing to publish papers on fuzzy set theory. Nowadays there are more than 25 journals with fuzzy in the title. To my humble opinion this is too much and will at the end negatively influence the quality of the papers accepted. In my talks to colleagues many times I heard them complaining with respect to the huge number of papers they had to referee. I myself have been a referee for 65 journals with around 80 papers per year.... Many times I experienced that I wrote a negative report on some paper that some months later I saw published unaltered in another journal.... A similar comment can be given with respect to the increasing number of conferences in our domain. Let's keep watchful with respect to the quality of our journals and conferences!

To conclude I can definitely say that I am very happy that 35 years ago I discovered Lotfi's paper. It enjoyed my life and I sincerely hope that more and more young people will become aware of the power of this gradual model and start to further develop it and apply it in all possible domains to facilitate our life.

Fig. 43.3. Lofti A. Zadeh at the first IFSA congress in Palma de Mallorca in 1985. One may also recognize Marialuisa McAllister and Ramon Lopez de Mantaras.

"Fuzzy" in Georgian is "aramkapio"

Tatiana Kiseliova

It was always interesting however not always understandable for me how professional historians estimate a past event. Even two persons observing the same scene can give it different characteristics. Not going deep into philosophic discussions, I would like to point out here, that my intend to write about "fuzzy + Lotfi Zadeh + Georgia" is considered only from my subjective point of view. Such courage to speak about this theme is based on my rather long stay in the fuzzy society, which integrates scientists from Georgia and many countries all over the world.

I had heard for the first time the word "fuzzy" from my PhD supervisor Academician Vova Chavchanidze [19] at the beginning of 90s. Chavchanidze was known in Georgia and in the former Soviet Union not only for his non-standard ideas in Cybernetics and Artificial Intelligence [3, 4], but also as a very wise manager and a inimitable toastmaster. Zadeh's postulates were accepted by Chavchanidze rather naturally, because Chavchanidze's philosophy in science and life was not restricted by white and black colours.

In that time there were not so many scientists in Georgia following fuzzy directions. But fortunately, just by chance I was lead to a professor of the Tbilisi State University, Dr. Tamaz Gachechiladze, physicist, who had investigated the fuzzy set theory from the position of practical applications. In particular, the most important results of Tamaz Gachechiladze concern the theory of fuzzy selective and semantic information [5, 6, 15]. On his innovative papers a whole generation of masters and doctors were brought up [13].

The time when I was writing my PhD thesis (the 90s) was very difficult for Georgia: there was no electricity, no public transportation, and there were problems with food. But supervisors did continue doing research despite all these negative events, and we, the graduate students, were together with them in this fascinating work.

As a result I prepared my PhD thesis concerning applications of fuzzy sets and fuzzy logic in medicine. Before the Thesis Defense it was necessary to convince 25 members of the scientific board (with classical mathematics-physics-engineering educations) that fuzzy sets had a right to exist.

From the end of the last century there were dozens of Theses' Defenses on fuzzy sets and fuzzy logic, such that the word "fuzzy" had already stopped to be indeterminate in Georgia [20]. The acceptance of a fuzzy philosophy was also impeded by the lack of up-to-date scientific papers and books at that time in Georgia. Thus, the first request from my colleagues when I (also by fortunate accident) started to work at the

R. Seising et al. (Eds.): *On Fuzziness: Volume 1*, STUDFUZZ 298, pp. 295–299.
DOI: 10.1007/978-3-642-35641-4_44

TU Dortmund, Germany, in the Department of Computer Science I (at that time, one of the leading centres of Fuzzy Sets and Fuzzy Logic in Germany [28]) was "send us papers!".

When the deficiency of scientific information in Georgia disappeared, scientists in Georgia got a possibility to compare their research and scientific ideas with results of scientists from other countries. Moreover, Georgian scientists got a possibility to attend international conferences and workshops [8, 11, 14]; to organise the same level of conferences in Georgia [18, 31]; as well as to publish their papers in the leading international scientific journals [7, 22, 24, 30].

Starting 2006 the Iv. Javakhishvili Tbilisi State University [21], (where most "fuzzy" scientists are currently working) has undertaken by numerous reforms, which are not yet finished. Almost every year new study programs are announced, but what is important for the fuzzy society in Georgia, is that the interest of students on fuzzy directions does not become weaker from one year to another. Therefore the subject "Fuzzy Sets and Fuzzy Logic" as a separate one or as a part of "Soft Computing" (or "Computational Intelligence") courses has entered several Bachelor, MA and PhD programs [21].

Students applied such fuzzy methodologies as fuzzy decision making, fuzzy preferences, fuzzy control, some others in their practical works. Students are actively taking part in the scientific projects, and attend conferences. During the last 10 years several Georgian and International Foundations have financed some 5 research projects, where the word "fuzzy" is present. These projects embraced the whole directions of current research of fuzzy sets and fuzzy logic in Georgia and the researchers that are working in these directions. Among them is Revaz Grigolia, who already in the 70th has constructed the algebraic semantics of n-valued Lukasiewizc-Tarski logical systems [17]. His current work is connected with many-valued logics and their algebraic models [9, 10]. Gia Sirbiladze deals with extremal fuzzy dynamic systems [29]. Teimuraz Tsabadze works on fuzzy relations and group decision making [32]. Anna Sikharulidze, Irina Khutsishvili, Bezhan Ghvaberidze, Temur Latsabidze, Bidzina Matsaberidze, Temuri Manjaparashvili have done an interesting work in fuzzy decision making [11, 16]. I am currently working on the problem of rare diseases and their suspicion on the base on fuzzy approaches [1, 23, 25, 26]. Active cooperation with the paediatrician Karaman Pagava (Tbilisi State Medical University) developed not only scientific publications, but also interesting practical results: a modern centre of rare diseases was founded in Tbilisi, Georgia, where software based on our algorithm is used.

The investigations go forth and in order not to stop this moving, new educated students should be involved in the research process. The interest in fuzzy direction should be awaken, a student should be motivated. To tell the truth, I do not know a more effective way than the personal example of a scientist,or a professor. For example, it is not disputable, that the charm of Lotfi Zadeh, his attentiveness to different people, an inimitable style to present his ideas and results of research – this is of course not a complete characteristics' list – have impact on the foundation and existence of the very active fuzzy society. Each meeting of the members of this society (there are scientists from all over the world) – conferences, symposiums, workshops, research visits – has a result, always brings new ideas, impacts for all participants and all this in a very friendly atmosphere.

Fig. 44.1. The author and Prof. Zadeh during the 7th Fuzzy Days Conference Dortmund, Germany, October 1-3, 2001

The sooner a student will get to such society, the better motivation would he/she obtain. "Join us, – always invites Lotfi Zadeh surrounded by many people at each meeting. – We are discussing an interesting problem". Following this advice, already several PhD students from the Tbilisi State University have presented their work at international conferences [2, 12, 27].

To the best of my knowledge, Lotfi Zadeh never visited Georgia, Tbilisi, in spite of the large area covered by his travels around the world; but his message has indeed reached Georgia. I also do not know, if there is a collection of words for "fuzzy" in all world languages. If there is no Georgian entry, it can be added : "aramkapio" means "fuzzy" in the language of Georgia.

References

1. Acampora, G., Kiseliova, T., Pagava, K., Vitiello, A.: Towards Application of FML in Suspicion of Non-Common Diseases. In: 2011 IEEE International Conference on Fuzzy Systems, Taipei, Taiwan, June 27-30, pp. 2073–2079 (2011)
2. Alshibaia, P.: On Finitely Valued Bimodal Symmetric Gödel Logics. In: 8th Panhellenic Logic Symposium, Ioannina, Greece, July 4-8, pp. 130–131 (2011)
3. Andrew, A.M.: Obituaries. Kybernetes 39(5), 838–841 (2010)
4. Chavchanidze, V.V.: Heuristico-conceptual Programming. In: International Joint Conference on Artificial Intelligence (IJCAI), p. 919 (1977)
5. Criado, F., Gachechiladze, T., Jorjiashvili, N., Mandjaparashvili, T., Meladze, H., Sirbiladze, G., Tsilossani, T., Tsertsvadze, G.: Fuzzy Analysis (Image Construction) of the Language Structure on a Finite Set of Insufficient Data. Journal of Quantitative Linguistics 11(1-2), 93–132 (2004)
6. Criado, F., Gachechiladze, T., Meladze, H., Tsertsvadze, G.: The Bag Model in Language Statistics. Information Science 147(1-4), 13–44 (2002)
7. di Nola, A., Grigolia, R.: MV-Algebras in Duality With Labelled Root Systems. Discrete Mathematics 243(1-3), 79–90 (2002)
8. di Nola, A., Grigolia, R.: Free MV-Algebras. In: Stepnicka, M., Novák, V., Bodenhofer, U. (eds.) Proceedings of the 5th EUSFLAT Conference on New Dimensions in Fuzzy Logic and Related Technologies, vol. 2, pp. 207–208. University of Ostrava (2007)
9. di Nola, A., Grigolia, R., Lettieri, A.: Projective MV-Algebras. International Journal of Approximate Reasoning 47(3), 323–332 (2008)
10. di Nola, A., Grigolia, R., Spada, L.: A Discrete Representation of Free MV-Algebras. Mathematical Logic Quarterly 56(3), 279–288 (2010)
11. Dvalishvili, P., Khutsishvili, I., Sirbiladze, G.: Decision Precising Fuzzy Technology to Evaluate the Credit Risks of Investment Projects. In: ISDA, pp. 103–108. IEEE (2010)
12. Fandoeva, M., Kiseliova, T., Sikharulidze, A.: Investigation of the Heat Index in Georgia Based on the Most Typical Fuzzy Expected Values. In: Sirbiladze, G., Sikharulidze, A., Khutsishvili, I., Lominadze, T. (eds.) Proceedings of the 3rd International Conference on Computing and Computational Intelligence, June 26-28. Electrical and Computer Engineering series. A Series of Reference Books and Textbooks, pp. 302–308. WSEAS Press, Tbilisi (2009)
13. Gachechiladze, J., Gachechiladze, T., Khutsishvili, I.: A New Approach to Analysing Fuzzy Data and Decision-making Regarding the Possibility of Earthquake Occurrence. In: Proceedings of the 3rd International Conference on European Computing Conference, ECC 2009, pp. 299–301. World Scientific and Engineering Academy and Society (WSEAS), Stevens Point (2009)
14. Gachechiladze, T., Kiseliova, T., Tsintsadze, M.: Fuzzy Set Membership Function Construction Probabilistic Model for Applications in WWW. In: Proceedings of the First International Conference on Web and Information Technologies (ICWIT 2008), Sidi Bel Abbes, Algeria, June 29-30, pp. 343–349 (2008)
15. Gachechiladze, T., Sirbiladze, G.: Restored Fuzzy Measures in Expert Decision-making. Information Science 169(1-2), 71–95 (2005)
16. Ghvaberidze, B., Latsabidze, T., Matsaberidze, B., Sirbiladze, G.: Using a Minimal Fuzzy Covering in Decision-making Problems. Information Science 179(12), 2022–2027 (2009)
17. Grigolia, R.: Algebraic Analysis of Łukasiewicz-Tarski n-valued Logical Systems. Selected Papers on Łukasiewicz Sentential Calculi, Wroclaw, pp. 81–91 (1977)

18. Revaz, G., Liparteliani, R. (eds.): Ordered Groups and Lattices in Algebraic Logic (OGLAL 2011), September 20-23. Tbilisi University Press, Tbilisi (2011)
19. `http://dic.academic.ru/` Chavchanidze Vladimir Valer'yanovich. In: Academic Dictionaries and Encyclopedias
20. The Ilia Chavchavadze National Parliamentary Library of Georgia, `http://www.nplg.gov.ge/`
21. Ivane Javakhishvili Tbilisi State University, `http://www.tsu.edu.ge/Ivane`
22. Kiseliova, T.: A Theoretical Comparison of Disco and CADIAG-II-like Systems for Medical Diagnoses. Kybernetika 42(6), 723–748 (2006)
23. Kiseliova, T.: Towards an Interpretation of the Connectivity Analysis in the Frame of Fuzzy Equivalence Relations for Medical Diagnoses. In: Proceedings of the IFSA World Congress and EUSFLAT Conference 2009, Lisbon, Portugal, July 20-24, pp. 1577–1582 (2009)
24. Kiseliova, T., Moraga, C.: Tatra Mountains Mathematical Publications, vol. 21, pp. 19–30. Tatra Mountains Mathematical Publications (2001)
25. Kiseliova, T., Pagava, K.: New Approach to Estimate Different Drugs and/or Other Medical Interventions Effectiveness Based on Fuzzy Logic Principles. Georgia Medical News 3(156), 65–68 (2008)
26. Kiseliova, T., Pagava, K., Korinteli, I., Korinteli, M.G., Shonvadze, D.N.: Purposefulness of Using Fuzzy Logic Approaches in the Rare Disease Clinical Trials. Georgian Medical News 193(4), 22–24 (2011)
27. Liparteliani, R.: MVn-Algebras With Constants. In: 8th Panhellenic Logic Symposium, Ioannina, Greece, July 4-8, pp. 121–122 (2011)
28. Mesiar, R., Novák, V.: Fifteen Years of Fuzzy Logic in Dortmund. In: Reusch, B. (ed.) Computational Intelligence, Theory and Applications. Advances in Soft Computing, pp. 127–138 (2006)
29. Sirbiladze, G.: Fuzzy Identification Problem for Continuous Extremal Fuzzy Dynamic System. Fuzzy Optimization and Decision Making 9(3), 233–274 (2010)
30. Gia, S., Sikharulidze, A.: Weighted Fuzzy Averages in Fuzzy Environment. Part I: Insufficient Expert Data and Fuzzy Averages. International Journal of Uncertainty, Fuzziness and Knowledge-Based Systems 11(2), 139–157 (2002)
31. Sirbiladze, G., Sikharulidze, A., Khutsishvili, I., Lominadze, T. (eds.): Proceedings of the 3rd International Conference on Computing and Computational Intelligence, June 26-28. Electrical and Computer Engineering series. A Series of Reference Books and Textbooks, pp. 302–308. WSEAS Press, Tbilisi (2009)
32. Tsabadze, T.: A Method for Fuzzy Aggregation Based on Group Expert Evaluations. Fuzzy Sets and Systems 157(10), 1346–1361 (2006)

45

Lotfi Zadeh, Fuzzy Sets, and I: A Personal Odyssey

George J. Klir

When I immigrated to the United States in 1966 (see [1] for details), I was fortunate to begin my academic career at UCLA. Although I taught courses in computer architecture and design, I was primarily interested at that time in developing a general theory of systems. This motivated me to arrange a meeting with Lotfi Zadeh in Berkeley shortly after I settled in Los Angeles. I knew and admired his work in system theory, and I wanted to exchange ideas with him in this area. The meeting went very well. When I was about to leave, he mentioned that he had just embarked on research in a radically new area – a theory of fuzzy sets. He gave me a signed reprint of his paper, published just a few months before my visit, in which he outlined basic ideas of this envisioned new area [2]. This was my first acquaintance with Lotfi Zadeh and with fuzzy sets, which influenced my academic life in a profound way. As is explained later, fuzzy set theory has eventually become one of my primary research interests, and Lotfi and I have developed a close relationship and friendship.

The mathematical ideas introduced in the paper I brought from Berkeley were so radically different from anything else in mathematics at that time that I had to read the paper several times to properly comprehend them. Gradually, I found them intuitively appealing, and that aroused my interest in learning more about this prospective new area of mathematics. I was fortunate to discover two excellent early papers by Joseph Goguen [3, 4], in which Zadeh's ideas were further developed, especially the idea of fuzzy logic, which in [2] was mentioned only casually in a footnote. Moreover, these papers contained references to other relevant papers, and some of the latter contained additional relevant references, etc. Through this process, I was able to identify almost all papers published in this emerging new area. I tried to collect as many of them as possible and I studied them carefully. I was especially fortunate that Lotfi was kind enough to routinely send me his new publications, virtually all dealing with various aspects of fuzzy set theory.

In the 1960s, I was fully preoccupied with research devoted to the development of a sound conceptual framework for describing and formalizing key categories of general systems. This led to my decision to move to SUNY-Binghamton in 1969, where the academic environment for this kind of research was considerably more favorable than at UCLA. My interest in collecting and studying publications on fuzzy set theory and fuzzy logic was at that time motivated solely by curiosity. However, the more I learned about these new areas, the more I recognized their relevance to

R. Seising et al. (Eds.): *On Fuzziness: Volume 1*, STUDFUZZ 298, pp. 301–309.
DOI: 10.1007/978-3-642-35641-4_45

my own work on systems. In 1975, I suggested to generalize the various categories of systems emerging from my conceptual framework to systems based on variables whose states were fuzzy sets. I presented this idea at two conferences (see, e.g. [5]). The response was quite positive and that encouraged me to further develop this idea, which at that time was still rather "half baked." However, an even greater encouragement came soon from a series of three important papers by Zadeh [6], in which he introduced the concept of a linguistic variable. These papers gave me not only confidence that I was on the right track, but also answers to some questions that I myself had not been able to answer.

Throughout the 1970s, I continued to follow the growing literature on fuzzy set theory quite diligently. While I increasingly recognized the significance of this emerging area, I also became aware that it had not been well received, by and large, within the academic community. Except for a small group of pioneering researchers who worked on it, fuzzy set theory was largely viewed with skepticism and sometimes even with an open hostility. Some influential scientists and mathematicians harshly criticized Zadeh's seminal paper [2], often in a hostile and emotional way, and he deserves credit for withstanding their criticism admirably. These circumstances urged me to take a more active role in supporting this young and promising area. I saw the first opportunity for doing that when I founded the International Journal of General Systems in 1974. I included fuzzy systems among the areas defining the scope the journal and I invited Lotfi Zadeh and Joseph Goguen to represent this area on the Editorial Board. Over the years, an impressive number of papers and eleven Special Issues on various aspects of fuzzy systems have been published in the journal. Another early opportunity came about when I was able to arrange, as editor of a book series, the publication of the very first book on fuzzy decision-making [7].

In the late 1970s, I was greatly inspired by Zadeh's idea of a theory of graded possibilities based on the notion of a fuzzy restriction [8], and I became interested in comparing possibilistic systems with probabilistic systems. I realized soon that the theory of graded possibilities was a natural generalization of classical possibility theory, while probability theory, contrary to common beliefs at that time, was not. One outcome of my study of possibilistic systems was a well-justified possibilistic measure of uncertainty (and the associated uncertainty-based information), very different from the well-established probabilistic measure of uncertainty and information – the Shannon entropy. This outcome stimulated me to study uncertainty and uncertainty-based information in various types of systems more comprehensively, which led eventually to my formulation of a broad research program known as Generalized Information Theory (GIT) [9].

I should mention at this point that Zadeh's 1978 paper on the theory of graded possibilities [8] was published in the inaugural issue of *Fuzzy Sets and Systems*, the first journal fully devoted to fuzzy set theory and its applications. I was invited to serve on Advisory Board of the journal, which increased my contact with the growing community of researchers working in this area.

Throughout the 1980s, fuzzy set theory and uncertainty gradually became the primary foci of my research. It was thus natural that I tried to participate actively as much as possible at conferences and other events pertaining to these areas. This was a decade filled not only with impressive advancements in fuzzy set theory, but also with many important events that were favorable to further progress in this area. In fact, there were too many of them to be covered in this short essay, but two of them stand out as milestones. One was the founding of the first professional society supporting fuzzy set theory in 1981 under the name *North American Fuzzy Information Processing Society* (NAFIPS). One year after its inauguration, NAFIPS began to organize Annual Meetings – high quality conferences on fuzzy set theory and fuzzy logic - and five years later, it began publishing its own journal – the *International Journal of Approximate Reasoning*. Another milestone was the founding of an international organization aimed at promoting fuzzy systems and related areas worldwide. Founded in 1984 under the name *International Fuzzy Systems Association* (IFSA), this organization adopted *Fuzzy Sets and Systems* as its official journal and has organized biennial World Congresses since 1985.

Through my active participation in conferences and other events devoted to fuzzy set theory and related areas, including those organized by NAFIPS and IFSA, I gradually became well acquainted with most of the key contributors to fuzzy set theory and related areas. Lotfi Zadeh participated, usually as a keynote speaker, in almost all conferences that I attended. His keynote speeches were always exciting and each contained some new ideas. It was even more exciting and enjoyable to talk with him personally during coffee breaks and other conference venues. I learned a lot from his wisdom via these frequent informal discussions. We eventually developed a close relationship and friendship. I nominated him for an Honorary Doctoral Degree at SUNY-Binghamton in 1988. The nomination was unanimously accepted and he was awarded the degree in 1989; see our joint photo in Figure 45.1, which was taken right after the ceremony.

In spite of the impressive developments in fuzzy set theory, fuzzy logic, and fuzzy systems in the 1980s, these areas were still little known within the academic community. I increasingly felt that the time was ripe to develop a graduate course in these areas and I proposed one. I was fortunate that, contrary to the situation at other universities at that time, the academic community at SUNY-Binghamton was supportive of my proposal. I started to teach the course as an elective in our graduate program in systems science in 1985. The course was well received and class enrollments were always impressively high. The content of the course and my class notes continuously evolved to capture new developments in this rapidly advancing field. In 1987, I received a rather unexpected offer from Prentice Hall to publish a textbook based on this course. I accepted the offer and converted and expanded my rather extensive class notes (with the help of one of my graduate students, Tina Folger, who had just completed the course) into a textbook that was published in 1988 [10]. The course and the textbook attracted to our program some very talented graduate students interested in studying fuzzy systems. A group of these students and myself are shown in Figure 45.2 with Lotfi Zadeh when he came to Binghamton in 1989 to receive his Honorary Doctoral Degree.

Fig. 45.1. Lotfi Zadeh and I right after he was awarded an Honorary Doctoral Degree from SUNY-Binghamton in 1989

During the 1980s, attitudes toward fuzzy set theory by the academic community had gradually changed. The earlier wholesale and usually rather emotional attacks on the theory were gradually replaced with more focused and rational debates, initiated largely by its opponents. All but one of these debates took place in various journals and each involved multiple correspondents discussing a particular position paper. One debate, in which I happened to be involved, was an oral debate between two persons. One person was Peter Cheesman, whose position was that probability theory is the only sensible description of uncertainty, which is adequate for all problems involving uncertainty. As the second person in this debate, I challenged this position. The debate took place at Cambridge University (in a historical building designed specifically for scientific debates) on the occasion of the Eighth Conference on Maximum Entropy and Bayesian Methods in 1988, and it was based upon an agreement made one year earlier. The debate took several hours and followed rigorously the traditional procedural protocol of Cambridge University. It is unfortunate that it was not recorded. However, a transcript of my presentation is available [11]. The debate was concluded by allowing each person in the audience to vote for one of the two positions. My position lost by only two votes out of more than 100 votes, which seems to me a surprisingly positive result at a conference devoted fully to classical probability theory.

Fig. 45.2. Lotfi Zadeh, I, and a group of Ph.D. students specializing on fuzzy systems at SUNY-Binghamton in 1989

Shortly after I returned home from the debate, I received a message informing me that I was nominated to become the next President of NAFIPS. This surprised me, but I accepted the nomination. I served as NAFIPS President until 1991. Two years later, at the IFSA Congress in Seoul, South Korea, I was elected by the IFSA Board of Directors as IFSA President for the period 1993-1995. There was a growing feeling at that time that IFSA should play the role of an international federation of national, regional, and other organizations supporting fuzzy systems and related areas. This required that the existing IFSA Constitution and Bylaws be radically changed. I took it as a challenge to do that. As it turned out, the challenge was much greater than I expected, primarily due to extensive and difficult negotiations among many players involved, whose mutually contradictory views had to be somehow reconciled. To make the long story short, the new IFSA Constitution and Bylaws were eventually approved at the Fifth IFSA World Congress in Sao Paulo, Brazil, in 1995, and IFSA began to operate as a Federation.

Let me mention an interesting episode involving Lotfi and me that occurred at two successive conferences on fuzzy theory and technology organized by Paul Wang in North Carolina in the 1990s. A prestigious "Lotfi A. Zadeh Best Paper Award" was inaugurated at the 1993 conference with the stipulation that the first recipient would be Zadeh himself for his seminal 1965 paper [2]. I was asked to present the award to him at the conference banquet. After the presentation (see Figure 45.3), the paper for the second award, selected by a special committee from papers presented at the conference, was announced. To my great surprise, the committee selected my paper [12]. According to the rules, I was supposed to receive the award at the next conference in 1994. I was again surprised when it was Lotfi Zadeh himself who presented the Second Lotfi A. Zadeh Best Paper Award to me at this time.

Fig. 45.3. My presentation of the First Lotfi A. Zadeh Best Paper Award to Lotfi for his 1965 seminal paper [2] at the Second International Conference on Fuzzy Theory and Technology, in Durham, North Carolina, in 1993

The 1990s are now generally recognized as a decade of some groundbreaking developments for fuzzy set theory and related areas. In the application domain, the most visible were the amazingly successful applications of fuzzy systems, especially fuzzy controllers, in Japan (see, for example, [13] and [14] for details). It is likely that these predominantly engineering applications led also to the IEEE endorsement of fuzzy systems by instituting annual IEEE International Conferences on Fuzzy Systems (FUZZ-IEEE) in 1992 and by publishing *IEEE Transactions on Fuzzy Systems* since 1993. In the theoretical domain, there were some breakthrough developments in foundational issues of fuzzy logic in the 1990s, as is exemplified by the work of Peter Hájek [15].

People who worked in the areas of fuzzy set theory, fuzzy logic, and fuzzy systems were suddenly in great demand throughout the 1990s. I remember visiting Japan at least twice each year for conferences and various other events. I almost always had the pleasure of seeing Lotfi during these visits and observing how much he was respected and admired in Japan. I got a sense of how much he was in demand there during one of my longer stays in Tokyo. Lotfi came to Tokyo for an interview on Japanese television at the beginning of my stay. After the interview, he returned to Berkeley, only to come to Tokyo again during the second week of my stay. He and I participated at a conference on fuzzy systems in Tokyo and travelled then together to Seoul, South Korea, for another conference devoted to fuzzy systems. After that, I returned home while Lotfi travelled once more to Japan for another engagement.

In spite of the heavy travel and my time-consuming service to IFSA, I pursued as much research as I could on issues in some subareas of fuzzy set theory, such as aggregation operations on fuzzy sets, fuzzy relational equations, fuzzy arithmetic, computing with granular probabilities, defuzzification, linguistic retranslation, and others. In 1995, my second and much more extensive book on fuzzy set theory and fuzzy logic, coauthored with Bo Yuan and containing some of our own results, was published again by Prentice Hall [16]. The book also contains an insightful and very generous Foreword by Lotfi Zadeh. When working on this book, we frequently consulted Zadeh's papers. In this regard, we found a collection of his papers on fuzzy sets published in 1987 [17] was very helpful. After our book was published, we felt that the time was ripe for another such collection and we actually prepared one [18] (see also Figure 4).

Since the beginning of the 21st century, literature on the theory and applications of mathematics based on fuzzy sets and fuzzy logic has tremendously increased. As a result, it has been increasingly difficult to keep track of all developments in this field. There is no doubt, however, that Lotfi Zadeh has continued to play a central role in advancing the field, primarily by expanding its frontiers. This is exemplified by his computational theory of perceptions [19] or his more recent generalized theory of uncertainty (GTU) [20].

One visible trend during the last decade or so has been a growing number of successful applications of fuzzy mathematics in various sciences. I became personally involved in geology (see [21], with a Foreword by Lotfi Zadeh) and, more recently, in a rather exploratory way, in the psychology of concepts [22]. However, my main focus in the 21st century has been on advancing GIT. Principal results are presented in my book [23] and in a more recent paper I wrote jointly with Andrey Bronevich [24]. Further progress in GIT is now largely contingent on the development of theories of monotone measures defined on fuzzy sets of various types. Although some initial results in this direction have already been obtained via collaboration with my former colleague, Zhenyuan Wang, and are presented in our recent book [25], much more research work is still needed in this regard.

Let me conclude this short essay by examining GIT with respect GTU. When I read Zadeh's first paper on GTU [20] several years ago, I soon recognized that GUT and GIT share a similar goal, but use very different approaches to achieving it. It is rather easy to see that GTU follows a top-down approach in which information is associated with statements describing perceptions in natural language. Each such statement is viewed as a constraint regarding some perceptual domain of concern. These constraints are usually complex and involve various modalities. GUT is a research program that begins with the characterization of the most general modalities, which are then further classified as needed. The overall aim is to develop an operational capability for dealing with information described in natural language. GIT, on the contrary, follows a bottom-up approach by which the very special and narrow, but well-developed classical theories of information (possibilistic and probablistic) are gradually generalized.

The distinction between the top-down and bottom-up approaches is certainly one of the main differences between GTU and GIT. However, it is not the only one or, perhaps, not even the principal one, as is so eloquently described by Lotfi in our recent e-mail correspondence. I quote from his e-mail message sent to me on March 21, 2012: "As I see it, in large measure GIT and GTU are complementary. The principal difference between GIT and GTU is the following. GIT is concerned, in the main, with measures of information, as in classical information theory. On another side, GTU is concerned, in the main, with the meaning of information. ... GIT is a significant generalization of classical information theory, but it stops short of consideration of semantic issues. In contrast, GTU stops short of exploration of issues which relate to measures of information." Although I was aware of this important difference between GTU and GIT, I took it for granted that this difference would disappear when the gap between the two theories is bridged. After receiving the insightful remark by Lotfi, I revised this rather naive attitude. I began to realize that there is no need to wait until the gap between the two theories is bridged, and that they both can be significantly enriched right now by borrowing the missing features from each other and applying them where appropriate and useful.

Fig. 45.4. Lotfi Zadeh, Bo Yuan and I, shortly after the second book of *Selected Papers by Lotfi A. Zadeh* [18] was published

References

1. Klir, G.J.: Systems Profile: The Emergence of Systems Science. Systems Research 5(2), 145–156 (1988)
2. Zadeh, L.A.: Fuzzy Sets. Information and Control 8(3), 338–353 (1965)
3. Goguen, J.A.: L-fuzzy sets. Journal of Mathematical Analysis and Applications 18(1), 145–174 (1967)
4. Goguen, J.A.: The Logic of Inexact Concepts. Synthese 19, 325–373 (1968-1969)
5. Klir, G.J.: Processing of fuzzy activities of neutral systems. In: Trappl, R., Pichler, F. (eds.) Progress in Cybernetics and Systems Research, Hemisphere, Washington, DC, pp. 21–24 (1975)
6. Zadeh, L.A.: The Concept of a Linguistic Variable and its Application to Approximate Reasoning. Information Sciences 8, 199–251, 301–357, 9, 43–80 (1975)
7. Kickert, W.J.M.: Fuzzy Theories on Decision-making. Martinus Nijhoff, Leiden (1978)
8. Zadeh, L.A.: Fuzzy Sets as a Basis for a Theory of Possibility. Fuzzy Sets and Systems 1(1), 3–28 (1978)
9. Klir, G.J.: Generalized Information Theory. Fuzzy Sets and Systems 40(1), 127–142 (1991)
10. Klir, G.J., Folger, T.A.: Fuzzy Sets, Uncertainty, and Information. Prentice Hall, Englewood Cliffs (1988)
11. Klir, G.J.: Is There More to Uncertainty than some Probability Theorists might have us Believe? International Journal of General Systems 15(4), 347–378 (1989)
12. Klir, G.J.: Multivalued Logics Versus Modal Logics: Alternative Frameworks for Uncertainty Modelling. In: Wang, P.P. (ed.) Advances in Fuzzy Theory and Technology, vol. II, pp. 3–47. Duke University, Durham (1994)
13. Schwartz, D.G., Klir, G.J.: Fuzzy Logic Flowers in Japan. IEEE Spectrum 19(2), 32–35 (1992)
14. Schwartz, D.G., Klir, G.J., Lewis, H.W., Ezawa, Y.: Applications of Fuzzy Sets and Approximate Reasoning. Proceedings of the IEEE 82(4), 482–498 (1994)
15. Hájek, P.: Metamathematics of Fuzzy Logic. Kluwer, Boston (1998)
16. Klir, G.J., Yuan, B.: Fuzzy Sets and Fuzzy Logic: Theory and Applications. Prentice Hall, Upper Saddle River (1995)
17. Yager, R.R., Ovchinnikov, S., Tong, R.M., Nguyen, H.T. (eds.): Fuzzy Sets and Applications: Selected Papers by L. A. Zadeh. John Wiley, New York (1987)
18. Klir, G.J., Yuan, B. (eds.): Fuzzy Sets, Fuzzy Logic, and Fuzzy Systems: Selected Papers by Lotfi A. Zadeh. World Scientific, Singapore (1996)
19. Zadeh, L.A.: Outline of a Computational Theory of Perceptions Based on Computing with Words. In: Sinha, N.K., Gupta, M.M. (eds.) Soft Computing and Intelligent Systems, pp. 3–22. Academic Press, San Diego (2000)
20. Zadeh, L.A.: Toward a Generalized Theory of Uncertainty (GTU) - an outline. Information Sciences 172(1), 1–40 (2005)
21. Demicco, R.V., Klir, G.J. (eds.): Fuzzy Logic in Geology. Academic Press, San Diego (2004)
22. Belohlavek, R., Klir, G.J.: Concepts and Fuzzy Logic. MIT Press, Cambridge (2010)
23. Klir, G.J.: Uncertainty and Information. John Wiley, Hoboken (2006)
24. Bronevich, A., Klir, G.J.: Measures of Uncertainty for Imprecise Probabilities: An Axiomatic Approach. International Journal of Approximate Reasoning 51(4), 365–390 (2010)
25. Wang, Z., Klir, G.J.: Generalized Measure Theory. Springer, New York (2009)

46

Fuzzy Rule Based Systems as Tools towards Solving the "Key Problem of Engineering"

Laszlo T. Koczy

46.1 Introduction

Key problem of engineering? Is there anything that is called so? In order to explain what is meant by this, let me start with some references to teaching Digital Design for B.Sc. Electrical/Electronics Engineering students in the first grade. The key part of the standard curriculum of this subject is a series of design algorithms for combinational and sequential circuits, even though it is obvious nowadays that the algorithmic design of LSI digital circuitry is a mathematically intractable problem that always leads to NP-hardness, and thus to insolvability in the classic sense. Learning the well established design approaches that have served well for SSI circuits in the early times of Digital Design nevertheless teaches a way of looking at the problems in a way that helps with finding *reasonably good*, or *almost* optimal solutions that can be very well used in the engineering practice and that might be applied in commercially feasible products, even in the case of more complex problems, up to the MSI level; while at the same time it teaches the students to think engineering optimization and design.

In my lectures on Fuzzy Systems at the beginning of the first class I usually try to explain the necessity of using fuzzy sets by presenting a few practical examples. Among those I like to mention the simple problem of driving a car, especially, how to do the simple maneuver of turning right at an intersection with elevated pavement. What is the essential point in this example? If we try to define the goal function to optimize at driving a car, the suggestions usually include the aspects of minimizing the elapsed time between start and destination, while the question of minimizing the consumed fuel is also an option. Let us construct now hypothetically a formal approach to this problem. Because both time and fuel consumption should be minimal, it is definitely necessary to determine the shortest trajectory of turning, such that none of the tires of the car touch the pavement thus causing damage to the rubber tire. Clearly, this is one of the restrictions that should always be observed, besides keeping speed limits, obeying traffic signals, etc. To deal with this question first it should be set what the minimal safety distance should be that must be kept between the wheels and the pavement. Let us say, it is 2 mm. Even though it sounds silly, assuming this safety distance the exact trajectory and the corresponding steering control can be theoretically calculated.

R. Seising et al. (Eds.): *On Fuzziness: Volume 1*, STUDFUZZ 298, pp. 311–323.
DOI: 10.1007/978-3-642-35641-4_46

It is obvious that this way of looking at the problem is silly to the degree of being ridiculous; nobody would consider a solution of the car turning problem unsolved if a trajectory would be followed by a driver that passes e.g. half a meter away from the pavement. Any good driver would prefer keeping a larger distance even though this way the car would use a *little more* fuel and would need a *little more* time for the maneuver, both amounts being negligible from the point of view of the resources used for the completion of the task, while doing all the calculations necessary to obtain the *exact optimum* (under the given restrictions) would need much more resources – if the exact model of the car and the road would be available, thus such a calculation would be possible at all.

The common point in both examples – the design of digital circuits and the planning of a simple car driving maneuver – an immense amount of resources would be necessary for finding the actual optimum. In both cases the optimal solution would cost much more (or in some cases would have virtually "infinite" cost), than any solution accepted in real life would allow. What is expected in reality is the *almost optimal* design of a circuit, or as a *reasonably optimal* turning trajectory for the car. In almost every engineering problem there are similar situations and problems to solve.

In real life there are always two essentially contradicting goals: One is obtaining the exact model of a given problem, in order to be able to determine the analytical optimum; while the other is reducing the "cost" of the solution (i.e. the consumption of resources, both in terms of time and space or equipment) to the minimum. As an example, in the context of circuit design, the cost includes the computational time needed for the design of a particular circuit as well as the actual amount of components used for the implementation of the circuit (obviously the costs of a single optimization having a different relative weight from the weight of the resources necessary for the implementation of the circuit itself). Let this be a simple case: The digital circuit to design is planned to be manufactured in a total of 1000 copies. If the result of the suboptimal design is containing in fact 1247 gate circuits, instead of the optimal 1245, i.e. two more compared to the optimum, the cost of the suboptimality of the design is 2000 units. Let us assume that every iteration of the circuit design procedure (algorithm) costs one unit. Let us also assume that by 5000 more iterations the procedure would reach the optimum. It is obvious then that having an optimal circuit costs two and a half times more than the fact that the manufacturer accepted a somewhat suboptimal solution. On the other hand, if the product were manufactured in a larger series, in this case in more than 2500 copies, the loss caused by the suboptimal design would be greater than the one by the further 5000 necessary repetitions in the optimization. If however, the size of the problem grows larger, soon a limit will be reached, when there is no more such reasonable number of repetitions that finds the exact optimum, thus the cost of searching for the optimum exceeds any cost caused by suboptimal design.

Similarly, in the turning car example, the calculations needed for determining the exact optimal steering control for the turning car would very likely cost orders of magnitude more than the unnecessary fuel consumed by the car and the surplus time needed when a wider safety margin is applied. A good engineer optimizes the total costs involved with the search for the solution and the quality of the solution itself together, the latter in the sense that any deviation from the optimum, any inaccuracy of the model results in a loss, i.e. it has costs.

In this sense *the key problem of engineering* is to find the complex optimum of efforts spent for solving a problem and losses caused by the acceptance of suboptimal solutions. In real life applications it is never the exact optimum that is sought for, instead of this the goal is to find the cheapest but acceptable approximate solution. In the next section it will be shown how fuzzy rule based systems serve this goal efficiently, and how the motivation behind proposing and evolving such systems is nothing else but to find "good enough" solutions for a wide class of problems.

Fig. 46.1. Laszlo Koczy and Lotfi A. Zadeh at the US Hungarian Joint Seminar on Pattern Recognition in the Hungarian Academy of Science, Budapest, June 1975

46.2 The Evolution of Rule Based Systems

46.2.1 Quantization and Expert Control in AI

Production rules and inference engines based on **if ... then ...** rules have been extensively used in early Artificial Intelligence, in expert systems, control and decision making as well. For a couple of relatively recent overviews of the topic see [6, 24]. The essential idea of this approach is that all knowledge available on the behavior of a system is quantized and for each quantum of the input state space the corresponding output state space value quantum is given in a particular production rule. There is a close analogy with the partial definition of a mathematical multivariable function by giving a table of input-output points, the input side determining input space vectors and the output side the corresponding output space values or vectors.

In information theory there are known laws concerning the necessary density of such samples, the Shannon Sampling Theorem or Cardinal Theorem of Interpolation Theory [16], which refers primarily to periodic signals but can be extended to the case of non-uniform samples as well. According to this theorem band limited functions, i.e. functions that can be decomposed into a (finite or infinite) set of periodical functions should be sampled with a greater frequency than the double of the maximal frequency occurring in the decomposition.

Quantization of the continuous physical reality is the essential point in transforming continuous (and often analytical) models into discrete models where individual values of the state space variables are clustered into interval type units within which the values cannot be differentiated from each other. Formally, such a sampled function corresponds to a step function that can be given by the production rules

$$\begin{array}{l}\textbf{If } x_1 \textbf{ is } a_{11} \textbf{ and } x_2 \textbf{ is } a_{21} \textbf{ and } \ldots \textbf{ and } x_k \textbf{ is } a_{k1} \textbf{ then } y_1 \textbf{ is } b_{11} \ldots \textbf{ and } y_m \textbf{ is } b_{m1} \\ \ldots \\ \textbf{If } x_1 \textbf{ is } a_{1r} \textbf{ and } x_2 \textbf{ is } a_{2r} \textbf{ and } \ldots \textbf{ and } x_k \textbf{ is } a_{kr} \textbf{ then } y_1 \textbf{ is } b_{1r} \ldots \textbf{ and } y_m \textbf{ is } b_{mr}\end{array} \tag{46.1}$$

This corresponds to the definition

$$y_1 = \begin{cases} b_{11} \textbf{ if } a_{11} \leq x_1 < a_{12} \ldots \textbf{ and } a_{k1} \leq x_k < a_{k2} \\ \ldots \\ b_{lr-1} \textbf{ if } a_{lr-1} \leq x_l < a_{lr} \ldots \textbf{ and } a_{kr-1} \leq x_k < a_{kr} \\ b_{lr} \textbf{ if } x_l = a_{lr} \ldots \textbf{ and } a_k = a_{kr} \end{cases}, \ldots y_m = \ldots \tag{46.2}$$

It is necessary that the intervals cover the input space (universe):

$$\begin{aligned} &a_{11} \leq x_1 < a_{12} \times \ldots \times a_{k1} \leq x_k < a_{k2} \cup a_{12} \leq x_1 < a_1 \times \ldots \\ &\ldots \times a_{k2} \leq x_k < a_{k3} \cup \ldots \cup a_{lr-1} \leq x_l < a_{lr} \times \ldots \\ &\ldots \times a_{kr-1} \leq x_k < a_{kr} \cup (a_{lr}, a_{2r}, \ldots a_{kr})^T = \\ &= X_1 \times \cdots \times X_k \end{aligned} \tag{46.3}$$

This defines a step function which is a rough approximation of the original f function $\mathbf{y} = f(\boldsymbol{x})$ (where $\boldsymbol{x} = (x_1, x_2, \ldots, x_k)^T$ and $\mathbf{y} = (y_1, y_2, \ldots, y_m)^T$). As stated above, f can be reconstructed completely (by interpolation) if the sampling theorem is satisfied.

There is however an alternative interpretation of production rule system (46.1), the logician's one. Here, each interval $a_{1i} \leq x_l < a_{1i+1} \times \ldots \times a_{ki} \leq x_k < a_{ki+1}$ is interpreted as a logic symbol, and the rule base itself is reinterpreted in the form

$$\begin{array}{c}\textbf{If } x \textbf{ is } A_1 \textbf{ then } y \textbf{ is } B_1 \\ \ldots \\ \textbf{If } x \textbf{ is } A_r \textbf{ then } y \textbf{ is } B_r\end{array} \tag{46.4}$$

where $A_1, A_2, \ldots, A_r, B_1, B_2, \ldots, B_r$ are all logic symbols.

This semantics means a loss of the original function view but has some other advantages: The classic inference schemes of logic, such as e.g. *Modus Ponens* may be straightforwardly applied. If namely "**If** x **is** A_2 **then** y **is** B_2" holds and "**If** x **is** A_2" is true, then "y **is** B_2" can be concluded without any further calculations. Other similar inference schemes, such as *Modus Tollens*, can also be applied.

What is the major problem with using this kind of rule based systems? The answer may be found in the computational complexity, i.e., the cost or need of resources involved with this approach. If for simplicity it is assumed that the quantization of the input space is equidistant in each dimension, a "grid" is set up, with T_1, $T_2, \ldots$, T_k intervals in X_1, $X_2, \ldots$, X_k, respectively. Here the total number of rules is $R = T_1 * T_2 * \ldots * T_k$. For estimating the order of this number, let us assume that $\forall i : T_i \leq T$, thus

$$R \leq T^k \tag{46.5}$$

There is no better estimation of R than $O\left(T^k\right)$, an exponential expression in terms of the input variables k, and this fact explains why there are no real applications of symbolic rule based expert control for systems where the input variables are more than about three! It does not matter whether the interpolative or the logic interpretation are used, this classic AI approach is more a toy than a real tool for engineers.

46.2.2 Fuzzy Sets and Fuzzy Rules

An essential change in the applicability of production rule systems came when Zadeh proposed the concept of *fuzzy sets* [27]. Fuzzy sets enabled covering the input state space (cf. 46.3 above) in a partial and overlapping way. While quantization required an exact cover, using fuzzy granulation with partly overlapping "intervals" in the condition parts of the rules, the number of "grid elements" in each dimension and thus the value of T in (46.5) could be reduced. The general concept of *modeling complex systems by applying a production rule system* (each rule representing a *fuzzy relation*, i.e. a fuzzy subset of the $X \times Y$ input – output space) was proposed by Zadeh himself some years later in [28]. Zadeh's fuzzy rules may be represented in a very similar way to (46.4), however the symbols occurring in parts of the rules have different interpretations:

$$\begin{array}{c}\textbf{If } x \textbf{ is } \tilde{A}_1 \textbf{ then } y \textbf{ is } \tilde{B}_1 \\ \ldots \\ \textbf{If } x \textbf{ is } \tilde{A}_r \textbf{ then } y \textbf{ is } \tilde{B}_r\end{array} \tag{46.6}$$

$\tilde{A}_i$ being fuzzy sets of $\boldsymbol{X}$ (defined by their membership functions $\mu\left(\tilde{A}_i\right): \boldsymbol{X} \rightarrow [0,1]$) and $\tilde{B}_j$ being fuzzy sets of $\boldsymbol{Y}$ (defined by their respective membership functions $\mu\left(\tilde{B}_j\right): Y \rightarrow [0,1]$). This paper of Zadeh, proposing the use of Compositional Rule of Inference (CRI) for obtaining conclusions or control actions from as a response to an observation started a revolution in non-conventional control and decision making, especially after a practical implementation was given to this new approach by Mamdani [15], where the fuzzy rules in (46.6) were orthogonally decomposed:

$$\textbf{If } x_1 \textbf{ is } \tilde{A}_{11} \textbf{ and } x_2 \textbf{ is } \tilde{A}21 \textbf{ and } \dots \textbf{ and } x_k \textbf{ is } \tilde{A}k1 \textbf{ then } y_1 \textbf{ is } \tilde{B}_{11} \dots \textbf{ and } y_m \textbf{ is } \tilde{B}_{m1}$$

$$\dots$$

$$\textbf{If } x_1 \textbf{ is } \tilde{A}_{1r} \textbf{ and } x_2 \textbf{ is } \tilde{A}_{2r} \textbf{ and } \dots \textbf{ and } x_k \textbf{ is } \tilde{A}_{kr} \textbf{ then } y_1 \textbf{ is } \tilde{B}_{1r} \dots \textbf{ and } y_m \textbf{ is } \tilde{B}_{mr} \tag{46.7}$$

The question is now, whether there is a change in the order of complexity (46.5) with these new types of rule bases? Fuzzy granulation instead of classic quantization modifies formula (46.3) describing the coverage of the input universe:

$$\tilde{A}_1 \cup \tilde{A}_2 \cup \ldots \cup \tilde{A}_r = ?\boldsymbol{X} \tag{46.8}$$

should hold according to (46.6), and

$$\tilde{A}_{11} \times \ldots \times \tilde{A}_{k1} \cup \tilde{A}_{12} \times \ldots \times \tilde{A}_{k2} \cup \ldots \cup \tilde{A}_{1r} \times \ldots \times \tilde{A}_{2r} = ?\,X_1 \times \ldots \times X_k \tag{46.9}$$

according to (46.7). As union was defined by Zadeh in [27] as $\max\left\{\mu\left(\tilde{A}_i\right), \mu\left(\tilde{A}_j\right)\right\}$, the equalities in (46.8) and (46.9) do definitely not hold in the classic sense apart from nonsense cases where the kernels cover the input space. Instead of this a fuzzy cover of degree α, simply a *fuzzy α-cover*, could be set as the necessary condition for the rule base, thus

$$\boldsymbol{X}: \mu\left(\tilde{A}_1 \cup \tilde{A}_2 \cup \ldots \cup \tilde{A}_r\right) \geq \alpha \tag{46.10}$$

should hold instead of (46.8) and the following instead of (46.9):

$$X_1 \times \ldots \times X_k : \mu(\tilde{A}_{11} \times \ldots \times \tilde{A}_{k1} \cup \times \ldots \times \tilde{A}_{1r} \times \ldots \times \tilde{A}_{2r}) \geq \alpha \tag{46.11}$$

Here usually $\alpha \geq 0.5$, so that the cover of the input space is reasonably “dense”, i.e., for any crisp singleton observation there is at least one rule whose antecedent matches with this observation in a degree of at least 0.5, i.e., “more yes than no”.

In (46.11) fuzzy sets $\tilde{A}_1, \tilde{A}_2, \ldots, \tilde{A}_r$ do overlap partially however their respective kernels do not, and thus the area covered in degree 1 (as in (46.3)) is necessarily a proper subset of $\boldsymbol{X}$ ($X_1 \ldots X_k$). What is the advantage of this compared to the quantization case? The number of terms in the fuzzy approach is *much smaller* than in the symbolic – crisp case, because in the areas between the kernels, there is no full coverage, nevertheless, an interpolation of the conclusions to be obtained from the surrounding rules (a combination weighted by the degrees of matching of the observation and the respective antecedents of the individual consequents) leads to an interpretable conclusion, i.e. (46.5) changes to

$$R \leq T'^k, \text{ where } T' << T, \text{ thus } R = O(T'^k). \tag{46.12}$$

This fact explains why there are quite many industrial applications of the dense fuzzy controller (Zadeh's CRI and derived methods) and inference machine with five to six, sometimes up to ten input variables. Using fuzzy systems reduces T to a smaller value.

It is relevant how many (and what kind of) rules are stored in the knowledge base and how many (and how complex) steps are needed for obtaining a conclusion from an observation. In a simplified approach where uniform complexity is taken, any step is considered as unit time and thus it is sufficient to consider only the size of the rule base, i.e., the number of rules taken into account in worst case for a single inference step.

The alternative Takagi-Sugeno controller using rule bases consisting of rules like

$$\textbf{If } \boldsymbol{x} \textbf{ is } \tilde{A}_1 \textbf{ then } \boldsymbol{y} = f_i(\boldsymbol{x}) \tag{46.13}$$

or

$$\textbf{If } x_1 \textbf{ is } \tilde{A}_{11} \textbf{ and } x_2 \textbf{ is } \tilde{A}_{21} \textbf{ and } \ldots \textbf{ and } x_k \textbf{ is } \tilde{A}_{kl} \textbf{ then } y = f_i(\boldsymbol{x}) \tag{46.14}$$

does not differ from the above in terms of coverage and complexity.

Many researchers have interpreted CRI type rule bases in the logician's way, considering production rules as implications, while looking for some kind of extended *Modus Ponens*. From the point of view of complexity and resources these approaches are comparable with the function type interpretations. For an early overview of this approach see [7].

46.2.3 The Next Step: Less Fuzzy Rules with Similar Efficiency, Interpolative Models

Expression (46.12) does not differ from (46.5) in the structure, only in the fact that for the same problem, T' is expectedly less than T. How can this exponential expression be further decreased in order to increase the value of k so that the model remains tractable? One way is to further decrease T', so that it reaches its minimum for a given model. This approach does not lead however to an essential reduction of the complexity in the sense that quotient k remains unchanged, thus it follows the trend of the fuzzy rule based approach compared to the symbolic – quantized rule base method.

The starting point of reducing T further is at the semantic interpretation of fuzzy rules. An important step towards more thorough understanding of fuzzy rules of the type **If** $\boldsymbol{x}$ **is** $\tilde{A}$ **then** $\boldsymbol{y}$ **is** $\tilde{B}$ can be found in [4], where the meaning is interpreted as "The more $\boldsymbol{x}$ is $\tilde{A}$, the more is $\boldsymbol{y}$ $\tilde{B}$". This interpretation assumes that the fuzzy labels $\tilde{A}$ and $\tilde{B}$ represent properties that have *graduality*, assuming that the relative degree of any $\tilde{A}'$ being similar to the "ideal" $\tilde{A}$ might be expressed as a grade, changing from "not at all" to "absolutely true", through "little", more or less", "very much", etc. Another interesting idea was proposed in [23] where a formal reasoning scheme allowed obtaining a conclusion by calculating the "distance", or dissimilarity between $\tilde{A}$ and $\tilde{A}'$, a single value $d\left(\tilde{A},\tilde{A}'\right)$. An even more algorithmic approach was proposed in [3] where a graphic construction method was shown for drawing the membership function of the conclusion B^* point by point from the membership functions of the observation A^*, and the antecedent and consequent, $\tilde{A}$ and $\tilde{B}$, resp. These approaches allowed relaxing the rule base in the sense that condition $\alpha \geq 0.5$ in (46.10) and (46.11) was removed and only $\alpha \geq \varepsilon > 0$ was requested, thus having a less dense cover of the input space by the antecedents. There is a common weak point of these approaches still: the condition of having at least some overlap between observation and antecedent, i.e. the fact that the space should be covered to a positive degree in any case.

In 1988 with K. Hirota we suggested to completely put aside the cover condition and we proposed a new family of inference algorithms: *Fuzzy Rule Interpolation.* While our primary motivation was to reduce the complexity of the rule base, to decrease the value of T, as a side effect, the new approach brought an additional advantage, the technical possibility to calculate a conclusion where there was no overlap between observation and any of the antecedents. Although the latter do not cover $\boldsymbol{X}$ in this case, they only "span" it by rules spread out in the whole state space, for rule bases that might be obtained by tuning from dense starting rule bases (dense in the sense of the antecedents forming an $\alpha \geq \varepsilon > 0$ cover) [2] conclusions could be still calculated.

The first such inference method was called *Linear Fuzzy Rule Interpolation* and it was based on the Fundamental Equation of Rule Interpolation (FERI), an equation deducted from the idea of gradual rule interpretation and extending the idea of analogical reasoning and revision principle based towards a general reasoning technique applicable for almost arbitrarily located rules – whenever the rules are spanning the space in a general Shannon sense. The basic approach and minimal necessary conditions for its applicability (the presence of metrics and at least partial ordering in the state space) were introduced and discussed in some detail in papers [8–10].

Soon it was shown that under certain rather general conditions there is no need to calculate the conclusion point by point: if both the original rule base and the observation are piecewise linear (e.g. trapezoidal), it is sufficient to calculate for the breakpoints only, i.e., no essential increase of computational complexity is brought in by the α-cut based approach, rather there is a decrease of complexity, because here the number of necessary antecedents and thus rules in the base can be decreased to a theoretical minimum that still contains sufficient information on the model. Some analyses of the conditions can be found in [14, 17, 18]. What is the essential improvement of the rule interpolation algorithms compared to the basic CRI- and Mamdani-algorithms? The complexity of the rule base and thus of the reasoning algorithm is now

$$R \leq O\left(T''^{k}\right), \text{ where } T'' << T' << T. \tag{46.15}$$

How sparse might the rule base be if interpolative reasoning is applied? At present no analytical answer to this question is available, thus the next open problem should be investigated in the future in order to firmly establish the exact mathematical foundations of fuzzy rule interpolation applied to rule based models.

Open Problem 1. Let extend the Shannon Theorem to fuzzy rules and rule based models representing fuzzy functions (mappings). This means the (local and global) denseness of the antecedents covering the input state space, necessary for the (maybe ε-good) reconstruction of the original fuzzy mapping in the form

$$\tilde{F} : \tilde{P}(\boldsymbol{X}) \to \tilde{P}(\boldsymbol{Y}) \tag{46.16}$$

where $\tilde{P}(\boldsymbol{X})$ and $\tilde{P}(\boldsymbol{Y})$ denote fuzzy power sets, i.e. the sets of all fuzzy subsets of the respective universes (maybe satisfying certain conditions, such as normality, convexity, or trapezoidal shape). The meaning of (46.16) is that for any observation $A^*\tilde{F}$ should deliver an unambiguous fuzzy conclusion $B^{*'} = \tilde{F}(A^*)$ correctly, so that

$$\forall y : \left| \mu\left(B^{*'}, y\right) - \mu\left(B^*, y\right) \right| \leq \varepsilon \in (0,1] \tag{46.17}$$

Some experimental results on examples showed that reconstruction accuracy is definitely not a monotonic function of the denseness of the rule base applied. In [12] I also showed that by using rule interpolation models it was possible to transform Mamdani-like and Takagi-Sugeno-like models into one another, even, these two were equivalent in limit. These two basic fuzzy rule based models offer equally good solutions for the "key problem".

After the initial fuzzy rule interpolation algorithms a series of new more efficient algorithms were proposed. In [1] a very generally applicable approach was given, where neither normality, nor convexity of the membership functions in the rules was requested. In [25] a transformation was proposed that allowed a very general approximation technique in the transformed space, including extrapolation. A good practical approach for general interpolation and extrapolation methods was proposed in [5] (the paper receiving an IEEE CIS Best Paper Award) and later, by the same author a new adaptive approach was given [26]. The original linear interpolation was extended by our team to multiple dimensions where we succeeded to prove the mathematical stability of the method, a very important step towards establishing a "tool kit" for solving the Key problem in the sense that stability guarantees low sensitivity of the model against noise and imprecision in the available measured or observed data, forming the starting base of the rule type model to be established [22]. This technique, by the way, might be considered a very efficient general interpolation technique, applicable for any problem, independently from fuzzy rule based modeling.

46.2.4 Going Beyond Decreasing T: Hierarchical Fuzzy Models and Interpolative Hierarchy

Even though using fuzzy rule interpolation might help with reducing T in (46.5), the problem of exponentiality is not eliminated, nor is the exponent decreased. While it is very probable that such problems can never be reduced to a complexity less than exponential, the decrease of k might essentially influence the practical computability and real time applicability of a given rule based model.

The first explicit and successful attempt to do so was done by Sugeno, who studied the behavior and reasoning strategy of a number of helicopter pilots and constructed a model that corresponded to the general conclusions drawn from his interviews: The main point being helicopter pilots always deciding first which of the numerous parameters and indicators should be used in a *particular situation or maneuver*, this way reducing the effective number of variables to deal with *at the same time*. The first successful helicopter autopilot experiments applying *hierarchical fuzzy rule* system were done by Sugeno [19].

In 1993 we attempted to formalize and extend the idea applied in Sugeno's approach [11]. We intended to find a model in the general case where no subspace dividing the unstructured model into a set of structured sub-models could be identified. The idea of this model is as follows: $\boldsymbol{X} = X_1 \times ... \times X_k$, where there is a Z_0 such that $\mathbf{Z_0} = X_1 \times ... \times X_{k_0}$, $k_0 < k$. Let Π be a partition in $\mathbf{Z_0}$ such that $\Pi = \{D_1, D_2, \ldots, D_s\}, \bigcup_{i=1}^{s} D_i = \Pi$, and let further R_i be sub-rule bases so that each R_i is valid in and only in D_1, for $i = 1, \ldots, s$. Then in $\mathbf{Z_0}$ we have the meta-rule base

$$\begin{aligned} R_0 : &\textbf{If } \boldsymbol{z} \textbf{ is } \tilde{A}_1 \textbf{ then } \boldsymbol{y} \textbf{ is } R_1 \\ &\ldots \\ &\textbf{If } \boldsymbol{z} \textbf{ is } \tilde{A}_r \textbf{ then } \boldsymbol{y} \textbf{ is } R_s. \end{aligned} \tag{46.18}$$

with symbolic conclusions R_i ,where R_i stands for sub-rule base i, $\boldsymbol{z} \in \mathbf{Z_0}$ and

$$\begin{aligned} \forall i : R_i : &\textbf{If } \boldsymbol{x_{i1}} \textbf{ is } \tilde{A}_{i1} \textbf{ then } \boldsymbol{y} \textbf{ is } \tilde{B}_{i1} \\ &\ldots \\ &\textbf{If } \boldsymbol{x_{iri}} \textbf{ is } \tilde{A}_{iki} \textbf{ then } \boldsymbol{y} \textbf{ is } \tilde{B}_{iri}. \end{aligned} \tag{46.19}$$

$\forall i : \{\boldsymbol{x_i}\} = \boldsymbol{X_i}$, containing $k_i < k - k_0$ variables from the set $\{k_{0+1} \ldots k\}$, the subspace $\boldsymbol{X_i}$ being k_i-dimensional, so that $k_0 + k_i < k$.

In most real cases however $\bigcup_{i=1}^{s} D_i \subset \Pi$ holds, D_i forming a proper subset of $\mathbf{Z_0}$, rather than a real partition. The validity of any R_i is *gradually* fading away when getting further from the set . Thus Π is just forming a fuzzy cover of $\mathbf{Z_0}$, so that in the area $\mathbf{Z_0} - \bigcup_{i=1}^{s} D_i$ there is no valid R_i. For such systems we introduced *hierarchical fuzzy interpolation* [13] with a new algorithm *interpolating sub-rule bases* rather than rules. In this algorithm the projection of every observation A^* is decomposed into orthogonal projections $A_0^* = A^* \perp \mathbf{Z_0}$ and $\forall i : A_i^* = A^* \perp \boldsymbol{X_i}$, further

$$\boldsymbol{X_r} = \lceil \boldsymbol{X_1} \times \ldots \times \boldsymbol{X_{k-k_0}} \rceil, \quad \mathbf{Z_0} \times \boldsymbol{X_r} = \boldsymbol{X}$$

$\lceil \Xi \rceil$ denoting the smallest containing superset of all elements in $\boldsymbol{\Xi}$. Then each sub-rule base should be separately evaluated using A_i^* as observation, thus obtaining sub-conclusions B_i^* from each sub-rule base (46.19), and then in (46.18) each symbol R_1 should be substituted by its respective sub-conclusion B_i^*, thus obtaining a fuzzy rule base

$$
\begin{aligned}
R_0 : \textbf{If } z \textbf{ is } \tilde{A}_1 \textbf{ then } y \textbf{ is } \tilde{B}_1^* \\
\dots \\
\textbf{If } z \textbf{ is } \tilde{A}_r \textbf{ then } y \textbf{ is } B_s^*.
\end{aligned} \tag{46.20}
$$

and applying the inference engine for A_0^* (interpolation, or CRI-derivative), obtaining B^* as resulting conclusion. By evaluating (46.19) using the orthogonal component of the observation, each sub-rule base is properly weighted in the final conclusion.

What is the advantage from the point of view of the computational complexity? Because in each R_i only $k_i < k - k_0$ variables are used, if $max\{k_i\} = K < k - k_0$, the resulting number of variables used in any reasoning cycle is $k' < K + k_0$, and thus the overall complexity of the new algorithm is

$$R \leq O(T''^{k'}) < \text{ or in “lucky” cases even } << O(T''^{k}) < O(T^k) \tag{46.21}$$

Thus in (46.5) the complexity has decreased essentially, by reducing not only the value of T but also of k! Applying hierarchical (and interpolative) models offers a further step towards having a good balance between computational cost and model adequacy.

It remains an open question whether such essential reduction is possible at all, and if yes, whether it changes the “goodness” or adequacy of the original model so that the solution thus offered is not an acceptable solution any more. This can be put in another form:

Open Problem 2. Under what condition can a suitable fuzzy cover $\bigcup_{i=1}^{s} D_i \subset \Pi$ always be determined so that the resulting hierarchical and interpolative fuzzy model is acceptable?

46.3 Conclusions and Further Work

It was shown that the evolution of production rule systems received an essential “push” forward when Zadeh introduced fuzzy sets and later fuzzy rule bases with CRI reasoning. In our past work we tried to continue this trend and proposed interpolative and hierarchical interpolative fuzzy systems. All these steps contributed to having better and better solutions of the Key problem of Engineering, i.e. better and better approximate models and cost efficient algorithms.

In our further work we investigated the application of clustering, and various evolutionary and memetic algorithms for determining “as good as possible” but cost efficient fuzzy rules from input-output data.

Acknowledgement. This paper was supported by the National Scientific Research Fund Grant OTKA K75711, a Széchenyi István University Main Research Direction Grant and TÁMOP 421 B.

Fig. 46.2. At the Third World Automation Congress (WAC 1998) in Anchorage, Alaska, 1998: fltr: Peter Varlaki, Annamaria Koczy Varkonyi, Laszlo Koczy, Lotfi Zadeh and Antal Penninger.

References

1. Baranyi, P., Koczy, L.T., Gedeon, T.D.: A Generalized Concept for Fuzzy Rule Interpolation. IEEE Transactions on Fuzzy Systems 12(6), 820–832 (2004)
2. Burkhardt, D.G., Bonissone, P.P.P.: Automated Fuzzy Knowledge Base Generation and Tuning. In: Proceedings of the FUZZ-IEEE 1992 – First IEEE International Conference Fuzzy Systems, San Diego, CA, pp. 179–196 (1992)
3. Ding, L., Shen, Z., Mukaidono, M.: Principle for Approximate Reasoning-based on Revision Principle. In: Fuzzy Logic Systems Institute (publ.): Proceedings of the 2nd International Conference on Fuzzy Logic and Neural Networks, IIZUKA 1992, Iizuka, Fukuoka, Japan, pp. 305–308 (1992)
4. Dubois, D., Prade, H.: Gradual Rules in Approximate Reasoning. Information Sciences 61(1-2), 103–122 (1992)
5. Huang, Z., Shen, Q.: Fuzzy Interpolation and Extrapolation: A Practical Approach. IEEE Transactions on Fuzzy Systems 16(1), 13–28 (2008)
6. Klahr, D., Langley, P.W., Neches, R.T.: Production System Models of Learning and Development. The MIT Press, Cambridge (1987)
7. Novák, V.: Fuzzy Sets and Their Applications. Adam Hilger, Bristol (1989)
8. Koczy, L.T., Hirota, K.: Ordering, Distance and Closeness of Fuzzy Sets. Fuzzy Sets and Systems 59(3), 281–293 (1993)

9. Koczy, L.T., Hirota, K.: Approximate Reasoning by Linear Rule Interpolation and General Approximation. International Journal of Approximate Reasoning 9(3), 197–225 (1993)
10. Koczy, L.T., Hirota, K.: Interpolative Reasoning with Insufficient Evidence in Sparse Fuzzy Rule Bases. Information Sciences 71(1-2), 169–201 (1993)
11. Koczy, L.T., Hirota, K.: Approximative Inference in Hierarchical Structured Rule Bases. In: 5th IFSA World Congress (IFSA 1993), Seoul, pp. 1262–1265 (1993)
12. Koczy, L.T.: Fuzzy if...then rule models and their transformation into one another. IEEE Transactions on Systems, Man, and Cybernetics, Part A: Systems and Humans 26(5), 621–637 (1996)
13. Koczy, L.T., Hirota, K., Muresan, L.: Interpolation in Hierarchical Fuzzy Rule Bases. International Journal of Intelligent Fuzzy Systems 1(2), 77–84 (1999)
14. Kovács, S., Koczy, L.T.: Shape of Fuzzy Conclusion Generated by Linear Interpolation of Trapezoidal if...then... Rules, vol. 6, pp. 83–93. Tatra Mountains Mathematical Publications (1995)
15. Mamdani, E.A., Assilian, S.: An Experiment in Linguistic Synthesis with a Fuzzy Logic Controller. International Journal of Man-Machine Studies 7(1), 1–13 (1975)
16. Marks II, R.J.: Introduction to Shannon Sampling and Interpolation Theory. Springer, New York (1991)
17. Qiao, W.Z., Mizumoto, M., Shi, Y.: An Improvement to Koczy and Hirota's Interpolative Reasoning in Sparse Fuzzy Rule Bases. International Journal of Approximate Reasoning 15, 195–201 (1996)
18. Shi, Y., Mizumoto, M.: On Koczy's Interpolative Reasoning Method in Sparse Fuzzy Rule Bases. In: Proceedings of the 10th Fuzzy Systems Symposium, Osaka, Japan, pp. 221–224 (1994)
19. Sugeno, M., Griffin, M.F., Bastian, A.: Fuzzy Hierarchical Control of an Unmanned Helicopter. In: Proceedings of the 5th IFSA World Congress, Seoul, Korea, pp. 179–182 (1993)
20. Sugeno, M., Yasukawa, T.: A Fuzzy-logic-based Approach to Qualitative Modeling. IEEE Transactions on Fuzzy Systems 1(1), 7–31 (1993)
21. Takagi, T., Sugeno, M.: Fuzzy Identification of Systems and its Applications to Modeling and Control. IEEE Transactions on Systems, Man, and Cybernetics SMC-15(1), 116–132 (1985)
22. Tikk, D., Joó, I., Koczy, L.T., Várlaki, P., Moser, B., Gedeon, T.D.: Stability of Interpolative Fuzzy KH Controllers. Fuzzy Sets and Systems 125(1), 105–119 (2002)
23. Türksen, I.B., Zhao, Z.-Q.: An Approximate Analogical Reasoning Approach Based on Similarity Measures. IEEE Transactions on Systems, Man, and Cybernetics 18(6), 1049–1056 (1988)
24. Waterman, D.A., Hayes-Roth, F.: Pattern-Directed Inference Systems. Academic Press, New York (1978)
25. Yam, Y., Koczy, L.T.: Representing Membership Functions as Points in High-dimensional Spaces for Fuzzy Interpolation and Extrapolation. IEEE Transactions on Fuzzy Systems 8(6), 761–772 (2000)
26. Yang, L., Shen, Q.: Adaptive Fuzzy Interpolation. IEEE Transactions on Fuzzy Systems 19(6), 1107–1126 (2011)
27. Zadeh, L.A.: Fuzzy Sets. Information and Control 8(3), 338–353 (1965)
28. Zadeh, L.A.: Outline of a New Approach to the Analysis of Complex Systems and Decision Processes. IEEE Transactions on Systems, Man, and Cybernetics SMC-3(1), 28–44 (1973)

47

Quest for Rigorous Combining Probabilistic and Fuzzy Logic Approaches for Computing with Words

Boris Kovalerchuk

47.1 Introduction

Lotfi Zadeh initiated three fundamental concepts: (1) the concept of a *linguistic variable*, (2) the concept of a *fuzzy set* (with a *membership function* (MF) for linguistic terms that gradually changes between 0 and 1), and (3) the concept a *matrix of linguistic rules* that connect linguistic variables. These concepts are outside of the main stream of concepts used in both the probability theory and the control theory. These concepts are critical for the area that Zadeh later denoted as Computing with Words (CWW) [17]. The *elegance* and *intuitiveness* of these three concepts deeply impressed me when I learned about them a long time ago. This stimulated my interest to contribute to this field. First I noticed from Zadeh's initial work on linguistic variables [15] that operations with fuzzy sets such as min, max, product and others were introduced just as illustrative examples/prototypes of *possible operations* with fuzzy sets. More work was needed to define operations that will be appropriate for CWW. I also noticed that people started to use these "sample" operations without much justification and critical analysis. Next I was impressed by the work of the Zimmermann's team [12], [21] were they analyzed appropriateness of these operations experimentally for linguistic terms metallic and container in German. Later on we conducted a similar experiment [4] in Russian for the same terms, but with different objects. Our results confirmed Zimmermann's negative results in spite of significant differences between these languages in the abilities to create new words by "concatenating" two words.

Using data in ([21], table 6.2) we also computed the average difference of 23.7% between humans' answers for the object to be a metallic container and the min(x,y) operation value. Here, x is evaluation for the object to be metallic and y to be a container asked separately. In the computation of 23.7% we removed outliers; the difference is greater, 37.76%, if outliers are not removed. For product operation x·y the difference was even higher. Such negative results led to developments of a large collection of other And operations that include compensatory operations [21] that exploited multiple t-norms.

This made fuzzy logic very different from the probability theory with only one And operation, $P(x\&y) = P(y/x) \cdot P(x)$. In fuzzy logic we need to pick up and justify an operation before using it. This is a difficult task and later a way around was

R. Seising et al. (Eds.): *On Fuzziness: Volume 1*, STUDFUZZ 298, pp. 325–336.
DOI: 10.1007/978-3-642-35641-4_47

found for fuzzy control. The *conceptual justification of the operation* was substituted by *tuning parameters* of the operation and values $x = m_1(a)$ and $y = m_2(a)$ of membership functions using training data, neural networks and other machine learning methods.

In natural language (NL) CWW getting *training data* is much more difficult especially due to less certainty about the *context* of the NL statements. Therefore, the progress in NL CWW was not as impressive as in fuzzy control.

Lotfi Zadeh posed a set of CWW test tasks and asked whether probability theory can solve these tasks. The recent discussion on relations between fuzzy logic and probability theory for CWW started at the BISC group with a "naïve" question from a student: *"What is the difference between fuzzy and probability?"* Numerous previous debates can be found in the literature (e.g., [1]; [16]. It continued at the uncertainty panel that B. Bouchon-Meunier organized at the World Conference on Soft Computing (WCSC) in May 2011 in San Francisco with L. Zadeh, B. Widrow, J. Kacprzyk, B. Kovalerchuk, and L. Perlovsky as panelists.

47.1.1 Context Challenges

Probability Theory, Fuzzy Logic, Dempster-Shafer theory, Rough Sets are oriented to somewhat *different contexts*. However, the appropriate context for a given application often is not clearly formulated, and thus it is very difficult to (*a priori*) select one of the approaches in favor of another. Fuzzy membership functions often are produced (measured) by using frequencies which is a probabilistic way to get MFs. The user of these MFs should get an answer for the question: "Why should we use T-norms and T-conorms with these 'probabilistic' MFs instead of probabilistic operations?" The same question is important from the theoretical viewpoint.

47.2 In What Sense Is Probability Theory Insufficient?

47.2.1 Extreme Positions and Real Challenges

The extreme position known in the fuzzy logic community is expressed by Von Altrock [13]. He stated that *lexical uncertainty* deals with the uncertainty of the *definition of the event* itself and that the probability theory *cannot* be used to model this, as the combination of subjective categories in human decision processes does not follow its axioms. Opposite positions are also exist for a long time [e.g., [1]. Table 47.1 summarizes both extreme views.

Another popular argument that CWW requires a conceptual framework that differs from the probabilistic framework is based on the differences in the nature of stochastic and lexical uncertainties [13]. However, counterexamples exist: water and air have different nature, but hydrodynamics and aerodynamics are modeled by very similar mathematical models.

Fig. 47.1. Lotfi Zadeh at the Computational Intelligence Conference in Honolulu, 08.17.2009. He was the keynote speaker on CWW invited by the author who was the Conference chair.

Zadeh [18], [19] (BISC 03/29/2012) stated *insufficiency of only the standard probability theory* (PT) to deal with CWW. The standard PT is defined by what is found "in textbooks and taught in the classroom," which is much smaller than the whole scope of the PT. This is an important conceptual difference. From our viewpoint this is not a claim of fundamental insufficiency of PT for CWW. It is a claim that PT models for CWW are not developed without claiming that they cannot be developed within probabilistic framework. We have at least three flavors of probability: (1) frequency-based, (2) subjective, and (2) axiomatic (Kolmogorov' axioms). The last one abstracts the first two interpretations and there maybe others yet unknown. Thus, thinking about probabilities only as frequencies of repeating events is a very narrow view of probability that should be avoided.

Table 47.1. Comparison of Extreme Probabilistic and Fuzzy Logic positions

Probabilistic Position	*Fuzzy Logic Position*
All kinds of uncertainty can be expressed with probability theory.	Stochastic and lexical uncertainties have different nature and require different mathematical models.
Probability theory can model stochastic uncertainty, that a certain event will take place.	Probability theory can model only stochastic uncertainty, that a certain event will take place.
Probability theory can model lexical uncertainty with the uncertainty of the definition of the event itself.	Probability theory cannot model lexical uncertainty with the uncertainty of the definition of the event itself.
Combination of subjective categories in human decision processes does not follow axioms of fuzzy logic theory.	Combination of subjective categories in human decision processes does not follow axioms of probability theory.

To fill deficiencies of the standard PT Zadeh proposed a *Perception-based Probability Theory* and *Generalized Theory of Uncertainty* (GTU) [18],[20], (BISC 03/29/2012) with solutions for CWW including that are consistent with the probabilistic framework. He also stated that he has not attempted to construct an axiomatic approach to GTU, believing that "it will be very hard, perhaps impossible, to do it". To support this statement Zadeh referenced his *Impossibility Principle*: "The closer you get to reality the more difficult it becomes to reconcile the quest for relevance and applicability with the quest for rigor and precision."

From my viewpoint the situation is not so hopeless. In several our works it was shown that scientific rigor, relevance, and applicability are reachable when Zadeh's linguistic variables, membership functions and probabilities combined in what we call a *linguistic context space* [5], [6], [8]. That creates a rigorous base for combination of fuzzy logic and probability concepts. The application of this approach is shown in the next section on one of Zadeh's test tasks.

Another popular idea is that fuzzy sets and probabilities are *complementary*. I fully agree with this, but for the reasons that differ from just pointing to success of applications where fuzzy sets and probability combined. The success in application is a reason to take a deeper look to discover the reason of success, but success itself is not sufficient to claim complementarity. We actually took a look at successes of fuzzy control and discovered compelling reasons for success that involves implicit use of probability spaces along with Zadeh's linguistic variables fuzzy sets, and interpolation [7].

47.2.2 Probability vs. Possibility

"What is missing in standard probability theory is the concept of possibility....The absence of this concept *limits* the problem-solving capability of standard probability theory" [20] (04/16/2012, 04/19/2012). Below we attempt to clarify the issue of limitation. Consider a midsize car with five seats.

Question 1: What is the *probability* that 9 people *are* in a midsize car? It is very low, say less than 0.01. To get this answer we watch midsize cars coming to the parking lot and count the number of people in each coming car during the day.

Question 2: What is the *possibility* that 9 people *are* in a midsize car? It is very high, say 0.95. To get this answer we can imagine that four people are sitting on the laps of four others excluding the driver or we can actually seat 9 people as described. This can happen in the case of emergency to be able to escape from a dangerous flood place. Another way to support 0.95 is to notice that Guinness World Record 2011 is 27 people in the 4-seat Mini car. It is obvious if 27 is possible then 9 is easily possible too with much higher possibility.

It seems that this example confirms Zadeh's statement that probability and possibility are different. Now consider another question:

Question 3: What is the *probability* that 9 people *can* be in a midsize car? It is very high, say 0.95. To get 0.95 we can select first randomly 100 people. Next we select 9 people from these 100 people randomly and test if they can sit in the car on laps of each other in the car. Then we repeat this random selection of 9 people multiple times and compute the frequency of success of putting 9 people in the car. Why do we expect that this result will produce a number close to 0.95? We assume that the percent of big people that will have difficulties to sit on the laps in the population is relatively small, say no more than 5%. Next the probability to pick up randomly 9 big people at the same time is small again. More accurate estimates would require knowing the actual share of big people in the whole population.

Now we ask the following questions. Is probability 0.95 as an answer for the Question 3 actually answering about the *possibility* of 9 people in the car? Do we answer Question 2 in this way? Is Question 3 within the probabilistic framework?

It seems that the answers for all these questions are positive, while the standard probability textbooks do not talk about questions like Question 3 as Zadeh pointed out. Note that Question 3 is about *probability of the modal statement* with word "can". In the 1980s, *P.* Cheeseman [1] discussed the probability *P* on statements that include the word possible: P(It is possible to put n passengers into Carole's car). It seems that the probability theory of such modal statements is not developed yet, while there are a few works on modal probability logic. In essence such a new theory would be in the same second-order realm as probabilities on probabilities, fuzzy sets on fuzzy sets, probabilities on possibilities, and possibilities on probabilities. Note that Question 3 is on *probabilities on possibilities* and is formulated in the probabilistic framework that satisfies Kolmogorov's axioms. This conversion of possibilistic Question 2 to Question 3 that is in the probabilistic framework shows that *translation between possibilistic and probabilistic languages is possible*. A particular language can be more convenient, more compact, more intuitive, faster to obtain, etc. However both languages should allow producing the same result with the same rigor. While there is a still active discussion about limits of modeling possibilities by the probability theory it will be good to generate more pro and con examples at this stage of discussion to avoid overgeneralized claims.

47.2.3 Mutual Exclusion

There is a popular idea voiced at BICS and multiple publications that concepts like old, young, short, and tall are imprecise overlapping concepts and, therefore need to be modeled by using fuzzy set membership functions not with the probability theory that deals with crisp disjoint (mutually exclusive) elementary events, e.g., die sides. This justification is incomplete and leads to the conceptual difficulties. It does not tell us how to construct these membership functions. A common way to get MF's values is using frequencies of subjective human answers. The fuzzy logic literature is full of such frequencies for computing MFs for ages Young, Old, Middle Age etc, e.g., [2]. This is a probabilistic way to get MFs, which contradicts the idea that PT fundamentally cannot capture such uncertain concepts.

47.2.4 Probability Theory and Linguistic Uncertainty

Table 47.1 contains a statement from the extreme fuzzy logic position: the probability theory can model *only* stochastic uncertainty that the event will take place, but *cannot* model lexical uncertainty of the definition of the *event*. While PT has an origin in stochastic not linguistic uncertainty as a theory of chances and frequencies back in the 18th century, after A. Kolmogorov published an axiomatic probability theory in 1933, the probability theory moved onto much more abstract level. In this axiomatic theory, elementary events can be elements of *any nature* from sides of dice to *words* such as young and old viewed just as *labels*. Note that word "young" differs from an uncertain real-world concept of being young. It seems that equating a word and an uncertain real-world concept often a reason of the claim that the mutual exclusion axiom of probability prevents modeling such concepts. It can be done via labeling [3].

We also need to have in mind that PT has two very different parts: *abstract* PT as a part of mathematical measure theory and *mathematical statistics* as an area dealing with stochastic uncertainty in the real world. Mathematical statistics matches stochastic uncertainties with abstract PT, but it does not prohibit *matching other linguistic* and *subjective uncertainties* with the abstract PT. This is done with the development of *subjective PT*, e.g., [14] and works on rigorous combination of *probabilistic concepts with linguistic variables* [3], [6], [8] inspired by a very productive concept of linguistic variables developed by Zadeh [15]. The last approach focuses on *formalizing contexts of lexical uncertainty*.

Zadeh's [18] work in CWW produced generic rules that can be specialized with possibilistic constraints and lead to possibility theory, probabilistic constraints that lead to probability theory; and random-set constraints that lead to the Dempster-Shafer theory of evidence. His perception-based theory of probabilistic reasoning with imprecise probabilities deals with tasks such as: given the perception: Usually Robert returns from work at about 6 p.m.; the question is: What is the probability that he is home at 6:30 p.m.?

47.2.5 Probability and Partial Truth

Another suggested dividing line between fuzzy logic and probability theory is a statement that PT cannot model partial truth. Zadeh offered the following example for the consideration [18], [20]: "Suppose that Robert is three-quarters German and one-quarter French. If he were characterized as German, the characterization would be *imprecise*, but *not uncertain*. Equivalently, if Robert stated that he is German, his statement would be *partially true*; more specifically, its truth value would be 0.75. Again, 0.75 has no relation to probability."

If we interpret "probability" in last statement as a *common natural language word* then this statement is very consistent with our understanding of this word. However, if we interpret it as a *term of the formal mathematical probability theory* then we may notice that 0.75 can be interpreted as probability because the mathematical term probability has a wider meaning. The axiomatic formal PT is special case of the mathematical measure theory, where 0.75 is just the value of the measure that may have multiple ways to get it. Moreover these ways are outside of the axiomatic theory.

47.3 Linguistic Context Space and Multiplicity of Solutions of Zadeh's Test Problems

Zadeh [18], [20] formulated a set of CWW test problems that include the following problems: (1) Usually Robert returns from work at about 6 p.m. What is the probability that he is home at 6:30 p.m.? (2) Probably John is tall. What is the probability that John is short? (3) What is the probability that my car may be stolen? (4) How long does it take to get from the hotel to the airport by taxi?

Zadeh's approach to such problems [18] is (i) representing uncertain concepts listed in the problem using fuzzy sets and/or probabilities, and then (ii) using these representations to the answers. In (1) uncertain concepts are "about" and "usually". In (2) uncertain concepts are "tall", "short", and "probably".

The *linguistic context space* method formally defined in [5]; [6] adds the *context* of uncertain concepts. Adding context is important because context can change the answer. In (2) the context sets up frameworks for the scope of words tall and short. In (2) and many other NL tasks context C is not expressed explicitly while it plays an important role to derive a conclusion. Do we have in mind only tall and short alternatives only or a wider context with more alternatives? In (2) it is intuitively clear that the answer "Probably John is short" is not correct. Thus, other words are needed to express the probability for John to be short. These words could be highly unlikely, more or less unlikely, fifty-fifty, probable, highly probable, or many others. The choice of words can change the answer, but it is not derivable from (2). It depends on the context.

Zadeh [19] (BISC, 8.17.2011) proposed three versions of this problem. "Given: *Probably John is tall*. Version 1: *What is the probability that John is short?* Version 2: What is the probability that John is *very short*? Version 3: What is the probability that John is *not very tall*? What can be assumed is that the imprecise terms tall, short, very short, not very tall and probable are labels of *fuzzy sets* with specified *membership functions*. Alternatively, the terms may be assumed to be labels of specified *probability distributions*. The answer should be a fuzzy probability."

Below we focus on version 1 of problem (2). Consider several sets of linguistic terms (linguistic variables) a part of different contexts:

Set $PJ1$: {improbable, probable}, or {unlikely, probable};
Set $PJ2$: {false, unlikely, probable, true};
Set $PJ3$: {false, possible, probable, true};
Set $PJ4$: {false, highly unlikely, more or less unlikely, fifty-fifty, probable, highly probable, true};
Set $H1$: {short, tall};
Set $H2$: {very short, short, medium, tall, very tall}.

The choice of these sets can change the solution. For context C_1 with sets unlikely, probable and short, tall a common sense answer is: Unlikely John is short (in context C_1). In a variation of this context the answer can be Improbable that John is short (in context C_1). For context C_2 with sets unlikely, probable and very short, short, medium, tall, very tall, a common sense answer is the same: It is unlikely that John is short (in C_2). Extra terms in the linguistic variable for the height did not chance the answer while they have expanded the context. For context E_3 with set PJ4 that contains terms "highly unlikely" and "more or less unlikely" and set short, tall we have *two alternative answers*: It is highly unlikely that John is short and it is more or less unlikely that John is short (in E_3).

How to get these results computationally? Assume that John's height is 180 cm with the probability $p(\text{tall}, 180) = 0.8$ and the probability that he is short $p(\text{short}, 180) = 0.2$. A voting experiment could give these numbers. Say, 100 people vote whether a man of height 180 cm is tall or short. Alternatively, one person can give these numbers as personal subjective probabilities. In Kolmogorov's terms here we have a probability space with two elementary events $\{\text{short } (180), \text{tall } (180)\}$ and probabilities of these elementary events $p(\text{short } (180)) = 0.8$ and $p(\text{tall } (180)) = 0.2$. Here the mutual exclusion follows from the fact that words short and tall are different.

The probability is defined on *linguistic labels* (short, tall) that are distinct, not on *the natural language concepts of short and tall people* that *have no sharp border*. For any other height h we can construct similar sets of elementary events $\{\text{short(h)}, \text{tall(h)}\}$ with probabilities of elementary events $p(\text{short}(h)) = x$ and $p(\text{tall}(h)) = 1 - x$. This is a very important distinction that we build a Kolmogorov's probability space on labels not fuzzy concepts "short" and "tall". As was already pointed out above the common critic of probability concept in the fuzzy logic community is that probability cannot be defined on the overlapping fuzzy concepts such as "short" and "tall". As we show it is not required to be able to solve this test problem. It is sufficient to build a probability space on a set of labels. Labels as different words are distinct and mutually exclusive.

Figure 47.2 (left) shows these probability spaces for each height h. This figure resembles a set of triangular membership functions commonly used to represent fuzzy linguistic variables. A probability space $S(180)$ is shown as a pair of circles on a vertical line at point h=180.As we can see from this figure, *just two membership functions serve as a compact representation of many simple probability spaces* described above. This is a *fundamental representational advantage of Zadeh's fuzzy linguistic variables vs. multiple small probability spaces*. In other words, *few fuzzy membership functions in a linguistic variable provide a quick way to build a huge set of simple probability spaces*. In this sense, fuzzy membership functions and probabilities are complimentary not contradictory. Thus, they are mutually beneficial by combining *fast model development and rigor*. More details are in [6].

Now we use a set $\{\text{unlikely}, \text{probable}\}$ and build a set of elementary events $\{\text{unlikely}(0.8), \text{probable}(0.8)\}$ for probability value 0.8 with, say, $P(\text{probable}, 0.8)$

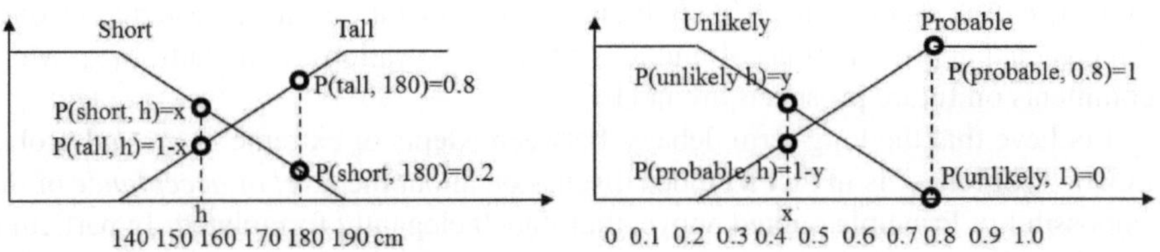

Fig. 47.2. Left: Sets of probability spaces $S(h)$ for elementary events $\{\text{short}(h), \text{tall(h)}\}$ for each height h. Right: Linguistic probabilities *unlikely*, *probable*.

$= 1$ and $P(\text{unlikely}, 0.8) = 0$. Other values of probabilities have own sets of elementary events, e.g., $\{\text{unlikely}(0.3), \text{probable}(0.3)\}$ with, say, $P(\text{probable}, 0.3) = 0.25, P(\text{unlikely}, 0.3) = 0.75$.

In general we define a probability space $\{\text{unlikely}(x), \text{probable}(x)\}$ with $P(\text{probable}, x) = 1 - y$ and $P(\text{unlikely}, x) = y$. All of these multiple probability spaces are shown visually in Figure 47.2 (right), and are also *compactly* represented by just two membership functions $\mu_{unlikely}(h)$ and $\mu_{probable}(h)$.

Now having $P(\text{tall}, 180) = 0.8$ and $P(\text{probable}, 0.8) = 1$, we convert a numeric $p(\text{tall}, 180) = 0.8$ into a linguistic answer $p(\text{tall}, 180) = \text{probable}$. Formally, it can be done by computing max $\{P(\text{probable}, 0.8), P(\text{unlikely}, 0.8)\}$ to identify a linguistic term that best fits the 0.8. Similarly having $P(\text{short}, 180) = 0.2$ and $P(\text{unlikely}, 0.2) = 1$ we convert a numeric $p(\text{short}, 180) = 0.2$ into a linguistic $p(\text{short}, 180) = \text{unlikely}$.

For a more general case of a set {false, unlikely, fifty-fifty, probable, true} and a set {short, tall} the logic of computations is the same. We will have more probabilities, say, $P(\text{false}, 0.9) = 0$, $P(\text{unlikely}, 0.9) = 0$, $P(\text{fifty-fifty}, 0.9) = 0.2$, $P(\text{probable}, 0.9) = 1$, $P(\text{true}, 0.9) = 0.9$, but with the same linguistic result of computation: it is unlikely that John is short. However the numeric value of this probability will differ from 0.2 obtained for a smaller set {unlikely, probable}.

47.4 Conclusion and Prospects for Future

As was shown in section 3, just two membership functions serve as a *compact representation* of many simple probability spaces. This is a *fundamental representational advantage of Zadeh's fuzzy linguistic variables vs. multiple small probability spaces*. These few (typically 5-7) fuzzy membership functions within a linguistic variable provide a quick way to build hundreds of simple subjective probability spaces. This is a way how fuzzy membership functions and probabilities become *complementary not contradictory* and mutually beneficial by combining *fast model development and rigor*.

There are also other emerging ways to meet the quest for rigor by developing *more general uncertainty theories* such as (1) by incorporating both rational and *irrational agents* that generate uncertain statements [11], and (2) by developing an *operation approximation theory* where fuzzy logic T-norm operations are considered as one-dimensional approximations of multidimensional operations in the lattice [9]. More comments on future prospects are in [10].

I believe that the long-term debates between adepts of extreme fuzzy and probabilistic "churches" is in fact a hidden discussion about the *level of acceptance* of the Impossibility Principle quoted above, that Zadeh elegantly formulated. In particular it is the difference in the *level of scientific rigor* and heuristics that are considered as acceptable to get a result relevant to real world challenges. The probability theory itself has got its rigorous foundation only in 1933 with Kolmogorov's axioms after over 200 years of existence with multiple impressive results and deficiencies. Fuzzy logic is much younger. The history of science has many other examples when new theories reached rigor much later than they produced impressive and useful results along with "results" that were rejected later under more rigorous foundations. Thus, I believe that the productive future in CWW and modeling uncertainty in general is in switching from arguing which extreme position is more wrong to *searching for ways of how to make fuzzy logic and its combination with the probability theory more rigorous, while being still relevant to real world challenges*. To do this we first need to come up to the common concept of what is not rigorous in fuzzy logic. The probability theory would not have gotten a rigorous foundation if (1) its deficiency had not been recognized and (2) the fields of mathematics that it is based on had not been developed before, such as the set theory and the measure theory. I believe that if we follow this constructive path then at some moment the "fight" between the "churches" will be over without any specific effort and the new generalized theories and practice will emerge.

References

1. Cheeseman, P.: Probabilistic vs. Fuzzy Reasoning. In: Kanal, L.N., Lemmer, J.F. (eds.) Uncertainty in Artificial Intelligence, pp. 85–102. Elsevier (North-Holland), Amsterdam (1986)
2. Hall, L.O., Szabo, S., Kandel, A.: On the Derivation of Memberships for Fuzzy Sets in Expert Systems. Information Sciences 40, 39–52 (1986)
3. Hisdal, E.: Logical Structures for Representation of Knowledge and Uncertainty. Springer (Physica), New York (1998)
4. Kovalerchuk, B., Talianski, V.: Comparison of Empirical and Computed Values of Fuzzy Conjunction. Fuzzy Sets and Systems 46, 49–53 (1992)
5. Kovalerchuk, B., Klir, G.J.: Linguistic Context Spaces and Modal Logic for Approximate Reasoning and Fuzzy Probability Comparison. In: Proceedings of the Third International Symposium on Uncertainty Modeling and Analysis and NAFIPS 1995, pp. A23–A28. IEEE Press (1995)
6. Kovalerchuk, B.: Context Spaces as Necessary Frames for Correct Approximate Reasoning. International Journal of General Systems 25(1), 61–80 (1996)
7. Kovalerchuk, B.: Second Interpolation for Fuzzy Control. In: Fifth IEEE International Conference on Fuzzy Systems, New Orleans, pp. 150–155 (1996)
8. Kovalerchuk, B., Vityaev, E.: Data Mining in Finance: Advances in Relational and Hybrid Methods (ch. 7 on fuzzy systems). Kluwer, Boston (2000)
9. Kovalerchuk, B.: Interpretable Fuzzy Systems: Analysis of T-norm interpretability. In: Proceedings of the WCCI World Congress on Computational Intelligence, Barcelona, pp. 20–35 (July 2010)
10. Kovalerchuk, B.: BISC-group posts (2010-2012), `http://mybisc.blogspot.com/`
11. Resconi, G., Kovalerchuk, B.: Agents' Model of uncertainty. Knowledge and Information Systems 18(2), 213–229 (2009)
12. Thole, U., Zimmermann, H.-J., Zysno, P.: On the Suitability of Minimum and Product Operations for the Intersection of Fuzzy Sets. Fuzzy Sets and Systems 2, 173–186 (1979)
13. Von Altrock, C.: Fuzzy Logic and Neuro Fuzzy Applications in Business and Finance. Prentice Hall, New Jersey (1997)
14. Wright, G., Ayton, P. (eds.): Subjective Probability. Wiley, Chichester (1994)

15. Zadeh, L.A.: The Concept of a Linguistic Variable and its Application to Approximate Reasoning. Information Sciences 8, 199–249 (1977)
16. Zadeh, L.A.: Dicussion: Probability Theory and Fuzzy Logic are Complementary Rather than Competitive. Technometrics 37(3), 271–276 (1996)
17. Zadeh, L.A.: From Computing with Numbers to Computing with Words – From Manipulation of Measurements to Manipulation of Perceptions. IEEE Transactions on Circuits Systems 45, 105–119 (1999)
18. Zadeh, L.A.: Toward a Perception-based Theory of Probabilistic Reasoning with Imprecise Probabilities. Journal of Statistical Planning and Inference 105, 233–264 (2002)
19. Zadeh, L.A.: A Note on Z-numbers. Information Sciences 181, 2923–2932 (2011)
20. Zadeh, L.A.: BISC-group posts, 8.17.2011,03/29/2012, 04.16.2012, 04.19.2012, http://mybisc.blogspot.com/
21. Zimmermann, H.-J.: Fuzzy Sets, Decision Making, and Expert Systems. Kluwer-Nijhoff, Boston (1987)

48

In the Beginning Was the Word, and the Word Was Fuzzy

Vladik Kreinovich

48.1 Fuzziness of Our Lives: A Personal Story

The World Is Awesome. The world is immense and complex, it is not easy to understand, not easy to change – but we humans have mastered it reasonably well. Because of the unstoppable progress of human knowledge, we live happier and longer lives, we travel faster, we recover faster from illnesses and accidents. During the millennia of our civilization, great geniuses provided breakthrough insights, and numerous scientists and engineers, geniuses and simply talented, translated these insights into practically useful ideas.

In the history of science, we can track many such insights – e.g., the idea of an atom. The history of ideas is fascinating and complex, but in a nutshell, each idea follows the same basic trajectory: First, we have a vague philosophical idea, then it is transformed into a more precise (but still somewhat vague) idea formulated in the language of natural sciences, and finally, the idea becomes described in the absolutely precise language – language of mathematics.

I have always been fascinated by the two extremal point of this process: the original philosophical insight and the final absolutely precise mathematical model. Because of this fascination, I decided to study Math – with the emphasis on its fundamental applications to science and engineering.

What Are the Main Objectives of Science and Engineering? Our ultimate objective is to improve the world. For that, first, we need to know how the world operates, what will happen if we perform a certain action (or if we do not do anything). Making such predictions is the main objective of natural sciences: physics, biology, etc.

Once we know how the world operates, once we know what are the possible consequences of different actions, of different decisions, we can start deciding which actions, which decisions are the most beneficial. This is the subject of optimization, engineering, decision making, and other related disciplines. To make a meaningful decision, we must know which outcome is more beneficial to us – and which outcome is less beneficial; for complex decision, this is not easy to decide.

Finally, once a general decision is made, once an engineering design is selected, we need to find the details of this design. In other words, we need to translate a general description (e.g., an abstract mathematical description) of the desired decision into an exact sequence of well-defined steps – i.e., into an algorithm.

R. Seising et al. (Eds.): *On Fuzziness: Volume 1*, STUDFUZZ 298, pp. 337–341.
DOI: 10.1007/978-3-642-35641-4_48

Surprisingly, Everything Is Fuzzy. Because of my interests, I started attending three research seminars: a seminar on mathematical aspects of physics and space-time geometry led by Revolt Pimenov, a seminar on decision making and game theory led by Nikolai Vorobiov, and a seminar on algorithmic (constructive) aspects on mathematics and corresponding mathematical logic led by Nikolai Shanin – all three leading Russian researchers in their areas. Since these were seminars organized by the Math department, I expected a lot of mathematical models and proofs, and there were a lot of them. But surprisingly, all three researchers emphasized the extreme importance of informal, vague ideas and of imprecise reasoning.

I was not that surprised that when we describe human decision making or human reasoning, we need to take into account human imprecision. However, I was really surprised to learn that theoretical physicists, even the most mathematically skilled ones, use informal reasoning and intuition to decide which terms in the corresponding complex equations are "small" and can therefore be ignored – without explicitly defining what "small" means. Moreover, physical equations are usually so complex that without such simplifying reasoning, it is not feasible to come up with any solutions. A convincing example comes from the history of General Relativity: a famous mathematician David Hilbert came up, in 1916, with the same equations as Einstein with a delay of only two weeks – *but* all Hilbert had was equations, while Einstein also had approximate solutions, solutions based on informal reasoning, solutions that could be (and were in 1919) experimentally checked.

From Hegel to Zadeh. To tell the truth, I should not have been that surprised, because in the former Soviet Union, we all studied philosophy, and one of the main messages – coming from Hegel, a beloved philosopher of Marx and Lenin – was that the traditional two-valued logic was not always adequate for describing human reasoning. First, real properties are not always absolutely true or absolutely false – they are only true to a degree. Second, human reasoning is dynamic, our opinions change with time, real properties change with time, while the traditional logic is static. This was part of what Hegel called dialectics.

And this was something we hated because it was coming from our brutal communist dictators, dictators who did not hesitate to throw a well-known professor in jail just for reading books published in the West and for expressing their opposition to the regime in private talks. One of such arrested professors was Revolt Pimenov. He got off easily: instead of a long term in a prison hard-labor camp (that he endured in the 1950s), he was sentenced to an internal exile to a far North town. I visited him there, and you know what he talked about? Hegel. Pimenov loved Hegel, he believed that Hegel's vague ideas had great potential. He was not deterred by the fact that Communists loved Hegel: they also loved the music of Tchaikovsky and Beethoven, but they are still great composers – as well as Wagner is a great composer irrespective of the fact that Hitler loved his music.

Coming from Pimenov, a person who was not allowed to leave the town and had to weekly report to the political police, this was convincing. I started reading all this seriously. And then I happened to read some papers by Lotfi Zadeh and realized that this is it, this is – finally – a precise mathematical presentation of the vague ideas about vagueness.

I published this connection in one of my reviews in Zentralblatt für Mathematik – a mathematical review journal. I described this connection as a report in my philosophy class – and not only I got an A+, I – a student of Jewish origin – was invited to a post-graduate program in philosophy of math, an invitation which at that time (of the official Soviet persecution of Jews) was almost unheard of. (This invitation did not work out, by the way :-)

From Theory to Practice. Fuzzy logic became one of my areas of interest. At first, I was mostly interested in mathematical, theoretical aspects of fuzzy techniques. But it so happened that in 1980, after defending my PhD (in space-time geometry), I started working at the Institute of Electrical Measurement Instruments, where we were not only developing theoretical foundations but also helping to solve practical problems related to measurements and measuring systems. When talking to scientists and engineers, we realized that in their practice, in addition to measurement results, they use their intuition, their imprecise knowledge that they cannot express in exact mathematical terms – only in terms of natural language words like "small" or "very small". Some researchers proposed to use fuzzy techniques to handle this knowledge. My boss Gennady Solopchenko asked me, as a professional mathematician, to help Leon Reznik, his doctoral student, to look into these papers and to see how fuzzy techniques can be applied to our problems. I was hooked. Mathematics was interesting and still simple enough to be useful, and practical consequences of taking this imprecise knowledge into account were impressive. Leon incorporated fuzzy techniques into an automated system for testing combustion and jet engines – a system that became a crowning point of his dissertation.

From Slavery to Freedom. Soon after that, I emigrated to the US. Now I was able to attend conferences; previously, as most Soviet scientists, I could not attend conferences outside USSR without KGB permission – and this permission was almost never given. Now I was able to submit papers to international journals – previously, I could not do it without KGB permission which was almost never given; I was once summoned to the KGB and threatened with jail for smuggling my math paper abroad.

I saw all the great people doing research in fuzzy, I saw Lotfi himself – and I was amazed to realized that not only he was a great researcher, he was also a tireless promoter of fuzzy techniques, a tireless helper to young people – in short, a true leader.

Fuzzy is one of my main research interests – the other is a related area of interval computations. I am happy. I am happy that my results and applications – as well as results and applications of others – help solve practical problems. Not everything is perfect in this world – to put it mildly – but I look optimistically into the future. Human ingenuity, human goodwill have overcome many crises, and I am sure that eventually, the future will be good.

What will be the role of fuzzy in this future?

48.2 Future of Fuzzy

Fuzzy Is – and Will Be – Ubiquitous. In the past, there was a lot of publicity about the use of fuzzy techniques in the cars, camcorders, trains, elevators. You do not see that many article about fuzzy in the popular press anymore. Does that mean that there are fewer applications of fuzzy? Not at all. For example, in his plenary talk at the 2011 NAFIPS conference, Dimitar Filev mentioned that many control systems in the cars use fuzzy control. Fuzzy techniques have become so natural and commonplace that the newspapers no longer consider it worth mentioning. After all, calculus is also used a lot in engineering practice – but there are not too many articles in the newspapers about the use of calculus (or the use of algebra, about the routine use of computers) – because this is now mainstream. Similarly, fuzzy has largely become mainstream.

This is exactly what Zadeh intended – to create a new tool that is often helpful, this is what fuzzy has largely become, and this is what it will be in the future.

Future of Fuzzy: Research Directions. The successes of fuzzy techniques do not mean that all the problems have been solved. Far from it. There are many technical problems. And there is also an important fundamental problems that still needs to be researched further.

Indeed, as we have mentioned earlier, according to Hegel, there are two main reasons why the traditional logic is not fully adequate to describe human reasoning: first, it is crisp, while the actual reasoning is often fuzzy; second, it is static in the sense that truth values do not change, while human reasoning is dynamic. There are a few articles about dynamic fuzzy logic [2–5] but this direction is still not very well developed, and this is where a lot of progress still has to be made.

New Application Areas. As of now, most successful applications of fuzzy are to engineering. However, as I mentioned, my interest in fuzzy started because I realized the importance of imprecise reasoning in physics. As of now, there are few applications of fuzzy to fundamental physics; see, e.g., [1]. This is the area where I expect most progress in the future, and I think that it will help physics a lot.

Instead of Conclusion: Future Is Fuzzy, and Fuzzy Is Future. With all this progress, fuzzy techniques – as part of a general scientific toolbox – will undoubtedly continue to excel.

Acknowledgement. This work was supported in part by the National Science Foundation grants HRD-0734825 and DUE-0926721, by Grant 1 T36 GM078000-01 from the National Institutes of Health. My most sincere thanks to Rudolf Seising, Enric Trillas, Settimo Termini, and Claudio Moraga for their invitation, to all fuzzy researchers for being such a great community, and, of course, to Lotfi Zadeh, who started this area, who inspired us all – and who continues to generate great ideas.

References

1. Koshelev, M., Kreinovich, V., Nguyen, H.T., Bouchon-Meunier, B.: Uncertainty Representation Explains and Helps Methodology of Physics and Science in General. In: Proceedings of the Vietnam-Japan Bilateral Symposium on Fuzzy Systems VJFUZZY 1998, HaLong Bay, Vietnam, September 30-October 2, pp. 577–585 (1998)
2. Kreinovich, V.: Dynamic Fuzzy Logic Leads to More Adequate "And" and "Or" Operations. In: Aliev, R. (ed.) Proceedings of the 6th International Conference on Soft Computing, Computing with Words and Perceptions in System Analysis, Decision and Control, ICSCCW 2011, Antalya, Turkey, September 1–2, pp. 21–29 (2011)
3. Perlovsky, L.I.: Neural Network with Fuzzy Dynamic Logic. In: Proceedings of the International IEEE and INNS Joint Conference on Neural Networks IJCNN 2005, Montreal, Quebec, Canada (2005)
4. Perlovsky, L.I.: Fuzzy Dynamic Logic. New Math. and Natural Computation 2(1), 43–55 (2006)
5. Perlovsky, L.I.: Neural Networks, Fuzzy Models and Dynamic Logic. In: Köhler, R., Mehler, A. (eds.) Aspects of Automatic Text Analysis: Festschrift in Honor of Burghard Rieger, pp. 363–386. Springer, Germany (2007)

49

On Fuzzy Data Analysis

Rudolf Kruse, Pascal Held, and Christian Moewes

Fuzzy systems can be found in nearly all industrial branches, e.g. automobile, control engineering, finance, medicine, logistics, telecommunications. Their advantage is their inherent simplicity. Fuzzy rule-based models often turn out to be useful and easily understandable in many real-world applications. In order to learn such models from data – may they be fuzzy or not – intelligent data analysis methods for learning and reasoning are necessary. Thus there is a need for fuzzy data analysis.

The aim of this paper is twofold: In the first part Rudolf Kruse presents some memoirs on early research in fuzzy data analysis and some anecdotes about Lotfi Zadeh in this context. The second part is donated to real-world applications of fuzzy methods and some thoughts about perspectives of fuzzy data analysis.

49.1 Memoirs on Early Research in Fuzzy Data Analysis

Until 1980 the *pioneers* in fuzzy research had no problems with mathematicians and statisticians because there were only a few researchers in this field. On the contrary, lots of people were just curious what type of research was hidden behind the funny name *fuzzy sets*. At that time I was a student of mathematics at the University of Braunschweig in Germany. I asked my supervisor Ernst Henze, an open minded, application-oriented statistician, whether he could recommend a challenging and new topic for my diploma thesis. Henze proposed to study the new field of fuzzy systems because he had found an interesting paper by Lotfi Zadeh [1]. I got interested in the application of fuzzy measures and fuzzy integrals and studied the papers of Michio Sugeno. My doctoral thesis was on fuzzy measures. I draw my attention onto the fuzzy random variables and finished my habilitation in 1984. These days the situation in the fuzzy systems field changed drastically since (1) the number of researchers in this field increased rapidly, (2) it became apparent that dealing with fuzzy data is a complicated research topic with no simple solutions, (3) some fuzzy researchers published papers where it turned out that they were not familiar with the respective standard methods, and (4) there were newspapers articles about potentials and industrial successes of fuzzy logic methods, especially in control engineering. In this situation some mathematicians and statisticians realized that there were interesting real-world applications and capacities for new scientific fields. Other researchers tended to fight against fuzzy sets because they considered its community as an

R. Seising et al. (Eds.): *On Fuzziness: Volume 1*, STUDFUZZ 298, pp. 343–347.
DOI: 10.1007/978-3-642-35641-4_49

academic rival that is weaker from a scientific point of view, but having successes in newspapers and by industrial applications, too.

I attended the first IFSA Congress in Palma de Mallorca [2], a large congress in incredibly hot rooms where I met some well-known fuzzy researchers for the first time. Lotfi introduced himself walking on the street from the hotel to the conference rooms by saying “My name is Lotfi. I just reviewed your paper [3]”. I was deeply impressed because the famous Lotfi Zadeh seemed to be a *normal* person. What a surprise! At this conference in Palma during the breakfast and after a long night at the beach in El'Arenal I was invited to write a book about the topic of fuzzy data analysis for Reidel Publishing Company. The editor of this series, Heinz Skala, was aware of the fact that my doctorate student Klaus-Dieter Meier and myself had developed useful fuzzy methods and a software tool for statistical applications for the Siemens AG. Nevertheless did he recommend avoiding the name *fuzzy* in the title. Thus the title of the book was *Statistics with vague data* [4].

These days some mathematicians were already fighting against fuzzy methods. During a panel discussion at the 8th International Congress of Cybernetics and Systems in June 1990 the famous mathematician Saunders MacLane had heavily criticized fuzzy set theory: “The fuzzy world is often full of fog. The ingenious notion of a fuzzy set was a notable novelty. Unfortunately it has now become a considerably inflated fashion [...]. This may account for the present sorry state of fuzzy statistics [...]. One text which I examined [4] had little to say about data. In spite of several serious attempts, I have yet to find a decisive application.”

As you can imagine I got somewhat unhappy by listening to his contribution after Ron Yager's party on top of an apartment house in New York City. Lotfi commented the situation by saying “Take it as a compliment. Now the people know your name”. I am still grateful for his encouraging comments. Of course MacLane was right in some of his remarks concerning fuzzy theories. But he should be blamed for not seeing the potentials of these new ideas. Before the conference he was not even aware that there were already lots of successful fuzzy applications at that time, e.g. in washing machines, photo cameras.

49.2 Real World Applications with Fuzzy Methods

In 1986 the fuzzy research group in Braunschweig has been established. The group had several industrial projects, mainly in the field of uncertainty handling. The group for example implemented the first Bayesian Network in Germany for Dornier in 1988.

The group was asked by a Volkswagen (VW) research leader to evaluate the usefulness of fuzzy logic control theory. There were no fuzzy researchers at VW and they knew that the group had background in academics as well as in industry. The reason for the VW activity was as follows: On the one hand there were lots of newspaper articles about spectacular new fuzzy applications in Japan. On the other hand there were lots of warnings by several control engineers in Germany about using fuzzy methods in control engineering. So, as a good research manager, he had to check both the potentials of fuzzy logic as a new technology and the use of the marketing potential of the innovative label *fuzzy logic*. The group and VW agreed on a similar project named "Idle speed control with fuzzy logic". This project was of great interest for the group because here they were asked to study real applications in which mathematical background in fuzzy theory could help. For the group it was an interesting challenge to cooperate with control engineers. This project was a big success from the scientific and the technology transfer point of view. It was realized that there are several semantics of fuzzy sets, e.g. uncertainty, similarity, preference, and that control engineers use a completely different interpretation than people working in the field of fuzzy logic – in the narrow sense of a multivalued logic [5]. It turned out that fuzzy control can be seen as a new kind of interpolation – the control engineers liked this view, because it explains fuzzy control in "their" scientific language. A paper on these results received the best paper award of the *IEEE Transaction on Fuzzy Systems* in 1995 [6]. The idle speed controller that students developed within that project turned out to be better than the series line controller. The classical control engineers were puzzled. As a result of this project fuzzy control methods were tolerated at Volkswagen and another student of the group was allowed to develop an automatic gearbox that uses *fuzzy logic* to adapt to the driver's style. This controller was used in the New Beetle series [7]. From a methodological point of view, this problem was considered as a fuzzy data analysis problem. The task was to analyze data from the car to classify the sportiness of the driver. So this was a classification, data analysis, and model learning problem. Lotfi sent a fax with a New York Times article about the New Beetle in which the fuzzy automatic gearbox was mentioned.

Lotfi accepted the invitation by the Technical University in Braunschweig on the occasion of its 250th anniversary to give a talk about fuzzy logic. The conference was a big event because of the plenary talks from the Secretary of State Henry Kissinger, the later German chancellor Gerhard Schröder, the physics hero Carl Friedrich von Weizsäcker, Volkswagen boss Ferdinand Piëch and other prominent speakers. Lotfi and the first author of this paper gave a tandem plenary talk – the former presented some ideas about fuzzy logic and soft computing whereas the latter the fuzzy automatic gearbox. Lotfi was treated in Braunschweig as a VIP (see Fig. 49.1). A female employee was responsible for him during his stay in Braunschweig. Later on she wrote the following lines in a short essay [8] about Lotfi's stay in Braunschweig: "I had to be really careful not to loose him. Scarcely I left him alone for a moment when countless scientists immediately bustled around him. [...] I became aware why Prof. Zadeh attracted so much attention on him. My admiration for him grew every day that I could take care of him. No, not only because he is a authority on his métier but also because I got to know him as an always accommodating, grounded, pleasant human being despite his international celebrity."

49.3 Perspectives of Fuzzy Data Analysis

One can find successful fuzzy systems in almost all industrial areas where optimization, learning and handling imprecise knowledge play a role, i.e. classification, prediction, planning, control, decision making – just to mention a few fruitful areas. We guess that in the near future these *classical* ones will remain the main areas for successful industrial applications of fuzzy systems. These fuzzy systems have always impressed by their simplicity. Fuzzy rule-based models (e.g. á la Mamdani or Sugeno) often turn out to be useful, understandable, not complex and easy to handle.

In a long term run we think that there will be more *intelligent* systems in the role of a *companion of humans*. We already see the trend in the automobile industry where lots of assistant systems are used or in health care for elderly people, which will be another huge market in the future. In order to develop such systems, fuzzy methods could be helpful. Definitely, improved methods for human computer interaction are necessary. So we are sure that, in cooperation with neuroscience, new brain-machine interaction methods will have to be developed. For our research areas we can say that to reach the aim of having more *intelligent* methods, we need much better learning and reasoning systems.

Fig. 49.1. Lotfi in Braunschweig

We think, that in order to reach this goal there is a need for fuzzy data analysis. We must differentiate between *fuzzy* data analysis and *fuzzy data* analysis. The former deals with the analysis of classical data using methods based on fuzzy set theory.

These methods, e.g. fuzzy clustering or fuzzy regression analysis, have been used successfully in lots of industrial applications. The second approach tries to analyze *fuzzy data* by using statistical methods. It seems that there are lots of fuzzy data in the *real* world, and that these data should be used in intelligent systems. This second approach is conceptionally much more difficult than the first approach, because it is often not clear, what a *fuzzy datum* actually means. There are lots of different semantics of a fuzzy datum: Often a fuzzy datum is considered as a "solid object", in other cases it is considered as a kind of "summary" of a more complex underlying phenomenon. The chosen semantics of the fuzzy data have to be taken into account in a serious statistical analysis. So we need models that are able to handle different type of phenomena, e.g. second-order uncertainty models. Such models are much more complicated than "classical" fuzzy methods, and industrial users often hesitate to use such complicated models that need further theoretical insides. Another problem with the second approach is that there are neither software tools available, nor databases for benchmarks. Nevertheless are there lots of challenging open real-world problems in which *fuzzy data* occur, and there is a need to evaluate such data, e.g. for decision making. So, we think that in the future, by using improved mathematical models that combine different qualitative and quantitative modeling approaches, and by increased computational power, fuzzy data analysis and related uncertainty handling technique can be successfully applied in lots of applications.

References

1. Zadeh, L.A.: Probability measures of fuzzy events. Journal of Mathematical Analysis and Applications 23(2), 421–427 (1968)
2. Kruse, R.: On a language and an interpreter for calculation and statistics on linguistic data. In: Proc. First IFSA Congress, Palma de Mallorca, Spain (1986)
3. Kruse, R.: On the variance of random sets. Journal of Mathematical Analysis and Applications 122(2), 469–473 (1987)
4. Kruse, R., Meyer, K.D.: Statistics with Vague Data. In: Mathematical and Statistical Methods. D. Reidel Publishing Company, Dordrecht (1987)
5. Kruse, R., Gebhardt, J., Klawonn, F.: Foundations of Fuzzy Systems. John Wiley & Sons Ltd., Chichester (1994)
6. Klawonn, F., Gebhardt, J., Kruse, R.: Fuzzy control on the basis of equality relations with an example from idle speed control. IEEE Transactions on Fuzzy Systems 3(3), 336–350 (1995)
7. Schröder, M., Petersen, R., Klawonn, F., Kruse, R.: Two paradigms of automotive fuzzy logic applications. In: Jamshidi, M., Titli, A., Zadeh, L., Boverie, S. (eds.) Applications of Fuzzy Logic: Towards High Machine Intelligence Quotient Systems. Environmental and Intelligent Manufacturing Systems Series, vol. 9, pp. 153–174. Prentice-Hall, Inc., Upper Saddle River (1997)
8. Syring, F.: Lotfi Zadeh in Braunschweig. In: Rebe, B., Schnieder, E., Grumbach-Raasch, E. (eds.) Moment eines Jubiläums: 250 Jahre Technische Universität Braunschweig, Technische Universität Braunschweig, pp. 331–332 (1996)

These methods, e.g. fuzzy clustering or fuzzy regression analysis, have been used successfully in [illegible] industrial applications. The [illegible] [illegible] [illegible] [illegible] [illegible] [illegible] methods [illegible] the [illegible] [illegible] [illegible] [illegible] of the fuzzy data[illegible] [illegible] [illegible] [illegible] [illegible] [illegible] models that are able to handle different types of [illegible] [illegible] second-order uncertainties. Such models [illegible] [illegible] [illegible] [illegible] [illegible] fuzzy methods, [illegible]. Nevertheless are there lots of challenging open real-world problems in which [illegible] [illegible] occur, and there is a need to evaluate such data, e.g. for [illegible] [illegible] [illegible] [illegible] [illegible] [illegible] [illegible] [illegible] [illegible] [illegible] [illegible] [illegible] [illegible] [illegible]

[illegible]

50

The Beauty of Vagueness

Mila Kwiatkowska

The words "vagueness" and "vague" are often used to describe a quality, a thought, or a statement which is "incomplete" or "lacking precision." The Oxford English Dictionary defines vagueness (the quality or condition of being vague) as "a lack of distinctness or preciseness; indefiniteness." Similarly, the word "vague," is defined as "not definitely or precisely expressed; deficient in details or particulars." Often in the scientific context, the expressions "lacking precision" and "deficient in details" bring negative connotations. We live in the era, in which scientific perspective and the utmost precision of digital computers are highly regarded. Thus, in a way, the "vagueness" of our thoughts, "fuzziness" of our linguistic expressions, "indefinability" of our feelings, and "intangibility" of our perceptions of the external world are undervalued. The transcendental aspects of our human existence are reduced by "precise" description and "quantitative" analytical methods. In the age of fast, exact, digital computers, precision is deemed a virtue. We strive to be precise; we attempt to create absolute statements about the perceived reality, and we try to explain with an utmost accuracy the surrounding world and our place in the universe. Nothing is inherently wrong with precision or definiteness; yet the world and our place in it, are, in many ways, vague and indefinite. We cannot unequivocally quantify our feelings, perceptions, and interpretations of stimuli because vagueness is inherently present in our language, and fuzziness is a part of our perception. Our interpretation of the world is context-dependent, time-dependent, and, often, contradictory. Moreover, the world itself is constantly changing, we as human beings are constantly changing, and the perceptions of us and the external world, too, are constantly changing.

To address the obvious impossibility of precise expression, we split our perception of the world: we insist on precision in science, and we delegate vagueness to arts, humanities, and social studies. However, precision is a matter of a degree and, in fact, every scientific measurement or statement displays certain level of imprecision. Thus, in fact, precision does not exist or, at least, cannot be achieved given our finite limitations. On the other hand, imprecision and vagueness are in the center of artistic expressions and allow us to experience what in Kant's tradition would be called the "transcendent." Poetry uses ambiguous, interpretable, vague language to describe, express, and evoke the undefined, imprecise, yet beautiful feelings and emotions. Paintings are created with blurred shapes, soft edges, and flickering light. Their visual and symbolic ambiguity allows us to create our own fuzzy or crisp perceptions and to freely interpret or not interpret them at the same time. *There is vagueness in beauty, and there is beauty in vagueness.*

R. Seising et al. (Eds.): *On Fuzziness: Volume 1*, STUDFUZZ 298, pp. 349–352.
DOI: 10.1007/978-3-642-35641-4_50

We exist in a beautifully imprecise reality, and we should integrate the precise quantitative approach with the fuzzy, yet mystical, qualitative approach. We constantly should connect and re-connect our detailed, precise, logical left-brain with the holistic intuitive, fuzzy right-brain. We need to address and continue to re-address the dichotomy between arts and science. We need to capture the beauty of vagueness and the vagueness of beauty.

My personal answer to this quest is the use of fuzzy logic for the modeling of vagueness in medical applications, particularly, applications in sleep medicine and psychiatry. Medicine is both an art and a science; therefore, it is based on scientific facts and, at the same time, it applies the scientifically-based reasoning in a humanistic way. The notion of imprecision (vagueness), missing or partial information, and degrees of uncertainty are specific to all medical data. Moreover, vagueness is intrinsic to many medical concepts. Concepts such as "quality of life," "mental health," "sleepiness," "sadness," and "depression" are difficult to define, measure, and quantify. Furthermore, many medical decisions must be made based on subjective, uncertain, and imprecise information. In particular, the diagnostic process in sleep medicine and psychiatry is based not only on objective data, but relies, in a large proportion, on subjective data. The subjective data are obtained from clinical interviews with patients and self-reporting instruments, such as questionnaires, standardized scales, patient's logs, and family reports. The objective data involve medical examinations, clinical tests (e.g., Electroencephalography, EEG, the recording of the electrical activity of the brain; EEG is used, for example, in the diagnosis of sleep disorders), lab results, and medical images (e.g., functional magnetic resonance imaging, fMRI, the images representing the brain activity associated with the changes in blood flow; fMRI is used in the diagnosis of neurological disorders). Thus, the computerized models in order to represent medical data and human decisions used in diagnosis, prognosis, and treatment, must explicitly represent vagueness and must provide reasoning methods which tolerate vagueness.

Therefore, the traditional approaches of hard computing operating on precise numbers and using categorical approaches of true and false values must be replaced by computational models and reasoning techniques allowing for degrees of imprecision, uncertainty and non-monotonic reasoning. In 1965, Lotfi Zadeh published his paper "Fuzzy sets," in which he introduced the term "fuzzy set", extended the fuzzy set theory, and created fuzzy logic as a new field of study. Lotfi Zadeh introduced the quantitatively-expressed measurement of vagueness, which allows representation of "fuzzy" concepts. As stated by Lotfi Zadeh "most of the concepts encountered in various domains of human knowledge are, in reality, much too complex to admit of simple or precise definition." The many clinically important, yet, imprecise, concepts such as "sleepiness," "high blood pressure" "feelings of depression," "level of physical activity," "obesity" can be defined using linguistic variables, fuzzy membership functions, and fuzzy inference systems. This fuzzy-logic based representation allows for the creation of computer-based systems to support diagnosis and treatment of disorders such as, for example, obstructive sleep apnea and clinical depression.

As it was emphasized by Zadeh, imperfections must be studied and accounted for in the models of reality. With the availability of large clinical data sets and electronic patient records, mismatched levels of precision (imprecision) have been recognized as a crucial issue in database systems, decision-support systems, data mining, machine learning, and information retrieval on the Web. In my research, I concentrate on imprecision, its definition, classification, and interpretation in context of medical data and medical decision making. Furthermore, I work on modeling and creation of clinical decision rules and clinical decision support systems (CDSS). These systems should model and manage all aspects of imprecision in data, information, and knowledge. Three questions are particularly important in the context of the research on imprecision: (1) how to model different types and levels of imprecision in data, information, and knowledge, (2) how to integrate data, information, and knowledge characterized by various levels of imprecision, and (3) how to model and manage the notion of acceptable and unacceptable imprecision in the context of decision-making process.

The motivation for my research in vagueness comes from medical and computational domains. From the medical perspective, there is a need to support the creation of diagnostic rules that could be applied in specialized clinics and primary care, as well as in medical research and education. Furthermore, with the increasing availability of patients' electronic records and electronically stored research data, there is a need for a conceptual framework capable of representing the complexity, varied granularity, heterogeneity, imprecision, and incompleteness of medical data. From a computational perspective, traditional computational models were designed for mechanical systems. Clinical systems are inherently qualitative, context-dependent, incomplete, and imprecise. Moreover, the clinicians expect the CDSS to be transparent, i.e. human-readable and updatable. Therefore, there is a need to create computational models that are appropriate for modeling of biomedical systems and sufficiently formalized for automation, yet comprehensible and interpretable by humans. In my research, I have applied a fuzzy-logic framework to practical medical problems. In collaboration with clinicians, Dr. Najib Ayas, Dr. Frank Ryan, Les Matthews, and Dr. Krzysztof Kielan, I have represented a number of medical concepts and diagnostic rules by the sets of fuzzy rules. These representations have been used by a fuzzy inference mechanism to evaluate the data and support the diagnostic process.

I am deeply indebted to the founder of fuzzy logic, Lotfi Zadeh, and the numerous fuzzy logic researchers involved in creating the link between what is precise and quantitative and what is imprecise and qualitative. I wish to thank all of them for the many opportunities for learning how to integrate our perception of reality and connect the world of science to the world of arts. Moreover, I would like to thank the many organizers of NAFIPS annual conferences, during which I have had the opportunity to present and discuss some of my and my students' findings. As a special memento, I cherish the picture below from NAFIPS 2008, where my student, Michelle Broadway, and I had an opportunity to present fuzzy systems for the evaluation of physical activity and for the evaluation of treatment methods for obstructive sleep apnea (a common and serious respiratory disorder caused by the repetitive collapse of the soft tissues in the throat as the result of the natural relaxation of muscles during sleep).

Fig. 50.1. NAFIPS 2008, New York. Dr. Lotfi Zadeh with the author of the paper (left) and TRU student, Michelle Broadway (right).

51

Flexible Concepts Are Fuzzy Concepts

Jonathan Lawry

51.1 Introduction

The last few decades have seen remarkable advances in Artificial Intelligence, with some form of intelligent system now embedded in a wide range of devices and software, from mobile phones to internet search engines. However, there are many aspects of intelligent behaviour that these systems still cannot replicate. For example, intelligent systems still cannot negotiate a contract online and they cannot promote a certain viewpoint or construct an informal argument. They cannot devise a political slogan or catchphrase. Indeed, the extent to which intelligent agents can take part in any but the most semantically simple dialogues is very limited. These tasks require, amongst other things, an ability to be imprecise or vague on appropriate occasions and in order to achieve certain goals. Furthermore, they require that intelligent systems be able to evolve their own semantic structures and to adapt conceptual models according to context and their current tasks and goals. In an age when intelligent systems must increasingly find patterns and structure in rapidly evolving data rich environments, new methodologies for embedding flexibility and representing imprecision and uncertainty in concept definitions can open the way to a new generation of uniquely robust and adaptable systems.

Symbols and concept labels are powerful representational tools which enable intelligent agents to communicate and reason at a reduced level of granularity, in a highly granular and complex environment. Grouping together different elements of the environment according to similarity or shared characteristics provides a mechanism for abstracting relevant general properties, rules and relations. Indeed, this mechanism lies at the heart of Zadeh's idea of *computing with words* [18]. However, these conceptual groupings should be inherently flexible so as to reduce discontinuities in decision making which may result from crossing the, partly arbitrary, boundaries between categories. A critical element of this flexibility should also be the explicit representation of order information. A category naturally induces an ordering on the underlying space representing the relative extent to which the category label can be appropriately applied. From this perspective some elements are more typical examples of a concept than others. Indeed certain characteristics may be viewed as prototypical for a concept even though they are not taken to be necessary and sufficient conditions for membership in the associated category. For example,

R. Seising et al. (Eds.): *On Fuzziness: Volume 1*, STUDFUZZ 298, pp. 353–358.
DOI: 10.1007/978-3-642-35641-4_51

the attribute of flight might be consider prototypical of the concept bird, perhaps to some extent explaining why robins are more typical members of the category of birds than penguins.

This richer approach to concept representation provides intelligent agents with a robust, effective and adaptable framework for communication, learning and reasoning at an appropriate level of information granularity. Concept flexibility is certainly important in any distributed system where there is no top-down mechanism for defining concepts but where meaning emerges and evolves through interactions between different agents, each aiming to convey useful and relevant information. If concepts are emergent phenomena then it is almost inevitable that different agents will have varying definitions of the same concept. Flexibility, conceptual ordering and an explicit representation of the uncertainty associated with concept boundaries is critical if undesirable errors and misunderstandings are to be avoided.

51.2 The Need for a More Flexible Notion of Concepts

The centrality of concepts and categories in cognition has been recognized by a succession of thinkers from Aristotle through to Locke, Frege, and Carnap. The philosophical consensus has converged on a classical bivalent understanding about the nature and representation of concepts well summarized in Carnap's [1] notions of intension and extension. Accordingly, intension is the 'meaning' or 'sense' [3] of a concept perhaps corresponding to a set of attributes or properties which must be satisfied by any object for it to be correctly denoted by the corresponding label or symbol. From the intension of a concept we can then identify its extension corresponding to those objects, belonging to some subset of the world, which satisfy the concept (e.g. the set of red objects in the room).

In the classical view the extension of a concept is bivalent, so that for any object in the world the intension results in binary classification of that object either as belonging to the extension or belonging to its complement. In fact, since the complement of the extension is also taken to be the extension of the negation of the concept then this is equivalent to adopting the law of the excluded middle. Immediately we see that this model of concepts allows for none of the vagueness characteristic of words and labels in natural language. There is no sense in which an instance can be inherently a borderline case of the concept (and its negation). Furthermore, the classical representation can capture only very limited information about how the concept is actually used in natural language communications. Instead, the intension of a concept is simply assumed to draw a boundary between its extension and that of its negation. From this perspective there is no reason to distinguish between any two examples of the concept so that, for example, all red objects are equal and none are redder than others. Clearly this has significant implications concerning what information can be inferred from assertions involving concepts modelled in the classical manner. For instance, suppose the police have a number of suspects for a crime and they learn, perhaps from a witness, that the actual criminal is *tall* [12]. Then a classical representation of tall will simply allow them to divide the suspects into two disjoint sets. They should then restrict all further investigations to those suspects belonging to the extension of tall. However, this strategy would have a number of difficulties if adopted in practice. For example, supposing the witness has a somewhat different definition of tall than the police. Perhaps they have grown up in a different cultural environment leading them to draw a different boundary between tall and not tall, or perhaps the difference simply results from natural variability in meaning resulting from the way in which language is learnt from experience. In either case this may result in the police permanently eliminating the true criminal from their inquiries. Furthermore, since classical concepts provide no order information then the witness statement would not suggest any particular ranking of those suspects which have been classified as tall.

This unstructured model of concepts has been questioned by both philosophers and psychologists. Wittgenstein [14] argues that there is no formal definition of the concept *game* and in fact there is no one property shared by all those things which can be described as games. Instead we see a 'complicated network of similarities overlapping and criss-crossing'. In such cases we cannot reasonably expect to identify necessary and sufficient conditions capturing a unified view, across native English speakers, of the exact definition of the concept. In other words, Wittgenstein would deny that there is an intentional aspect to the concept game of the form proposed by Carnap. Following the ideas of Wittgenstein and motivated by psychological studies into natural categories Rosch [11] proposed a radically different model of concepts based around the notions of prototype and typicality, subsequently developed and extended by Lakoff [6]. The central tenet of prototype theory is that concepts, rather than being defined by formal rules or mappings, are represented by prototypes and that categorization is based on similarity to these prototypes. By taking typicality to be a decreasing function of distance from prototypes, this approach would naturally explain the fact that some instances are seen as more typical exemplars of a concept than others. For example, robins are more typical examples of birds than penguins, since the latter have certain atypical characteristics such as the inability to fly. Through this notion of typicality, prototype theory naturally embeds a notion of order into concept definitions, so that, for example, some people are taller than others in that they are more typical instances of the concept. Such an ordering could provide the police with a mechanism for ranking all of the suspects on the basis of their new information. In addition, this approach could reduce the risk of mistakenly eliminating the actual criminal early in the investigation because of a difference in concept boundaries between the police and the witness. Also, as argued by Van Deemter [12], this ranking approach may, on average, reduce the search time for finding the criminal amongst the suspects.

In addition to typicality another important aspect of natural concepts missing from the classical model is semantic uncertainty. Given that language learning is, to a significant extent, an inductive process based on evidence from interactions and communications between individuals, we would expect such individuals to exhibit significant uncertainty concerning the definition of concepts and consequently the applicability of the associated labels. In particular, we would expect there to be significant uncertainty regarding the location of concept boundaries so that it is unlikely that an individual will be able to identify a precise extension set for a concept. Instead an agent would at best be able to identify a probability distribution over possible concept models. In our view, semantic uncertainty about concepts is epistemic in the sense that it results from a process of inductive learning whereby individuals attempt to learn the underlying categorization conventions of the language. This is close to the epistemic theory of vagueness expounded by Williamson [13] but with a subtle though important distinction. The epistemic theory would seem to assume the existence of some objectively correct, but unknown, set of criteria for determining whether or not a given instance satisfies a vague concept. In other words, that each concept has an objectively correct intension about which individuals are uncertain due to their lack of information on how the concept can be used in practice. However, in practice the rules and conventions of language use are not imposed by some outside authority but, rather, are represented as a distributed body of knowledge, shared across a population, and emerging as the result of interactions and communications between individuals. From this perspective the idea that there are objectively correct concept intensions is rather hard to justify. Instead, it is more realistic to assume that individuals, when faced with decision problems concerning categorization, find it useful as part of a decision making strategy, to *assume* that there is a set of language rules specifying an intension of each concept. In other words, in deciding what to assert individuals adopt an *epistemic stance* [8] and behave *as if* the epistemic view of vagueness is correct.

51.3 Fuzziness in Flexible Concepts

In recent work [9] my colleague Yongchuan Tang and I have proposed a concept representation model which combines both prototype theory and random set theory. This model assumes that there is an underlying metric space of attributes, (Ω, d), relevant to the concept definition and in which typicality is proportional to the distance from a region of prototypical values in this space. Here we follow Gardenfors [4] in highlighting the importance of such *conceptual spaces* in which concepts are represented as convex regions. However, we also argue that given semantic uncertainty the boundaries of such regions must be inherently uncertain. This is a degree based approach which incorporates both typicality and semantic uncertainty by quantifying the degree to which any given instance can be *appropriately* described by a concept label, and as such is a special case of the *label semantics* framework [7]. The underlying idea is essentially as follows: A concept label L is defined by a prototype P, corresponding to a subset of the conceptual space Ω, together with an uncertain distance parameter ε. L is then said to be *appropriate* to describe an element x from Ω if $d(x,P) \leq \varepsilon$ i.e. x lies within a distance ε of the prototype for the concept. Given the inherent semantic uncertainty as represented by a probability density on the threshold ε, the degree of appropriateness of L as a description of x, is then determined by the probability that $\varepsilon \geq d(x,P)$. Clearly this is an inherently probabilistic random set based model and consequently is not fully truth-functional in the way that fuzzy set theory is. However, the calculus is naturally functional in a weaker sense and coincides with Zadeh's [15] original min-max model for a significant restricted class of logical combinations. Furthermore, it has at its heart two of the best known semantics for fuzzy membership functions proposed in the literature i.e. similarity to prototypes, and random set theory [2].

I have been fascinated by the random set theory interpretation of fuzzy sets [5], [10] since 1994 when I first came to Bristol University to work with Jim Baldwin and Trevor Martin. In my view it holds out the best prospect of an integrated theory linking semantic and epistemic uncertainty [1] in a principled way. Now I do not believe that Zadeh is enthusiastic about the random set theory or other probabilistic interpretations of fuzzy sets [17]. Furthermore, the relationship between fuzzy sets and prototype theory which he outlines in [16] would seem to be based more on the idea of a fuzzy intensional model from which fuzzy prototypes are then derived, rather than on interpreting membership functions as similarity to a prototype. Hence, most probably he would not entirely approve of the flexible concept models I have proposed, and of course I would disagree. However, for this current volume such differences are rather beside the point. What matters is that Lotfi Zadeh's seminal work on Fuzzy Set Theory has opened the door to the study of more flexible models of natural concepts and categories, which may yet prove to be the key to true artificial intelligence. While I may not agree with all of the details of Lotfi's work I have always found it totally inspirational, often driving my own work in directions I would not otherwise have dreamed of.

References

1. Carnap, R.: Meaning and Necessity. University of Chicago Press, Chicago (1947)
2. Dubois, D., Prade, H.: The Three Semantics of Fuzzy Sets. Fuzzy Sets and Systems 90(2), 141–150 (1997)
3. Frege, G.: Über Sinn und Bedeutung. Zeitschrift für Philosophie und philosophische Kritik 100, pp. 25–50 (1892); English translation as 'On Sinn and Bedeutung'. In: Beaney, M.(ed.) The Frege Reader, pp. 151–171. Blackwell, Oxford (1997)
4. Gärdenfors, P.: Conceptual Spaces: The Geometry of Thought. MIT Press, Cambridge (2000)
5. Goodman, I.R.: Fuzzy Sets as Equivalence Classes of Random Sets. In: Yager, R.R. (ed.) Fuzzy Set and Possibility Theory. Recent Developments, pp. 327–342. Pergamon Press, New York (1982)
6. Lakoff, G.: Women Fire, and Dangerous Things. University of Chicago Press, Chicago (1987)
7. Lawry, J.: Modelling and Reasoning with Vague Concepts. SCI, vol. 12. Springer, Heidelberg (2006)
8. Lawry, J.: Appropriateness Measures: An uncertainty model for vague concepts. Synthese 161(2), 255–269 (2008)
9. Lawry, J., Tang, Y.: Uncertainty Modelling for Vague Concepts: A Prototype Theory Approach. Artificial Intelligence 173(18), 1539–1558 (2010)

[1] Such as you find in the statement 'Ethel is tall' when Ethel's height is also unknown.

10. Nguyen, H.T.: On Modeling of Linguistic Information using Random Sets. Information Science 34, 265–274 (1984)
11. Rosch, E.H.: Natural Categories. Cognitive Psychology 4, 328–350 (1973)
12. van Deemter, K.: Not Exactly: In Praise of Vagueness. Oxford University Press, Oxford (2010)
13. Williamson, T.: Vagueness. Routledge, London (1994)
14. Wittgenstein, L.: Philosophical Investigations. Blackwell, Oxford (1953)
15. Zadeh, L.A.: Fuzzy Sets. Information and Control 8(3), 338–353 (1965)
16. Zadeh, L.A.: A Note on Prototype Theory and Fuzzy Sets. Cognition 12(1), 291–297 (1982)
17. Zadeh, L.A.: Probability Theory and Fuzzy Logic are Complementary Rather than Competitive. Technometrics 37(3), 271–276 (1995)
18. Zadeh, L.A.: Fuzzy Logic = Computing with Words. IEEE Transactions on Fuzzy Systems 4, 103–111 (1996)

52

Fuzzy Ontologies for the Game of Go

Chang-Shing Lee, Mei-Hui Wang, and Olivier Teytaud

Abstract. This chapter presents a developed fuzzy ontology model for computer Go applications. Unlike previous research, this chapter employs features derived from professional Go players' domain knowledge to transform them into the opening book sequence and to represent them by a fuzzy ontology for the game of Go. Afterward, the domain experts validate the built fuzzy ontology. The developed fuzzy ontology has been verified through the invited games for Go programs playing against human Go players. The results show that the fuzzy ontology can work for computer Go application.

52.1 From Ontology to Fuzzy Ontology

Ontology is an explicit specification of a conceptualization and a formal specification of a shared conceptualization. It is also a good knowledge representation and communication model for intelligent agents. However, it is widely pointed out that classical ontology is not sufficient to deal with imprecise and vague knowledge for some real world applications. The fuzzy ontology is an extension of the domain ontology that is more suitable to describe the domain knowledge for solving the uncertainty reasoning problems [5, 7]. As a result, fuzzy ontology can effectively help to handle and process uncertain data and knowledge.

52.2 History in Computer Go Development

Go is a board game that is much more complex than chess. However, despite several decades of artificial intelligence and computational intelligence, there are still no Go programs that can challenge a professional player in 19×19 games without handicap. This is because Go is a problem with high uncertainty, especially for big board games. Each Go player has his own way of thinking to play with his opponent, and each top professional Go player will take different strategies even though they face the same situation. Thus, in 1997, the IBM's Deep Blue Supercomputer beat the World Chess Champion, Garry Kasparov, while the game of Go is still one of the last board games where the strongest humans are still able to win easily against computers in big board games [4, 6, 8–10]. In 1998, Martin Müller won against *Many Faces of Go*, one of the top programs at that time, in spite of 29 handicap stones, an incredibly big handicap, so big that it does not make sense for human

R. Seising et al. (Eds.): *On Fuzziness: Volume 1*, STUDFUZZ 298, pp. 359–364.
DOI: 10.1007/978-3-642-35641-4_52

players. Ten years later, in 2008, *MoGo* and *CrazyStone* won Myung-Wan Kim (8p) and Kaori Aoba (4p) in 19×19 games with handicap 9 and 7 stones, respectively [6]. Both programs were using databases of patterns (based on confidence and support of rules for *MoGoTW*), which can be considered as fuzzy ontologies for the game of Go. Since 2008, IEEE Computational Intelligence Society (CIS), National University of Tainan (NUTN) in Taiwan, and other academic organizations have co-hosted or co-organized several human vs. computer Go-related events, including the *2008 Computational Intelligence Forum & World* 9×9 *Computer Go Championship* held in September 2008 [6], and 2009 *Invited Games for MoGo vs. Taiwan Professional Go Players* (*Taiwan Open 2009*) held in February 2009 (Fig. 52.1 (a)) [10]. In 2008, human won most of the games; however, in 2009, the Go program *MoGo* made two new world records by winning a 19×19 game with 7 handicap stones against the 9P professional Go player (Chun-Hsun Chou) and a 19×19 game with 6 handicap stones against the 1P professional Go player (Li-Chen Chien) in *Taiwan Open 2009*.

(a) (b)

Fig. 52.1. (a): Competition @ Taiwan Open 2009; (b) Competition @ FUZZ-IEEE 2009

52.3 Human vs. Computer Go Competition in IEEE CIS

The flag conference of IEEE CIS, 2009 International Conference on Fuzzy System (FUZZ-IEEE 2009) held in August 2009, started to formally support the *Human vs. Computer Go Competition* (Fig. 52.1 (b)) [10]. During FUZZ-IEEE 2009, there was the first win of a computer program (*Fuego*) against a 9P player (Chou-Hsun Chou) in 9×9 as white. On the other hand, none of the programs could win against in 19×19, in spite of the handicap 7, showing that winning with handicap 7 against a top level player is still almost impossible for computers, in spite of the win by *MoGo* with handicap 7 in the *Taiwan Open 2009* [10]. Also, during FUZZ-IEEE 2009, no computer program could win as black in 9×9 Go with komi 7.5 against the top professional Go player [10]. The only wins in 9×9 games as black against a professional Go player were realized by *MoGo/MoGoTW* against Catalin Taranu (5P) in Rennes, France in 2009 and the win against Chun-Hsun Chou (9P) in Taipei, Taiwan in 2009. The *Human vs. Computer Go Competition*, organized by IEEE

CIS, 2010 IEEE World Congress on Computational Intelligence (IEEE WCCI 2010), IEEE CIS Emergent Technologies Technical Committee (ETTC), and NUTN, was held in Barcelona, Spain on July 20, 2010 (Figs. 52.2 (a) and (b). A main novelty is the presence of 13×13 games, and the Go programs *MoGo* and *Many Faces of Go* even won against human (6D) in 13×13 Go with handicap 2 [9]. From the games results at the competition, we know that the Go programs won 9 out of the total 22 games. The average performance of the computer Go programs is approaching to the professional level, with Zen and CrazyStone ranking 4 Dan on KGS (5 Dan in blitz games). On the full 19×19 board, programs have racked up a number of wins (but still a lot more losses) on 6 and 7 handicap stones against top professional Go players [9, 10]. In 2011, hosted by IEEE CIS together with NUTN, INRIA team TAO of France, and Grid5000, the *Human vs. Computer Go Competition* was held at the 2011 IEEE Symposium Series on Computational Intelligence (IEEE SSCI 2011) in Paris, France (Fig. 52.2 (c)), and FUZZ-IEEE 2011 in Taipei, Taiwan (Fig. 52.2 (d)).

(a) (b)

(c) (d)

Fig. 52.2. (a): Opening Ceremony @ IEEE WCCI 2010. (b): Competition @ IEEE WCCI 2010. (c): Competition @ IEEE SSCI 2011. (d): Competition @ FUZZ-IEEE 2011.

Blind 9×9 Go games between human and Go programs were first held at the IEEE SSCI 2011. *MoGoTW* broke a new world record by winning the first 13×13 game against the professional Go player Ping-Chiang Chou (5P), with handicap 3 and reversed komi of 3.5. *MoGoTW* also won 3 out of 4 games of 9×9 blind Go and Pachi won one 19×19 game with handicap 7 against Chun-Hsun Chou (9P). In the three-day competition, held at FUZZ-IEEE 2011, the Go program Zen from Japan won each competition even playing 19×19 game with Chun-Hsun Chou (9*P*) with handicap 6, showing that the level of Go programs in 19×19 game is estimated at 4D. *MoGoTW* also won all of twenty 7×7 games under a specific komi, that is, setting komi 9.5 and 8.5 as *MoGoTW* is white and black, respectively, suggesting that in 7×7 perfect play is a draw with komi 9. Importantly, major successes in small board Go ($7 \times 7, 9 \times 9$) are all based on opening books, handcrafted for Fuego, mixing handcrafted expertise and automatic building for *MoGoTW*; these opening books of uncertain and variable knowledge can be considered as ontologies.

52.4 Fuzzy Ontology Model for Go Opening Book

The structure of constructing the ontology to express the knowledge of opening book for game of Go is shown in Fig. 52.3 (a). The first step is to invite Go players to play against Go programs via the Go-playing graphic interface such as Kiseido Go Server (KGS) or Go Graphical User Interface (GoGui). Once the game is started, the records of board games are stored by following the Smart-Go Format (SGF). The records of the Go games are stored into the SGF files repository. Then, the linguistic descriptions, including *very good* (*VG*) move, *good* (*G*) move, *uncertain* (*U*) move, *bad* (*B*) move, and *very bad* (*VB*) move, on each move and alternative branches are given by the invited Go players via MultiGo software or talking to the side assistant.

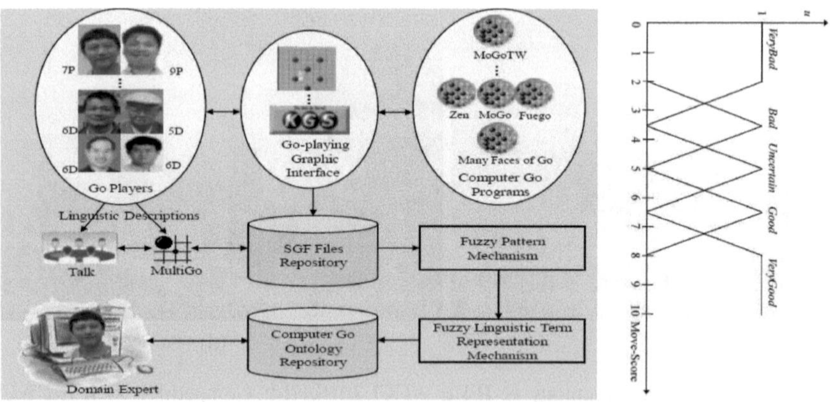

(a) (b)

Fig. 52.3. (a): Constructing the ontology to express some knowledge on the game of Go [4]; (b) Example of the fuzzy sets for fuzzy variable Move-Score [4]

Fig. 52.3 (b) shows an example of the fuzzy sets for fuzzy variable *Move-Score* = *VeryBad*, *Bad*, *Uncertain*, *Good*, *VeryGood*, which indicates that there are five fuzzy sets, including *VeryBad*, *Bad*, *Uncertain*, *Good*, and *VeryGood*, to describe the score of the move [4].

The opening book sequences are extracted based on the SGF files storing in the SGF files repository and obtained through the fuzzy pattern mechanism. The fuzzy pattern is used to present the opening book sequence for the fuzzy ontology model. The fuzzy linguistic term is used to represent the degree of goodness for each opening book sequence via the fuzzy linguistic term representation mechanism. Different Go player maybe give different linguistic description for the same opening book sequence. Finally, the computer Go ontology can be built by integrating fuzzy pattern and fuzzy linguistic term, and the domain experts validate and verify the correctness of the constructed computer Go ontology.

52.5 Conclusion

The computer Go advances in the recent years relied on the followings: (1) simulation-based methodologies which do not rely on evaluation functions make the Monte-Carlo Tree Search methodologies relevant in many games where such evaluation functions do not exist, (2) good compromise between exploration and exploitation, which can be used for many other areas far from games, e.g., planning, and (3) heavy parallelization could be used in clusters of multi-core machines or grids to precompute the important parts of the games, such as the opening books [1]. In addition to Go games, the generality of the approach makes it suitable for wide application fields such as (1) energy mangement applications [3], (2) other games (Havannah, Lines Of Action [13]), including clear breakthroughs in widely played card games [2, 12], (3) non-linear expensive optimization, and (4) active learning [11].

If the Go player, no matter a human or a computer, is able to do an excellent opening, the chance of winning will be increased, especially when playing on a small 9×9 board. As a result, it is very important for a computer Go to construct an excellent opening book, if a computer Go would like to challenge top human Go players. On the other hand, except for *VeryGood* moves and *Good* moves, the other types of moves, namely *Uncertain* move, *Bad* move, and *VeryBad* move, are also necessary to develop a computer Go assessment system in the future. If there is such a fuzzy ontology existing to represent the above-mentioned information, then Go programs will learn quickly and understand easily the opening sequences recommended by domain experts.

Acknowledgement. The authors would like to thank the support from National Science Council of Taiwan (NSC 99-2923-E-024-003-MY3 and NSC 100-2811-E-024-001).

References

1. Bourki, A., Chaslot, G., Coulm, M., Danjean, V., Doghmen, H., Hoock, J.-B., Hérault, T., Rimmel, A., Teytaud, F., Teytaud, O., Vayssière, P., Yu, Z.: Scalability and Parallelization of Monte-Carlo Tree Search. In: van den Herik, H.J., Iida, H., Plaat, A. (eds.) CG 2010. LNCS, vol. 6515, pp. 48–58. Springer, Heidelberg (2011)
2. Chou, C.-W., Teytaud, O., Yen, S.-J.: Revisiting Monte-Carlo Tree Search on a Normal Form Game: NoGo. In: Di Chio, C., Cagnoni, S., Cotta, C., Ebner, M., Ekárt, A., Esparcia-Alcázar, A.I., Merelo, J.J., Neri, F., Preuss, M., Richter, H., Togelius, J., Yannakakis, G.N. (eds.) EvoApplications 2011, Part I. LNCS, vol. 6624, pp. 73–82. Springer, Heidelberg (2011)
3. Couëtoux, A., Hoock, J.-B., Sokolovska, N., Teytaud, O., Bonnard, N.: Continuous Upper Confidence Trees. In: Coello, C.A.C. (ed.) LION 2011. LNCS, vol. 6683, pp. 433–445. Springer, Heidelberg (2011)
4. Hoock, J.-B., Lee, C.-S., Rimmel, A., Teytaud, F., Wang, M.-H., Teytaud, O.: Intelligent Agents for the Game of Go. IEEE Computational Intelligence Magazine 5(4), 28–42 (2010)
5. Lee, C.-S., Jian, Z.-W., Huang, L.-K.: A Fuzzy Ontology and its Application to News Summarization. IEEE Transactions on Systems, Man and Cybernetics Part B: Cybernetics 35(5), 859–880 (2005)
6. Lee, C.-S., Wang, M.-H., Chaslot, G., Hoock, J.-B., Rimmel, A., Teytaud, O., Tsai, S.-R., Hsu, S.-C., Hong, T.-P.: The Computational Intelligence of MoGo Revealed in Taiwan's Computer Go Tournaments. IEEE Transactions on Computational Intelligence and AI in Games 1(1), 73–89 (2009)
7. Lee, C.-S., Wang, M.-H., Hagras, H.: A type-2 Fuzzy Ontology and its Application to Personal Diabetic-diet Recommendation. IEEE Transactions on Fuzzy Systems 18(2), 374–395 (2010)
8. Lee, C.-S., Müller, M., Teytaud, O.: Special Issue on Monte Carlo Techniques and Computer Go. IEEE Transactions on Computational Intelligence and AI in Games 2(4), 225–228 (2010)
9. Lee, C.-S., Wang, M.-H., Teytaud, O., Wang, Y.-L.: The Game of Go @ IEEE WCCI 2010. IEEE Computational Intelligence Magazine 5(4), 6–7 (2010)
10. Rimmel, A., Teytaud, O., Lee, C.-S., Yen, S.-J., Wang, M.-H., Tsai, S.-R.: Current Frontiers in Computer Go. IEEE Transactions on Computational Intelligence and AI in Games 2(4), 229–238 (2010)
11. Rolet, P., Sebag, M., Teytaud, O.: Boosting Active Learning to Optimality: A Tractable Monte-Carlo, Billiard-Based Algorithm. In: Buntine, W., Grobelnik, M., Mladenić, D., Shawe-Taylor, J. (eds.) ECML PKDD 2009, Part II. LNCS, vol. 5782, pp. 302–317. Springer, Heidelberg (2009)
12. Teytaud, O., Flory, S.: Upper Confidence Trees with Short Term Partial Information. In: 3rd European Event on Bio-inspired Algorithms in Games (EvoGames 2011), Torino, Italy, April 27-29, pp. 153–162 (2011)
13. Winands, M.H.M., Bjornsson, Y., Saito, J.-T.: Monte Carlo Tree Search in Lines of Action. IEEE Transactions on Computational Intelligence and AI in Games 2(4), 239–250 (2010)

53

Fuzzy Objects

Jonathan Lee and Nien-Lin Hsueh

Abstract. As informal requirements are usually imprecise, extending object-oriented modeling to fuzzy logic for capturing and analyzing the informal requirements was proposed in the past years. In this article, we will introduce the related work in this area and give a brief introduction to the approach of Fuzzy Object Oriented Model (FOOM). FOOM is an approach based on fuzzy logic to formulate imprecise requirements along four dimensions: (1) to extend a class by grouping objects with similar properties into a fuzzy class, (2) to encapsulate fuzzy rules in a fuzzy class to describe the relationship between attributes, (3) to evaluate the membership function of a fuzzy class by considering both static and dynamic properties, and (4) to model uncertain fuzzy associations between classes.

53.1 Introduction

As Zadeh pointed out in [15], it is evident that almost all concepts in or about natural languages are almost fuzzy in nature. Rumbaugh and his colleagues [13], [15] have argued that OOM is a way of thinking about problems using models organized around real-world concepts that are usually expressed in natural languages. A number of researchers have reported progress towards the successful integration of fuzzy logic and object-oriented modeling in the Fuzzy Logic and Software Engineering literature, which can be classified into three categories based on their intended modeling purposes: knowledge representation for AI systems, data modeling for database systems and object-oriented modeling for conventional software systems.

Knowledge Representation for AI Systems

Lano has proposed to combine fuzzy reasoning and object-oriented representation for the real-world information [9]. A knowledge base is organized as a class hierarchy for representing concept categories, each class corresponds to a fuzzy set, whose membership functions is the proximity metric defined for the class. To support an approximate reasoning in systems based on prototypical knowledge representation, Terwilliger and Cambell have defined formalism for the representations and a general evaluation mechanism to deal with the form of knowledge [5], [14]. Each frame has three kinds of weighted attributes: necessary, sufficient and supplementary. The evaluation mechanism is based on fuzzy logic: the fuzzy match between prototypical description and sets of data is based on possibility theory and the relevance measure

R. Seising et al. (Eds.): *On Fuzziness: Volume 1*, STUDFUZZ 298, pp. 365–370.
DOI: 10.1007/978-3-642-35641-4_53

of each slot. To handle vagueness and imprecision in an expert system, Leung and Wong have integrated fuzzy concepts into object oriented knowledge representation [12]. An approach to querying fuzzy objects and the fuzzy relations between classes is also proposed.

Data Modeling for Database Systems

In [5], Dubois and Prade have advocated that classes can be intentionally described in terms of attributes that are distinguished between the range of allowed values and the range of typical values. The degree of inclusion between a class C_1 and a subclass C_2 is computed by comparing the ranges or the typical ranges of C_1 with the ranges or the typical ranges of C_2. In [8], the problem of object recognition is viewed as a classification problem, which is characterized by an objected-oriented knowledge representation and control strategies based on fuzzy pattern matching procedures. Bordogna et al. [1] propose a Fuzzy Object Oriented model for management of crisp and fuzzy data. Their work develops a fuzzy graph-based data model, which intends to generalize a graph model so that imprecision and uncertainty can be managed at different levels.

Object-Oriented Modeling for Conventional Software Systems

George et al. have utilized the ranges of fuzzy values of classes and objects for computing the degree of inclusion and membership, respectively [6]. To measure the class memberships, a similarity metric is formulated to measure the nearness between attributes' values in a superclass and its subclasses. Graham [7] has focused on the derivation of unknown values of attributes through the use of a-kind-of relation (AKO), generalized modus ponen and defuzzification techniques. In Graham's work, the notion of an object is extended to that of a fuzzy object in two ways: (1) attributes' values may be fuzzy, and (2) AKO is a matter of degrees.

A fuzzy object-oriented modeling technique (FOOM) is proposed by Lee [10], [11] to capture and analyze imprecise requirements through the following two steps: (1) to identify the possible types of fuzziness involved in the modeling of imprecise requirements, and (2) to investigate the potential impacts of incorporating the notion of fuzziness on the features of object orientation. FOOM is more general an approach than non-fuzzy ones in that FOOM can model both crisp and imprecise requirements. The details of FOOM will be introduced in next section.

53.2 Fuzzy Object-Oriented Modeling

Fuzzy object-oriented modeling technique (FOOM) is a modeling approach for requirements engineers to model and analyze imprecise requirements. FOOM extends the traditional OOM along several dimensions: (1) to extend a class to a fuzzy class which classifies objects with similar properties, (2) to encapsulate fuzzy rules in a class to describe the relationship between attributes, (3) to evaluate fuzzy class memberships by considering both static and dynamic properties, and (4) to model uncertain fuzzy associations between classes.

53.2.1 Inside a Fuzzy Class

Traditionally, a class is used to describe a crisp set of objects with common attributes, common operations and common relationships. In order to model the impreciseness rooted in user requirements, we extend a class to describe a fuzzy set of objects (called a fuzzy class), in which objects may have similar attributes, similar operations and similar relationships, for example, a set of interesting books or a class of clever students. In the meeting scheduler system, the class *ImportantParticipant* is modeled as a fuzzy class, that is, a participant may be an important one to a degree.

Since a fuzzy class is a group of objects with similar static properties (i.e., attributes, operations) and similar dynamic properties (i.e., relationships and rules), the membership degree of an instance to a fuzzy class is dependent on the properties, especially the values of attributes and the values of link attributes. In our example, the degree that a person belongs to the class *ImportantParticipant* depends on his status and his role in the meeting he attends.

Attributes with Fuzzy Ranges

The domain of an attribute is the set of all values the attribute may take, irrespective to the class it falls into; whereas, the range of an attribute in a class is defined as the set of allowed values that a member of a class may take for the attribute. In FOOM, the fuzziness in the range of an attribute in a class may be due to either a linguistic term or a typical value. A class may be fuzzy for the linguistic values its attributes can take. For example, the class *YoungMan* has a fuzzy range for the attribute age, since a person may take young or very young as values for his age. The range of an attribute is fuzzy because some of its values are deemed as atypical (i.e. less possible than other values), therefore, each value the attribute may take is associated with a typical degree. In our example, the class ImportantParticipant has a fuzzy range *student/0.4, staff/0.7, faculty/1* for the attribute status, which means that a faculty is typically an important participant, and a student is an important participant with a typical degree of 0.4.

Fuzzy Rules

Incorporating fuzzy rules in object-oriented analysis can help enrich the semantics of analysis models. Using fuzzy rules is one way to deal with imprecision where a rule's conditional part and/or the conclusions part contain linguistic variables. Fuzzy rules are an optional feature for a fuzzy class in FOOM and are thus classified as a dynamic property. Fuzzy rules are used to describe the internal relationship or external relationship. In the former, fuzzy rules describe the relationship between attributes inside a class. For example, in Figure 53.1, a rule "if the role is a staff, the participant importance is less important" describes the relationship between the attributes role and participant importance. In the latter, fuzzy rules are used to describe the relationship between two different classes.

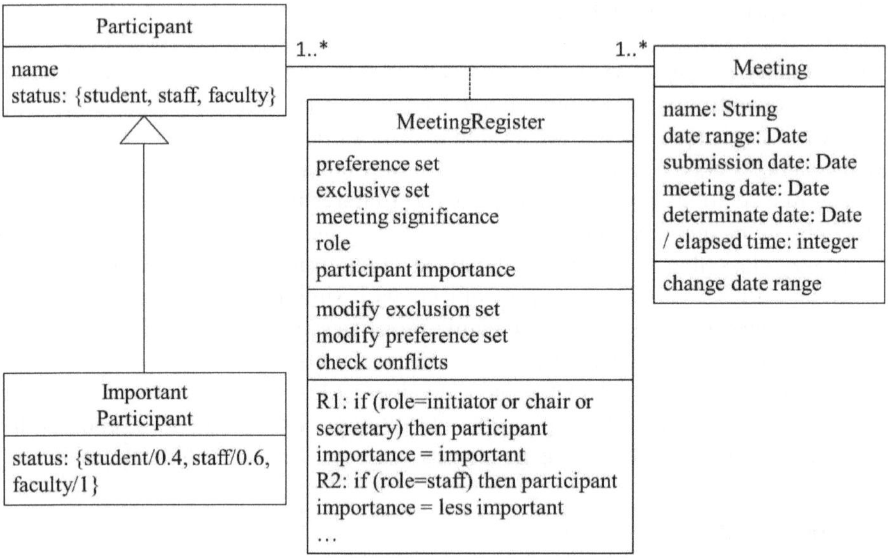

Fig. 53.1. An Example of a Fuzzy Class

53.2.2 Fuzzy Classification

Perceptual fuzziness refers to the compatibility between a class and an object (i.e. ISA), and the class membership between a class and its subclass (i.e. AKO). In FOOM, we extend crisp class memberships to fuzzy class memberships by allowing the existence of perceptual fuzziness. In this section, the notion of inheritance and how it affects the perceptual fuzziness are elaborated.

Perceptual Fuzziness between Classes and Subclasses

Traditionally, the AKO relationship between a class and its subclass is crisp, that is, an instance of a subclass is also an instance of its superclass. In FOOM, an instance of the subclass may be an instance of its superclass to some extent, i.e., the AKO degree ranges from 0 to 1. As the weak form of substitutability is maintained in FOOM, a subclass is constructed through extending new operations, redefining the inherited operations, adding new attributes or modifying the inherited attribute ranges. The perceptual fuzziness of an object to a class or a subclass to its superclass is calculated by evaluating both the static properties and dynamic properties. It is also important to note that not all attributes are necessarily related to the perceptual fuzziness. Referring to our example, the membership degree of a person to the class *ImportantParticipant* can be obtained by checking his status and his participant importance in the meeting he attends. An attribute that may affect a perceptual fuzziness is called a Focus Of Attention (FOA) attribute. Therefore, the attribute status is classified as a static FOA attribute, and participant importance a dynamic FOA attribute. The criticality of an FOA attribute indicates the relevance of the attribute to a perceptual fuzziness.

Perceptual Fuzziness between Classes and Objects

The class membership between an object and a class is crisp, that is, the ISA degree of an object to a class is either 1 or 0. In FOOM, a perceptual fuzziness between an object and a class is allowed. An object may belong to a class to a degree. In the meeting scheduler system, a person may belong to the class *ImportantParticipant* to some extent.

53.2.3 Uncertain Fuzzy Associations

Links and associations are means for establishing relationship among objects and classes. A link is a physical or conceptual connection between object instances. For example, "John *work-for* Simplex company". An association describes a group of links with common structure and common semantics. For example, "a person *work-for* a company". In traditional object-oriented approaches, only crisp associations are introduced, namely, an object either participates in an association or not. Usually, certain and precise knowledge about an association is not always available in the user requirements; furthermore, users' observations are sometimes uncertain and imprecise. Therefore, an adequate management of uncertainty and imprecision in the phase of requirements analysis is an important issue. The distinction between imprecise and uncertain information can be best explained by Dubois and Prade [3]: imprecision implies the absence of a sharp boundary of the value of an attribute; whereas, uncertainty is an indication of our reliance about the fuzzy information. An uncertain fuzzy association is allowed in FOOM. The imprecision of an association implies that an object can participate in the association to some extent, whereas uncertainty is referred to the confidence degree about the association. To represent the imprecision of an association, a special link attribute is introduced in FOOM to indicate the intensity that objects participate in an association. Fuzzy truth value, such as true, fairly true and very true, is used to serve as the representation of uncertainty for its capability to express the possibility of the degree of truth.

53.3 Conclusion

As was pointed by Borgida et al. [2], a good requirement modeling approach should take the problem of describing natural kinds into account; furthermore, Zadeh has indicated that almost all concepts in or about natural languages are almost fuzzy in nature [15]. In this article, we have proposed an approach to incorporating fuzzy concepts into object-oriented systems for modeling imprecise requirements. Several kinds of fuzziness involved in user requirements are identified: fuzzy classes, fuzzy rules, and fuzzy ranges of attributes, perceptual fuzziness, and uncertain fuzzy associations. FOOM offers an important benefit: to extend traditional object-oriented techniques to manage different kinds of fuzziness that are rooted in user requirements.

References

1. Bordogna, G., Lucarella, D., Pasi, G.: A Fuzzy Object Oriented Data Model. In: Proceedings of the Third IEEE Conference on Fuzzy Systems, pp. 313–318 (1994)
2. Borgida, A., Greenspan, S., Mylopoulos, J.: Knowledge Representation as the Basis for Requirements Specifications. IEEE Computer 18(4), 82–91 (1985)
3. Dubois, D., Prade, H.M.: Possibility Theory: An Approach to Computerized Processing of Uncertainty. Plenum Press, New York (1988)
4. Dubois, D., Prade, H.M., Rossazza, J.-P.: Vagueness, Typicality, and Uncertainty in Class Hierarchies. International Journal of Intelligent System 6(2), 167–183 (1991)
5. Dubois, D., Prade, H.M.: Fuzzy Sets and Systems: Theory and Applications. Academic Press, New York (1980)
6. George, R., Srikanth, R., Petry, F.E.: Uncertainty Management Issues in the Object-oriented Data model. IEEE Transactions on Fuzzy Systems 4(2), 179–192 (1996)
7. Graham, I.: Fuzzy objects: Inheritance under Uncertainty. Object Oriented Methods, pp. 403–433. Addison-Wesley, New York (1994) (Appendix I)
8. Granger, C.: An Application of Possibility Theory to Object Recognition. Fuzzy Sets and Systems 28(3), 351–362 (1988)
9. Lano, K.: Combining Object-oriented Representations of Knowledge With Proximity to Conceptual Prototypes. In: Dewilde, P., Vandewalle, J. (eds.) Proceedings of Computer Systems and Software Engineering (CompEuro 1992), pp. 442–446 (1992)
10. Lee, J., Kuo, J.-Y., Xue, N.-L.: A Note on Current Approaches to Extending Fuzzy Logic to Object-oriented modeling. International Journal of Intelligent Systems 16(7), 807–820 (2001)
11. Lee, J., Xue, N.-L., Hsu, K.-H., Yang, S.J.: Modeling Imprecise Requirements with Fuzzy Objects. Information Sciences 118(1-4), 101–119 (1999)
12. Leung, K.-S., Wong, M.-H.: Fuzzy Concepts in an Object Oriented Expert System Shell. International Journal of Intelligent Systems 7(2), 171–192 (1992)
13. Rumbaugh, J., Blaha, M., Premerlani, W., Eddy, F., Lorenson, W.: Object-oriented Modeling and Design. Prentice-Hall, London (1991)
14. Terwilliger, R.B., Campbell, R.H.: PLEASE: Executable Specifications for Incremental Software Development. Journal of Systems and Software 10(2), 97–112 (1989)
15. Zadeh, L.A.: Test-score Semantics as a Basis for a Computational Approach to the Representation of Meaning. Literary and Linguistic Computing 1(1), 24–35 (1986)

54

Humanistic Fuzzy Systems

E. Stanley Lee

I was fortunate to have the opportunity to contact with Professor Zadeh and fuzziness in the early 1970's. The frequent contacts, encouragements as well as the directions personally from Professor Zadeh became a continuous inspiration to my work. It is difficult to summarize and to re-call all the contacts, developments and meetings on fuzziness through the years but, at this remarkable point of time, I would like to congratulate the successful developments and applications on fuzziness during these past nearly 50 years and celebrate on Professor Zadeh's 90th anniversary.

54.1 Developments and Expectations of Fuzziness

I have my first contact with fuzziness and Professor Zadeh at the early 1970's. At that early time, it appears that fuzzy sets theory should be a very fruitful approach to modeling humanistic systems or systems that can only be described linguistically. As we know now, there exist very few applications of this theory in humanistic or social systems and the most fruitful developments so far is in the modeling and control of mechanistic systems. This surprising development can be explained from two different standpoints, namely, from the standpoints of the characteristics of the mechanistic systems and the difficulties in the handling of linguistic variables.

First, any large *practical* mechanistic system frequently have humanistic or social variables and thus it is very useful to be able to model the complete system with the special treatment of qualitative or linguistic variables by fuzzy sets. Secondly, even systems with no linguistic variables, it is usually very large and complex that a complete understanding of the system is practically impossible and thus some approximation by the use of fuzzy sets is very desirable. Thirdly, even for small systems with complete understanding – one example is in the focusing operation control in picture taking — some approximation is desirable or accuracy is not needed, furthermore, this approximation can make the approach realistic and considerably simple.

The second explanation can be obtained from the basic characteristics of the linguistic variables. There are two basic difficulties with these humanistic variables: first, they are difficult to represent mathematically on computer and second, even if they are represented, how to aggregate or "lumping" them together to obtain useful answers is a very difficult problem. These two difficulties were solved by professor Zadeh by representing linguistic variables by fuzzy numbers, granularity and fuzzy graphs, the mapping is obtained by the use of extension principle. Fuzzy logic and

R. Seising et al. (Eds.): *On Fuzziness: Volume 1*, STUDFUZZ 298, pp. 371–375.
DOI: 10.1007/978-3-642-35641-4_54

fuzzy if-then rules are ideally suited for representation, manipulation and aggregation of this fuzzy system.

Another ingredient is needed in the modeling and representation of social systems. This is because of the uncertainty, or the vagueness nature of the spoken language. In order to improve the model, a learning component should form an integral part of the modeling system. This learning component can continuous improve or update the already established model by the use of the newly available data. In other words, in addition to the problems of representation and aggregation, a fuzzy approach needs a learning component. Neural networks and support vector machines are ideal techniques for this purpose.

With the successful formation of these three components, namely, representation, aggregation and learning, more developments in modeling and control of social systems should be or hopefully expected in the future.

54.2 Initial Contact with Professor Zadeh and Fuzziness

My first contact with Professor Lotfi A. Zadeh and fuzziness was over forty years ago, while I was on sabbatical – Sabbatical leave from Kansas State University during 1972 – working with Professor Richard Bellman at University of Southern California. We were doing research work on the possibility of establishing global models representing the interactions between countries. The problem involves many quantitative as well as qualitative variables. Although at that early time simulation approach appears feasible, the problem is the handling of the qualitative variables.

Mathematical model making is an art. If the model is too large, not only the computation and analysis are difficult, the available data and understanding of the system may also be over used, resulting in a meaningless model. On the other hand, if the model is too small, although computation and analysis are easier, the results may be useless for being too simple. Furthermore, it appears that most practical large models cannot avoid qualitative, "humanistic" or "soft" variables – variables which are vague, difficult to define, and frequently subjective.

Professors Zadeh and Bellman are good friends, and I have the great opportunity to meet Professor Zadeh through Professor Bellman while I was at University of Southern California. Through the encouragements and directions from Professor Zadeh, I was able to do some research on fuzziness systems. In order to handle humanistic variables, we turned to fuzzy systems, which appear ideally suited for treating qualitative variables encountered in large practical models. During these investigations, it becomes increasingly clear that although computer is a very powerful tool, it is almost useless for modeling social systems whose variables are frequently linguistically represented. Furthermore, these linguistically represented variables – soft variables in soft computing as is well known now – have completely different characteristics: It is subjective instead of objective as required in scientific research. It is domain dependent and, philosophically, it is frequently desirable to be somewhat approximate instead of exact. Because of the desirability of some approximation, Professor Zadeh proposed the use of granularity, which can be represented by fuzzy graphs. Fuzzy graph of a function is a fuzzy relation. As suggested by Professor Zadeh, the mapping is obtained by the use of extension principle. Fuzzy logic and fuzzy if-then rules are ideally suited for the representation, manipulation and aggregation of this mapping fuzzy system.

54.3 Collaborations and Lectures on Fuzziness

During the last 20 years of the twentieth century, with Professor Zadeh's encouragement and collaboration, I give many lectures and also organized conferences in China, Taiwan and Singapore on fuzzy systems and fuzzy-neural network. For example, the *First Asian Fuzzy Systems Conference* was held in Singapore in November, 1993. And, with the support of Yuan-Ze University, Professor Zadeh was invited to give a series of lectures in Taiwan during December 1994. As a result of these activities, I met many colleagues doing research in this area and thus have the opportunity to continuously collaborating with these colleagues until now.

One particular area we worked on is in fuzzy relational inequalities with Professor P. Z. Wang and several of his students. At the time, Professor Wang was working at the Beijing Normal University and latter move to the University of Singapore. I am still collaborating with several colleagues met at that time. I was invited to give a series of lectures at the Chung Cheng Institute of Technology, Taiwan, on fuzzy set theory during May, 1991. Over 200 attended this lecture and the lecture was summarized in a monograph. Later on in 1993, I was appointed as a Chaired Professor at Yuan-Ze University, and on leave from Kansas State University. During this time, I collaborated with many colleagues on the use and development of fuzzy systems.

54.4 Collaborations and Developments on Direct Modeling of Engineering and Social Systems

One of the widely used learning algorithms is the neural network and, more recently, the support vector machine. Several investigators proposed different combined fuzzy-neural or fuzzy support vector machine networks. Using Takagi-Sugeno inference model, a fuzzy adaptive network (FAN) was constructed. The network of FAN provides a comprehensive visualization and adaptability system by retaining both the representation ability of fuzzy systems and the learning ability of neural network.FAN is a five-layered feed forward network. Each node in FAN performs a particular mode function on the incoming signals, which is characterized by a set of parameters. In order to reflect different adaptive capabilities, the nodes are represented by circles and squares. Circle nodes represent fixed nodes without parameters, while square nodes are adaptive nodes with parameters or fuzzy numbers to be adjusted. The basic difference from neural network is the presence of the adaptive nodes. In order to use back propagation, Gaussian membership functions are used. Both layers 1 and 4 are adaptive nodes. FAN is a powerful approximation tool for fuzzy systems, whose objective is to infer an association between specific input-output pairs. These input-output pairs are usually referred to as training data that characterizes the system to be identified. The training procedure is actually a sequence of adjusting the parameters in the network, including both consequence and premise parameters. The learning for the premise parameters is achieved by the use of back propagation. The error obtained in layer 5 is back propagated to layer 1. The consequence parameters set represents the coefficients of the linear functions in the fuzzy if-then rule. Since the resulting equations are similar to the fuzzy regression equations obtained by Professor Tanaka. Tanaka's approach, namely, by the use of linear programming to solve the learning problem is used.

The combined use of fuzzy inference systems and support vector machines forms another very useful learning network. Compared to neural network, support vector machine has two basic advantages: support vector machine is based on statistical learning theory and thus has a very good generalization characteristics and the learning problem can be solved by the use of quadratic programming instead of back propagation. Several fuzzy adaptive networks based on support vector machines were constructed and the results are compared with FAN.

Many modeling problems in engineering applications have linguistic or qualitative variables. With the collaboration of colleagues and PhD students, some of the systems studied by the use of fuzzy numbers or fuzzy inference networks are: water resources systems where both the quality and quantity are important with the consideration of pollution and flooding, the dynamic modeling of crop growth and irrigation with the consideration of historic rain fall, fuzzy clustering in parts coding and cell formation in group technology and cellular manufacturing, the influence of geometry and shape of welding bead in welding operations, and the fuzzification of the modeling of coal processing, liquefaction and gasification.

Another area we investigated is the modeling of the machining process with rotary ultrasonic machines for machining difficult and brittle material such as ceramic, silicon, titanium and its alloys. A particular interesting is in the manufacturing of silicon wafers where the dimensional error and the surface roughness must be very small. There are many unknown variables - due to the influence of thermal conductivity, heat problems and chemical reactivity – in this machining process and usually is carried out by experienced operators or experts. It appears ideally suited to use fuzzy if-then rules. We also used fuzzy numbers, fuzzy if-then rules, neural-fuzzy systems and fuzzy support vector machines to model social systems. One example is the credit rating problem in financial systems, where both qualitative and quantitative variables are present. Another problem is the heating or cooling of a facility with the consideration of thermal comfort. We also did some initial modeling on US presidential election. Some of the ergonomic problems investigated are the human stress and fatigue under various different working environments and the VDT legality, or in the viewing ability of a video display terminal.

54.5 Collaborations and Incorporating Fuzziness into Existing Modeling and Decision Making Approaches

Instead of directly modeling the engineering and social systems, many existing modeling, optimization and decision making techniques can be considerably improved by incorporating fuzziness. This is due to the fact that many systems are inherently uncertain. Many researchers have contributed in this area. The fuzzification of neural networks and support vector machines, as discussed above, are two examples in this direction. Other examples are the fuzzification of de novo programming, regression analysis algorithms, optimization or decision making algorithms and the use of fuzzy interactive decision-making to overcome some computational difficulties in multi-level decision making problems. Again, all of these approaches are taking the advantages of the inherent fuzziness of the problem. A monograph on fuzzy interactive computational approach for multi-hierarchy or multi-level system was writing. Another area is the multi-attribute decision making problems, where linguistic variables and not well defined variables are always present. In collaboration with Professor C. L. Hwang, we formed various versions of fuzzy multi-attribute decision making algorithms. Other areas investigated are the fuzzification of spatial statistics used in geography, mining and petroleum explorations and the fuzzification of queueing equations.

One of the most obvious fuzzification approaches is directly fuzzify the existing crisp equations or algorithms by using Zadeh's extension principle. We used this approach to fuzzify the equations or algorithms to obtain fuzzy queues and fuzzy de novo programming. However, the problem with this approach is that not only the resulting equations are very complicated due to the use of inverse functions in the mapping; the basic fuzzy inherent nature of the system is also not explored. Thus, almost all the fuzzifications were carried out by indirect approaches. In other words, instead of fuzzify the existing equations, new fuzzy equations were formulated by exploring the basic inherent fussy nature of the system.

54.6 Collaborations and Investigations on the Basic Theory of Fuzziness

Several different areas were investigated concerning the basic aspects of fuzziness, one is on the developments and relationships between fuzziness (or, possibility), evidence reasoning (due to Dempster and Schafer) and probability theory, the other is in the convexity and concavity of fuzzy sets from the standpoint of developing basic theories of fuzzy optimization, the third area is in the fuzzy relational equations, inequalities and constraints based on t-norms and other relational axioms, other areas investigated are interval arithmetic and its use in fuzzy arithmetic, interval fuzzy sets, and the comparison and ranking of fuzzy numbers.

55

ABC Intelligence on Fuzziness

Chin-Teng Lin

Three research areas, Artificial Intelligence (AI), Brain-like Intelligence (BI) and Computational Intelligence (CI) (denoted as ABC Intelligence), intertwine throughout my career. In my early days, I have shown strong interest in biology, psychology, and later on engineering. When I was an undergraduate student from 1982 to 1986, several hands-on courses on robotics covering automatic control, computer vision, AI, and microprocessor inspired my devotion of machine intelligence. When I arrived in West Lafayette of India for my graduate studies at Purdue University in 1988, I was given the book, Parallel Distributed Processing: Explorations in the Microstructure of Cognition by James L. McClelland, David E. Rumelhart and the PDP Research Group [1]. This book has opened my eyes on the "Connectionism", representing a set of approaches in the fields of artificial intelligence, cognitive psychology, cognitive science, neuroscience, and philosophy of mind. I have been deeply intrigued by the book due to its cross-disciplinary nature. My Ph.D. advisor, Professor George Lee, has leaded my ways to CI fields in fuzzy systems and neural networks. We published a paper on Fuzzy Neural Networks (FNN) in IEEE Trans. on Computers in 1991 [2], the early era of this field. We then wrote a FNN textbook together [3], published by Prentice Hall in 1996. During this period, I also received the enthusiastic support of Prof. Lofti Zadeh in FNN researches [4] which will be mentioned in details later. After receiving my doctoral degree, I returned to Taiwan to start tenure-track faculty position and research efforts in CI and its applications at NCTU. When I was appointed as the Director of NCTU Brain Research Centre at 2003, I branched into BI in computational neuroscience; my researches have embraced brain dynamics and the pursuit for Brain-inspired CI. What benefited me the most during this period is the international cooperation between NCTU and universities abroad. The interdisciplinary cooperation at home and abroad plays a pivot role at each transition of my research areas.

I was fortune to work with prominent researchers in AI, BI, and CI areas. Among them, Professor Zadeh's seminal work on Fuzzy Logic played the most important role in interconnecting these three "I"s. The linguistic representation and processing power of fuzzy logic is a unique tool to bridge symbolic intelligence and numerical intelligence gracefully. In my Ph.D. study in Purdue University around 1990, my research on the synergism of fuzzy logic and neural networks — fuzzy neural networks, the early era of this field, were majorly inspired by the "Soft computing" promoted by Professor Zadeh. He emphasized that the future would involve fuzzy logic, neural networks and genetic algorithms, and he lumped all these under the rubric of

R. Seising et al. (Eds.): *On Fuzziness: Volume 1*, STUDFUZZ 298, pp. 377–381.
DOI: 10.1007/978-3-642-35641-4_55

"soft computing". Professor Zadeh encouraged us to have the inclination and capability to become competent in all three of these areas. Moreover, I had the chance to receive his direct guidance when I visited his Lab in UC Berkeley as a BISC visiting scholar in 2004. During this period, I received valuable instructions from him on the linkage of fuzzy neural networks to biological neural networks, especially on the mapping of fuzzy automata and brain dynamics. Later on, the pioneering work of Bart Ksoko on Fuzzy Cognitive Map [5] and Jim Bezdek on Fuzzy C-means [6] further inspired me to seek for the synergism of low-level learning (e.g., neural networks) and high-level human-like thinking (e.g., fuzzy systems) into a functional unit with higher machine IQ. These have motivated me to concentrate on CI in general and on hybrid CI such as FNN in particular.

The inspiration from Professor Zadeh for me is not only in academics but also in the personal role model he set up. As a worldwide recognized "Father of Fuzzy Logic", Professor Zadeh's academic and social impacts to the world cannot be overemphasized. He also fully devoted himself to promote fuzzy logic around the earth. I am very impressed by his generosity in accepting my invitation to deliver keynote speeches in Taiwan several times during the past 20 years, taking many long flights from San Francisco to Taiwan. The photo (figure 55.1) was a treasured moment at IEEE SMC 2006. Even for the situation when he cannot make it due to time-conflict, Professor Zadeh still tried the best to deliver his talks in some ways. For example, in FUZZ-IEEE2011 held in Taiwan, he taped his talk and had it delivered to all the attendees in the conference on June 29th, 2011 in Taipei. His visits and talks directly contributed to the prosperous development of fuzzy logic researches in Taiwan and neighborhood regions. Moreover, what impressed me mostly is that, every time when he visited us, he concerned not only on the academics aspect but also on the economic/high-tech development aspects of Taiwan. I remembered vividly that when he visited us in 2005 for ICONIP2005, he told me some key numbers about Taiwan's economic situation when I received him in the Hall Lobby and elevator; the numbers that I were even not aware of! He also asked me to check out some related figures for him, including Taiwan government budget for technology R&D, National IT-related R&D projects funded by government and industries, IT industries export, Indexes of Knowledge Economics Development such as GPD percentage for industrial research, Revenue and Growing Rate of Top Uprising Industries in Taiwan, etc. He also told me clearly a very good development direction for Taiwan hi-tech industries based on his very knowledgeable understanding of Taiwan — Intelligent embedded software for IT systems/devices. This prediction has been proved to be totally true nowadays!

CI has formed a paradigm shift in the computation domain in the world. With CI's goal for addressing some of the most challenging real-world issues with biologically motivated computational paradigms, it serves as the most critical bridge to link natural intelligence and artificial intelligence for the scientific/engineering applications. Many profound breakthroughs and impacts on today's intelligent-technology world have emerged from the research and development covered by the fields of interest of CI. Among them, Fuzzy Set and Systems (FSS) deals with the theory, design or applications of fuzzy systems/models ranging from hardware to software, addressing the issues with not only biologically but also linguistically motivated computational paradigms. Four aspects are highly expected for the future development of CI in general and FSS in particular:

Fig. 55.1. From left to right: At IEEE SMC2006 in Grand Hotel of Taipei, Dr. De-Jeng, Liu (National Chiao Tung University); Prof. Lan-Da, Van (National Chiao Tung University, Taiwan); me (National Chiao Tung University, Taiwan); Prof. Hiroshi Tsuji (Osaka Prefecture University, Japan); Prof. Lofti Zadeh (University of California, Berkeley, USA); Prof. Yo-Ping Huang (National Taipei University of Technology, Taiwan); Prof. Ozer Ciftcioglu (Delft University of Technology, Netherlands); Prof. Chang-Shing Lee (National University of Tainan, Taiwan).

1. Scope Broadening of FSS. To reflect the rapid growing of Brain researches and Smart living technology nowadays [7], we will consider to expand the scope spectrum of FSS at both ends: at the basic research end, FSS can explore the interface of cognitive-neuroscience and fuzzy logic; at the application end, FSS can encourage the full-span applications of fuzzy logic in smart living technologies from house-hold devices to urban planning. This not only can keep FSS in the main stream of next-generation intelligent systems including "IIT (intelligent information technology)", but also can efficiently enlarge the domains of our authors and readers, and provide a platform for interdisciplinary cooperation/interactions of intelligent systems.
2. FSS-based interdisciplinary researches. One growing area is "biologically inspired information technique", which cover the aspects of sensation, perception,

reasoning, decision-making, and learning of biology as well as machine intelligence. Representative areas include fuzzy neural networks with structure and parameter learning, fuzzy term understanding in natural language processing, robotics and intelligent sensing, brain-computer interface, and NBIC [8] (Nano-Bio-Information technologies and Cognitive science). Some examples in the real-life applications of these basic researches are: (1) Smart City and Home - developing smart living technologies to improve health and save energy; (2) Intelligent Transportation Systems - developing vision-based intelligent technologies to improve safety and efficiency of vehicle transportation; and (3) Cognition and Neuroergonomics - developing devices to enhance human behavioral decision making under several forms of stress and cognitive fatigue. Another related and potential research is to develop and demonstrate fundamental transitional principles of operational neuroergonomics to enrich the CI realm. By gaining a better understanding of how human brain, body, and sensory systems work together to accomplish tasks in daily operational environments, we could develop basic principles for translation of basic brain reasoning and neuroscientific knowledge into optimal design of fuzzy human-system interfaces for complex operational settings. This, so-called "transitional brain and neuroscience", could enrich the basics and applications of CI realm not only on fuzzy systems, but also on neural networks and evolutionary computation. The impact applies to individuals, groups and society.
3. FSS Basic Researches with Breakthroughs and Innovations. While focusing on new technologies, we should continue to provide the highest recognition to the research works on relevant fundamental fuzzy theories. Laying the necessary biology/neuroscience and mathematical/physical foundations has been a source of pride and has brought great respect to our society.
4. Real-Life Applications: Starting in the era of 1990, Professor Zadeh's fuzzy logic has overwhelming impacts on the consumer products in the whole world. Around 1994, the Chinese University of Hong Kong conducted a survey to determine which consumer products were using Fuzzy Logic. The result was a thick report; some 150-200 pages long-washing machines, camcorders, microwave ovens, etc. What interested us weren't the particular applications so much as the breadth of applications — so many products were incorporating Fuzzy Logic. In the era of 2010 nowadays, there are rapidly growing needs of smart living technology, smart energy/carbon management, and translational neuroscience. And we have observed the full-span applications of soft computing in these areas including smart-home devices, smart meters, and homecare health/medical equipments, etc. Soft computing in general and fuzzy logic in particular have been becoming the backbone of next-generation iIT — intelligent IT industries! This phenomenon has also been reflected in the IEEE Transactions on Fuzzy Systems, where a new paper category for real-world applications of fuzzy logic was created to house the bounty application achievements of fuzzy logic.

All in all, I found myself enjoying very much in working in this area. Walking with the growing of the fuzzy theory and application, I am fortunate enough to be guided by the maestro of this generation, to be supported by numerous excellent coworkers and to be pleased to see the prospered developments and broad applications of the FSS.

Also, the innovative applications of the iIT industries are attracting much attention the present day. With the smart technology developing, I believe that it will bring a brand new life style to the modern people. And I am happy to realize that perhaps some contributions I am able to make to be a part of it.

Acknowledgement. This work was supported in part by the UST-UCSD International Center of Excellence in Advanced Bio-engineering sponsored by the Taiwan National Science Council I-RiCE Program under Grant Number: NSC-100-2911-I-009-101. This work was also supported in part by the Aiming for the Top University Plan of National Chiao Tung University, the Ministry of Education, Taiwan, under Contract 100W9633.

References

1. McClelland, J.L., Rumelhart, D.E., The PDP Research Group: Parallel Distributed Processing: Explorations in the Microstructure of Cognition: Foundations. The MIT Press, Cambridge (1986)
2. Lin, C.-T., George Lee, C.S.: Neural-Network-Based Fuzzy Logic Control and Decision System. IEEE Transactions on Computers 40(12), 1320–1336 (1991)
3. Lin, C.-T., George Lee, C.S.: Neural Fuzzy Systems: A Neuro-Fuzzy Synergism to Intelligent System. Prentice-Hall, Englewood Cliffs (1996)
4. Zadeh, L.A.: Fuzzy Logic, Neural Networks, and Soft Computing. Communications of the ACM 37(3), 77–84 (1994)
5. Kosko, B.: Fuzzy Cognitive Maps. International Journal of Man-Machine Studies 24(1), 65–75 (1986)
6. Bezdek, J.C., Ehrlich, R., Full, W.: FCM: The Fuzzy C-means Clustering Algorithm. Computers & Geosciences 10(2-3), 191–203 (1984)
7. Liao, L.-D., Lin, C.-T., McDowell, K., Wickenden, A., Gramann, K., Jung, T.-P., Ko, L.-W., Chang, J.-Y.: Biosensor Technologies for the Augmented Brain-Computer Interface in the Next Decades. Proceedings of the IEEE (accepted, 2012)
8. Roco, M.C., Bainbridge, W.S.: Converging Technologies for Improving Human Performance: Nanotechnology, Biotechnology, Information Technology and Cognitive Science. Kluwer Academic Publishers, National Science Foundation (2003)

Fuzzy Set and Possibility Theory in Optimization: L. Zadeh's Contributions

Weldon A. Lodwick and K. David Jamison

Abstract. The thesis of this article is that after the Bellman/Zadeh seminal 1970 fuzzy optimization paper [1], two new powerful optimization types have emerged over the last four decades: (1) Flexible optimization and (2) Optimization under generalized uncertainty. Flexible optimization arises when it is necessary to relax the meaning of the mathematical relation of set belonging. Optimization under generalized uncertainty arises when it is necessary to represent parameters whose values are known only partially or incompletely.

Our objective is to highlight the neglected power of what Bellman/Zadeh [1] introduced in 1970 which has led to two new, powerful, and distinct types of optimization, flexible optimization and optimization under generalized uncertainty. In the literature these are called fuzzy optimization and possibilistic optimization and subsumed under one classification, fuzzy optimization. This has led to some confusion and we clarify and argue that indeed, most applied optimization problems should consider the power and effectiveness of using fuzzy set and possibility theory as mathematical languages to state and solve real optimization problems. Both fuzzy set theory [16] and possibility theory [17] were developed by L. Zadeh. He was also the first to apply fuzzy set theory to optimization. The appearance of the 1988 Dubois/Prade book [3] pointed out that possibility theory is distinct from (though at times related to) fuzzy set theory and that possibility theory was not complete without its dual, necessity.

The chronology of fuzzy optimization, as we understand it is that in 1965 [16], L. Zadeh developed fuzzy set theory. In 1970 [1], the first application of fuzzy set theory to optimization was made by Bellman and Zadeh. In 1978 [17], L. Zadeh developed possibility theory. In 1986, M. Luhandjula [9] applied possibility theory to optimization problems. In 1988 [3], Dubois and Prade clearly separated possibility theory as distinct from fuzzy set theory adding to the canon of possibility its dual, necessity. Once necessity was clearly delineated as giving a more complete description of possibility theory, it eventual led to a more complete possibilistic optimization which as a *dual pair* has only recently been exploited [13], [14].

However, in fuzzy optimization, there is still confusion between how to use fuzzy sets distinct from using possibility/necessity distributions. We maintain that the only connection between the two distinct types of optimization is that they most often use "fuzzy intervals" as parameters. Fuzzy intervals are able to encode transitional set

R. Seising et al. (Eds.): *On Fuzziness: Volume 1*, STUDFUZZ 298, pp. 383–387.
DOI: 10.1007/978-3-642-35641-4_56

belonging (fuzzy sets) and information deficiency (possibility). To clearly delineate the differences, we propose two different names that will encompass these two distinct types, flexible optimization which uses fuzzy set membership functions (fuzzy intervals being one type of membership function) and optimization under generalized uncertainty, one type uses possibility and necessity distributions. Both flexible and generalized uncertainty optimization are broader than just fuzzy optimization and possibilistic optimization though both are one type in their respective categories. Moreover, we state that not only is there a semantic and theoretical difference between flexible optimization and optimization under generalized uncertainty, there is a significant algorithmic difference.

Flexibility, as modeled by fuzzy set theory and used here, pertains to the relationship "belonging" or "is an element of," that is, the relationship $\in$ in the context of a constraint set. When belonging takes on the meaning of "come as close as possible but do not exceed" then there is flexibility. For example, one may have a deterministic constraint of a fixed upper bound on the hours of labor based on the number of employees a company has. However, it is often possible and no doubt wise to have a pool of overtime and temporary laborers that the company can draw on at a perhaps increased cost. This is what we mean by flexibility. Flexibility is typically given by a concave function over a compact support and includes fuzzy numbers which are typically used, they are a particular type of membership function. Fuzzy sets have a semantic of transitional set belonging so that it is a natural mathematical language for problems in which flexibility is an inherent part of the modeling and analysis.

The meaning of generalized uncertainty as used here pertains to parameters of an optimization problem whose values are known to be incomplete or partially known and are distributions or whose distributions lie within an envelope bounded above and below by two distributions. That is, some or all the input data of the model is not a real number nor a probability density function, but nevertheless uncertain, not precise. Possibility theory is a natural mathematical language in which information deficiency is stated and analyzed.

Possibility theory grew out of fuzzy set theory and in fact, the title of the first paper on possibility theory [17] is "Fuzzy sets as a basis for a theory of possibility." So what distinguishes fuzzy sets and possibilities? First, the semantics of fuzzy sets and possibilities are distinct. Fuzzy sets are associated with transitional set belonging and possibilities are associated with information deficiencies. Second, the measures they induce are distinct (see [11]). That is, their underlying axiomatics are distinct. The confusion between fuzzy set theory and possibility theory arises most especially in optimization since both fuzzy optimization (flexible optimization) and possibility optimization (optimization under generalized uncertainty) use fuzzy intervals (numbers) which gives the impression that there is no distinction between fuzzy sets and possibility. In particular, fuzzy intervals (numbers) may mathematically encode transitional set belonging and fuzzy intervals may encode information deficiency or partially known values or non-specificity. To insure that distinctions are maintained, one must understand the semantic of the underlying parameters and relationships, that is, the model. Secondly, the data from which the parameters and relationships arose will dictate its semantics. If the model encodes relaxed set belonging, it represents flexibility and its language is fuzzy set theory. If the data arises from partial, ambiguous, or incompletely specified fountains, then it is a type of generalized uncertainty. Possibility theory is a mathematical language that may be used to represent and manipulate this type of uncertainty depending on the underlying structure.

One interpretation of possibility is simply as an ordering of a collection of possible outcomes of an uncertain event with $p(x) > p(y)$ implying that outcome x is considered more possible than outcome y. In this setting $p(x)$ is akin to a belief function. The interpretation we adopt in our setting of optimization is that of an upper probability bound. Thus, for a random outcome X,

$$\{X \mid Pos(A) \geq prob(X \text{ in } A) \geq Nec(A)\}$$

gives a bound on an unknown probability measure. This interpretation of possibility theory provides a method for combining possibility, intervals, probability, and interval-valued probability into a single mathematical framework (see [5], [8], [10], [15]).

A real-valued (deterministic) optimization model, as is well-known, is a *normative* mathematical model whose underlying system is most often *constrained* and its general (deterministic) form is:

$$z = \min f(x,c) \tag{56.1}$$

$$\text{subject to } g_i(x,a) \leq b_i \;\; i = 1,...,M_1 \tag{56.2}$$

$$h_j(x,d) = e_j \;\; j = 1,...,M_2. \tag{56.3}$$

We denote the constraint set by $\Omega = \{x \mid g_i(x,a) \leq b_i \;\; i = 1,...,M_1, h_j(x,d) = e_j \;\; j = 1,...,M_2\}$. It is assumed that $\Omega \neq \emptyset$. The values of a,b,c,d, e, $\leq,=$, and *opt* (min / max) are input (data), parameters, relationships, and operations. Our general model can be formulated as

$$z = \min f(x,c) \tag{56.4}$$

$$x \in \Omega(a,b,d,e), \tag{56.5}$$

where we denote the constraint set as a function of the input parameters for emphasis. When the $\in$ of "element of Ω" is relaxed, then we have flexible optimization. Note that from this point of view, all that has been called "soft constraints," including in meaning of "minimize" as it relates to the objective function, can be considered as a relaxation of set belonging. Thus, we will distinguish two broad types of optimization problems - *flexible optimization* and *optimization under generalized uncertainty*.

Two key ideas in the context of optimization for what L. Zadeh introduced are:

- Optimization models of real systems are very often *satisficing* (see [12]) in which case fuzzy and possibility methods are key approaches to satisficing optimization models. Satisficing is defined by Herbert Simon (see [12]) to mean that decision makers rarely work with the deterministically "best" solution or are even able to obtain "the best" solution to a real problem, but seek to obtain solutions that are satisfying. Solutions that are satisficing are inherently flexible in their values. Clearly, the usual deterministic models, if used to model decision processes described by Simon, need to be modified. Fuzzy and possibility optimization are able to model satisficing in a natural and direct way. They are mathematical languages that can and do encode satisficing problems.

- Many satisficing optimization models are *epistemic*. That is, models that are epistemic are those which we, as humans, construct from knowledge about a system rather than models that are constructed from the system itself. For example, an automatic pilot of an airplane models the system physics. A fuzzy logic chip that controls a rice cooker is a model of what we know about cooking rice rather than the physics of rice cooking.

We summarize what has been articulated using, additionally, insights from [2].

- Fuzzy optimization (what we call here *flexible optimization*), offers a bridge between numerical (deterministic) approaches and the linguistic or qualitative ones. The thrust of these approaches are to provide the analyst with what is the uncertainty in the results of a decision process.
- Fuzzy set theory and its mathematical environment of aggregation operators (*"and"*, *t-norms*), interval analysis, constraint interval analysis [6], fuzzy interval analysis, constraint fuzzy interval analysis [6], gradual numbers [4], and preference modeling, provide a general framework for posing decision problems in a more open way and provides a unification of existing techniques and theories.
- Fuzzy set theory has the capacity of translating linguistic variables into quantitative terms in a flexible and useful way.
- Possibility theory explicitly accounts for lack of information, avoiding the use of unique, often uniform, probability distributions.

Both flexible optimization and generalized uncertainty optimization problems begin with (56.1), (56.2) and (56.3) in the presence of data and relationships $\{a,b,c,d,e,\leq,=,\in\}$ that are either all or a mixture of real, interval, interval-valued probability, possibility with at least one parameter being one of these types.

Given our delineation, we follow the development in [7] who distinguish solution methods associated with each of our two optimization types.

1. *Fuzzy Decision Making:* Given the set of real-valued (crisp) decisions, Ω, and fuzzy sets, $\{\tilde{F}_i \mid i = 1 \text{ to } n\}$, find the optimal decision in the set Ω. That is,
$$\sup_{x\in\Omega} h\left(\tilde{F}_1(x),\dots,\tilde{F}_n(x)\right), \tag{56.6}$$
where $h : [0,1]^n \to [0,1]$ is an aggregation operator often taken to be the *min* operator, and $\tilde{F}_i(x) \in [0,1]$ is the fuzzy membership of x in fuzzy set $\tilde{F}_i$.
2. *Possibility Decision Making:* Given the set of real-valued (crisp) decisions, Ω, and the set of possibility distributions representing the uncertain outcomes from selecting decision $\mathbf{x} = (x_1,\dots,x_n)^T$ denoted $\Psi_x = \{\hat{F}^i_x, i = 1,\dots,n\}$, find the optimal decision that produces the best set of possible outcomes with respect to an ordering U of the outcomes. That is,

$$\sup_{\Psi_x \in \Psi} U(\Psi_x), \tag{56.7}$$

where $U(\Psi_x)$ represents an evaluation function of the set of distributions of possible outcomes $\Psi = \{\Psi_x | x \in \Omega\}$.

Our exposition has emphasized and distinguished the differences between fuzzy set theory and possibility and the implications to optimization models that use these entities. We indicated how the semantics are important in determining which of the various approaches must be used in seeking a solution to associated models.

References

1. Bellman, R.E., Zadeh, L.A.: Decision-making in a Fuzzy Environment. Management Science, Serial B 17, 141–164 (1970)
2. Dubois, D.: The Role of Fuzzy Sets in Decision Sciences: Old Techniques and New Directions. Fuzzy Sets and Systems 184(1), 3–28 (2011)
3. Dubois, D., Prade, H.: Possibility Theory. Plenum Press, New York (1988)
4. Fortin, J., Dubois, D., Fargier, H.: Gradual Numbers and Their Application to Fuzzy Interval Analysis. IEEE Transaction on Fuzzy Systems 16, 388–402 (2008)
5. Jamison, K.D., Lodwick, W.A.: The Construction of Consistent Possibility and Necessity Measures. Fuzzy Sets and Systems 132, 1–10 (2002)
6. Lodwick, W.A.: Constrained Interval Arithmetic. CCM Report, vol. 138 (February 1999)
7. Lodwick, W.A., David Jamison, K.: Theory and Semantics for Fuzzy and Possibility Optimization. Fuzzy Sets and Systems 158(17), 1861–1871 (2007)
8. Lodwick, W.A., Jamison, K.D., David Jamison, K.: Interval-valued Probability in the Analysis of Problems Containing a Mixture of Possibility, Probabilistic, and Interval Uncertainty. Fuzzy Sets and Systems 159(1), 2845–2858 (2008)
9. Luhandjula, M.K.: On Possibility Linear Programming. Fuzzy Sets and Systems 18, 15–30 (1986)
10. Neumaier, A.: Structure of Clouds (2005), submitted - downloadable: `http://www.mat.univie.ac.at/~neum/papers.html`
11. Puri, M.L., Ralescu, A.D.: Fuzzy Measures are not Possibility Measures. Fuzzy Sets and Systems 7, 311–313 (1982)
12. Simon, H.A.: Models of Bounded Rationality. MIT Press, Cambridge (1997)
13. Thipwiwatpotjana, P.: Linear Programming Problems for Generalized Uncertainty. Ph.D. Thesis, University of Colorado, Department of Mathematical and Statistical Sciences (2010)
14. Thipwiwatpotjana, P., Lodwick, W.A.: Pessimistic, Optimistic, and Minimax Regret Approach for Linear Programs Under Uncertainty (submitted, 2012)
15. Weichselberger, K.: The Theory of Interval-probability as a Unifying Concept for Uncertainty. International Journal of Approximate Reasoning 24, 149–170 (2000)
16. Zadeh, L.A.: Fuzzy Sets. Information and Control 8, 338–353 (1965)
17. Zadeh, L.A.: Fuzzy Sets as a Basis for a Theory of Possibility. Fuzzy Sets and Systems 1, 3–28 (1978)

57

Optimization under Fuzziness

Monga Kalonda Luhandjula

Abstract. Fuzzy set theory has a strong track record of success in the field of Optimization under uncertainty. It offers a proper framework for coming to grips with situations where imprecision and complexity are in the state of affairs in an Optimization setting. This paper presents my personal views on the descriptive and prescriptive power of Fuzzy set Theory in letting informational and intrinsic imprecision be taken into account in an Optimization model. The paper is jam-packed with information on how and why I started doing research in this field, along with encouragements and inspiration I got from Professors L. A. Zadeh and H. J. Zimmermann.

57.1 First Steps in Fuzzy Set Theory

My interest on Fuzzy set Theory dates back to early 1980's when I was completing a PhD at the Free University of Brussels. I was then fascinated by Zadeh's theory [1],[2] that challenges traditional reliance on two-valued logic and classical set theory as a basis for scientific inquiry. I quickly realized that Fuzzy set Theory was offering opportunities both as a language and a tool to cope with imprecision and complexity. I then was extremely keen to apply this theory to decision situations where imprecision plays a pivotal role. It was also clear to me that although probabilistic theories [3], [4],[5] claim to model decision making under uncertainty, there was a qualitatively different kind of imprecision which was not covered by these apparatus, that is: inexactness, ill-definedness, vagueness.

Correctness of statements and judgements, degrees of credibility, plausibility have little to do with occurence of events, the back-bone of Probability Theory. I was comforted on the above mentioned views by Prof H.-J. Zimmermann, the first Principal Editor of the Journal *Fuzzy Sets and Systems.* He provided me with guidance and intuition on when Fuzzy set Theory venue may be most appropriate and ultimately successful in the field of Mathematical Programming under uncertainty. Fruitful exchanges on many aspects related to Fuzzy Mathematical Programming namely, Flexible programming [6], Mathematical Programming with fuzzy number coefficients [7], Duality and sensitivity analysis, were catalysts that kept me continuing work on this field.

Early 1990's, I had a golden opportunity to visit Prof Zadeh at University of California, Berkely. I was impressed by his breadth of knowledge about all facets of his theory. I then took advantage of this visit to learn more about Fuzzy Set Theory, its

R. Seising et al. (Eds.): *On Fuzziness: Volume 1*, STUDFUZZ 298, pp. 389–393.
DOI: 10.1007/978-3-642-35641-4_57

relation with Black's work on vagueness [8] and some philosophical related issues ranging from ontological to application levels via epistemological one [9].

57.2 From Optimization to Fuzzy Optimization

Optimization is a very old and classical area which is of high concern to many disciplines. Engineering as well as Management, Politics as well as Medicine, Artificial Intelligence as well as Operations Research and many other fields are in one way or another concerned with Optimization of designs, decisions, structures or information processes. In a deterministic environment using a single well-defined criterion for evaluating potential alternatives, the optimal decision can be obtained through user-friendly mathematical programming software. Optimization procedure is, in this case a batch-type process assuming a closed model in which all information is available and in which the Decision Maker could provide and process all information simultaneously. In a turbulent environment involving intrinsic or informational imprecision, the Optimization process is not that simple. Fuzzy set Theory has proven to be of great help in representing and treating such imprecise information.

57.3 Fuzzy Mathematical Programming

The language of Fuzzy set Theory has been exploited with good reasons [10] to tolerate some leeways in the formulation of goals and constraints of a mathematical program. The soft goals and constraints of the mathematical program are, in this case represented by appropriate fuzzy sets reflecting the viewpoints of the Decision maker.

Making use of the Bellman-Zadeh's confluence principle [11] one may single out a satisfying solution in this context. Several variants of Zimmerman's model may be found in the literature [12] , [13], [14]. In [15], [16], Zimmermann's approach has been carefully adapted to deal with multi-objective linear and fractional problems.

Another situation where Fuzzy Set Theory enters into the dance in a mathematical programming setting is when considered parameters are not well-known and are modelized by fuzzy numbers. Several methods are described to solve this problem, see e.g. [17] and references therein. The general principle behind these methods is to convert the original problem into deterministic terms, by sticking as well as possible to uncertainty principles. Existing transformation strategies are based either on rules for comparing fuzzy numbers or on exploration of possibility and necessity measures. It is worth pointing out that a large amount of applications of Fuzzy mathematical programming exists supporting the efficiency and the effectiveness of Fuzzy Optimization techniques. An interested reader is referred to [18],[19] for a sample of these applications.

57.4 Fuzzy Stochastic Optimization

In some real-life problems one has to base decision on information which is both fuzzily imprecise and probabilistically uncertain. Although consistency indexes providing a union nexus between imprecision of possibilistic nature and uncertainty of stochastic type exist, there are no reliable ways of transforming one to another. This calls for new paradigms for incorporating simultaneously the two kinds of undeterminacy into mathematical programming models. Fuzzy stochastic Optimization is an attempt to fulfill this need.

The general methodological approach for Fuzzy Stochastic Optimization problems consists of crafting an uncertainty processing that suits the particular characteristics of the problem at hand, exploiting to a great extent the available structure [20]. This processing should embody a device that interprets the original Fuzzy Stochastic Optimization problem from the probabilistic and possibilistic lenses and perform good conversions from both the standpoints of effectiveness and efficiency. This methodological approach has given rise to different methods for solving Fuzzy Stochastic Optimization problems, see. e.g. [16], [21], [22], [23]. These ideas have been successfully applied to Portfolio selection [24] and to many other concrete real-life problems [25].

57.5 Concluding Remarks

Mathematical programming under fuzziness provides a corpus of scientific knowledge that permits to cope effectively with vagueness instead of merely thwarting, suppressing or downplaying it.

Freud told us that the history of science is the history of an alienation. Since Copernicus we no longer live at the centre of the universe; since Darwin man is no longer different from other animals and since Freud himself, conscience is just the emerged part of a complex reality hidden from us.

Paraphrasing Freud, we can say that since Zadeh we are no longer forced to approzimate real problems of the more-or-less type by yes-or-no type models. This is crucial in this post-modern era characterized by fragmentation of the truth and ascendancy of approximate reasoning.

Among lines for further development in the field of mathematical programming under fuzziness we may mention the following.

- Extension of the Bellman principle to the fuzzy and fuzzy stochastic cases so as to develop Fuzzy and Fuzzy stochastic cases so as to develop Fuzzy and Fuzzy Stochastic dynamic programming.
- Deep comparison of Fuzzy and Fuzzy stochastic Optimization techniques. This may help to design a user friendly Decision Support System able to help a Decision maker confronted with a problem of optimization under fuzziness.

References

1. Zadeh, L.A.: Fuzzy Sets. Information and Control 8, 338–353 (1965)
2. Zadeh, L.A.: The Concept of Linguistic Variable and its Application to Approximate Reasoning. Parts 1, 2, 3. Information Science 8, 199–249; 8, 301–357; 9, 43–80 (1975)
3. Karlin, S., Taylor, H.M.: A First Course in Stochastic Processes. Academic Press, New York (1975)
4. Winkler, R.L.: Statistics: Probability, Inference and Decision. Holt, Rinehart and Winston, Chicago (1971)
5. Winkler, R.L.: Introduction to Bayesian Inference and Decision. Holt, Rinehart and Winston, New York (1972)
6. Zimmermann, H.-J.: Description and Optimization of Fuzzy Systems. International Journal of General Systems 2, 209–215 (1976)
7. Ramik, J., Rimanek, J.: Inequality Relation Between Fuzzy Numbers and its Use in Fuzzy Optimization. Fuzzy Sets and Systems 16, 123–138 (1985)
8. Black, M.: Vagueness: An Exercise in Logical Analysis. Philosophy of Science 4, 427–455 (1937)
9. Kubinski, T.: An Attempt to Bring Logic Nearer to Colloquial Language. Studia Logica 10, 61–72 (1960)
10. Zimmermann, H.-J.: Fuzzy Programming and Linear Programming With Several Objective Functions. Fuzzy Sets and Systems 1, 46–55 (1978)
11. Bellman, R.E., Zadeh, L.A.: Decision Making in a Fuzzy Environment. Management Science 17, 141–164 (1970)
12. Verdegay, J.L.: A Dual Approach to Solve the Fuzzy Linear Programming Problem. Fuzzy Sets and Systems 14, 131–141 (1984)
13. Rommelfanger, H.: A General Concept for Solving Linear Multicriteria Programming Problems With Crisp, Fuzzy or Stochastic Values. Fuzzy Sets and Systems 158, 1892–1904 (2007)
14. Luhandjula, M.K.: Fuzzy Mathematical Programming: Theory, Applications and Extensions. Journal of Uncertain Systems 1, 123–136 (2007)
15. Leberling, H.: On Finding Compromise Solutions in Multicriteria Problems Using the Min Operator. Fuzzy Sets and Systems 6, 105–118 (1981)
16. Luhandjula, M.K.: Fuzzy Approaches for Multiple Objective Linear Fractional Optimization. Fuzzy Sets and Systems 13, 11–24 (1984)
17. Buckley, J.J.: Possibility Programming. Fuzzy Sets and Systems 26, 135–138 (1988)
18. Sommer, G., Pollatschek, M.A.: A Fuzzy Programming to an Air Pollution Regulation Problem. European Journal of Operations Research 10, 303–313 (1978)
19. Zmeskel, Z.: Application of the Fuzzy-stochastic Methodology to Appraising the Firm Value as a European Call Option. European Journal of Operations Research 135, 303–310 (2001)
20. Luhandjula, M.K.: Fuzzy Stochastic Linear Programming: Survey and Future Research Directions. European Journal of Operations Research 174, 1353–1367 (2006)
21. Van Hop, N.: Solvind Linear Programming Problems Under Fuzziness and Randomness Environment Using Attainment Values. Information Sciences 177, 2971–2984 (2007)
22. Liu, B.: Fuzzy Random Dependent-chance Programming. IEEE Transactions on Fuzzy Systems 9(5), 721–726 (2001)

23. Mohan, C., Nguyen, H.T.: An Interactive Satisficing Method for Solving Multi-objective Mixed Fuzzy-stochastic Programming Problems. Fuzzy Sets and Systems 117, 61–79 (2001)
24. Zmeskel, Z.: Value at Risk Methodology Under Soft Condition Approach. European Journal of Operations Research 161, 337–347 (2005)
25. Weber, K., Cromme, L.: Decision Support System in Marketing by Means of Fuzzy Stochastic Optimization. In: 75th Annual Meeting of the Gesellschaft für Angewandte Athematik and Mechanik (GAMM), Dresden, March 21-27 (2004)

58

How Much "Fuzzy" Has Been (and Is) My Life? A Few Impressions from a Physicist Debated between Experiments and Data Analysis of High-Energy Astrophysics

Maria Concetta Maccarone

My approach to fuzzy sets and possibility theory goes back many years ago when, at the beginning of Eighties' with Vito Di Gesù we started our studies on pattern recognition, features selection, image analysis and mathematical morphology.

Nowadays my scientific activity is a bit different and my memory may be blurred; nevertheless I'll try to remember the intriguing "atmosphere" we lived at that time.

In 1981 officially born the Institute which I belong, early named IFCAI, Institute of Cosmic Physics and Computer Science Applications. The Institute, established in Palermo and headed by Prof. Livio Scarsi, comprised mainly researchers in astrophysics and some external collaborators as Vito Di Gesù, professor at the Department of Mathematics. The denomination of the new Institute was not casual: Livio, my director, believed in the importance of interdisciplinary world and in its evolution; and Vito, with the same feeling, had the opportunity to define at IFCAI the group of "informatics applied to astrophysics" in which Vito and I, we worked together for a long time, about 20 years.

Our interests were soon addressed to new methods in the field of data and image analysis, aiming to obtain valid solutions alternative with respect to the classical ones. The starting point was that the physics science try to build exact models of phenomena, by analyzing experimental data, and to use these models to make predictions. It is obvious that in this process a very crucial point is the correct analysis and evaluation of the experimental data. Probability theory and algebra are surely essential to carry out this process as well as the human experience which may play a fundamental role whenever inexactness and vagueness are inherent to the data, or when cloudy quantities exist that are not describable in terms of probability distributions. The development of the physics is based on claims supported by results and models which must be confirmed or not by new experimental results. Physics is only able to hold temporary truths and this uncertainty is its beauty: main effort of the scientists is to convince the community, with subjective reasoning, about the majority of their claims. Based on this "human reasoning" we devoted our attention to the young fuzzy sets theory [12]; its application to image analysis seemed to be very promising when the probability model was not available or difficult to issue.

R. Seising et al. (Eds.): *On Fuzziness: Volume 1*, STUDFUZZ 298, pp. 395–399.
DOI: 10.1007/978-3-642-35641-4_58

Nevertheless, in analyzing our data we maintained the combination of probability and possibility theory, as a matter of fact that complementary information comes out from both of them.

One of our first works in this field was devoted to the classification of shapes in binary sparse images. In general, as Zadeh expressed in its Principle of Incompatibility ([12], [13], p. 28), complexity and precision bear an inverse relation to one another in the sense that, when the complexity of a problem increases, the possibility of analyzing it in precise terms diminishes. Thus 'fuzzy thinking' may not be deplorable, after all, if it makes possible the solution of problems which are too much complex for precise analysis. A complex problem may therefore be the recognition and classification of shapes where there are some unfavorable conditions for a classical analysis: the evaluation of the local points density is affected by the high data fluctuations; the inherent meaning of the measurements may be affected by the experimental environment which introduces dummy effects and need the knowledge of some a-priori artifacts; the data space dimensionality is high; the parameters dependence is not linear; and so on. These conditions make difficult to operate with only probabilistic methods, which are objective, and therefore we used also a fuzzy model where a membership degree is subjective.

So, starting from the theory developed by Zadeh and making use of the 'fuzzy entropy' defined by De Luca and Termini [4], we entered in the field of features selection. As first point we considered the case of sparse images characterized by data with low statistics and describing shapes of astronomical interest. By combining cluster analysis and uniformity test based on the statistical properties of the Minimum Spanning Forest, hierarchical decision criteria, and possibility functions, our works were firstly presented at the astronomical community [1], at the Pattern Recognition ICPR Conferences (1984, 1986) and at the first IFSA Congress, held in Palma de Mallorca, 1985. A detailed treatment of our method [2] was published in 1986; in brief, it made use of both cluster analysis and possibility theory to select the most significant features in multidimensional data.

I must stress that at the time, as always, new methods found no easy way into the astronomical community: statistics and probability theory remained primary and predominant with respect to the young fuzzy sets and possibility theories. It was difficult to convince the community of the applicability and effectiveness of these methods, but this gave us the impetus to continue even deeper into their study and application. Proof of this was that in 1988, at the 3rd Data Analysis in Astronomy Workshop, directors Livio Scarsi and Vito di Gesù, Professor Lofti Zadeh was invited to give a talk on "Fuzzy approach to random image analysis" (see Fig. 58.1). Unfortunately, due to something unexpected, I don't remember what, Zadeh was not able to attend the workshop but some authors presented fuzzy sets applications in astronomical cases, so activating a fruitful discussion with other scientists who were themselves defined against the world "fuzzy".

Vito and I, we continued to apply fuzzy logic concepts and tools in several different frameworks, always related to the astronomical data, as in the management of pictorial database and in the development of knowledge-based systems to analyze images [3], [5], [6]. Nevertheless, the most funniest application was the work we carried out during the Nineties, i.e. the extension of the mathematical morphology to gray-level images making use of fuzzy sets and possibility theories. The formation of an astronomical image is often accompanied by degradation; a pre-processing phase can be needed to enhance the significant parts without loss of relevant information before to physically describe and interpret the image data. A powerful framework in this case comes from the classical mathematical morphology theory which, by adopting the concept of umbra or multi-set, studies the transformations of an image when it interacts with a matching pattern (the so called structuring element) through well-defined local operators (erosion, dilation, ...).

Differently from the classical one, our approach consider the gray-level image as a fuzzy set to be processed via morphological and logical operators re-defined in the fuzzy space [7], [8]. In brief, firstly we modeled our gray-level images in fuzzy sets. In fact, if the gray levels of an image are properly scaled, we can regard the intensity of a pixel as its membership degree to the set of high-valued pixels; thus a gray level image can be regarded as a fuzzy set. To perform the scaling, we identified specific membership functions, or fuzzifiers, depending on the nature of the problem to be solved and on the available a-priori knowledge; for example, it could be convenient to fuzzify the image by using a simple linear scaling or applying a contrast intensifier or enhancing specific intensity bands. Then we defined the morphological operators in the fuzzy space as an extension of the classical Boolean ones. The application of such fuzzy morphological operators (fuzzy erosion, ...) or their combination (fuzzy closing, ...) produces always a fuzzy set. Finally, we defined fuzzy logical operators, mainly based on the "min", "max", and "fuzzy entropy" functions, which allowed us to reach several goals of low and medium level of image analysis such as cleaning, noise reduction, edge detection, skeletonizing [9], [10].

I'm very close and affectionate to this work that was one of the first in the field and involved me for several years; at that time, the use of the morphological framework in the domain of astronomical imaging was relatively limited yet, and this fact contributed to further stimulate our activity. A positive result was also the inclusion of fuzzy thinking among the topics of the European CCMA scientific network, Converging Computing Methodologies in Astronomy [11], activated and chaired by me in the period 1995-1997, and born as natural corollary of the IAPR-TC13, Pattern Recognition in Astronomy and Astrophysics Technical Committee, activated by Vito some years before.

But, life is always changing and, at the beginning of the new millennium my Institute moved its denomination in IASF, Institute of Astrophysics and Cosmic Physics, partially losing its early Computer Science component and assigning me to new responsibilities and commitments.

Anyway, even today and not only in my research activity, I always take into account fuzziness, fuzzy reasoning and its related: from my point of view, fuzzy sets and possibility theory reflect the human processes but they cannot be separated from statistics and probability. The two theories are not competitive each other; rather, they are compatible and complementary, as in the case of the clustering method based on the collaborative use of fuzzy sets and probability. In the application of the mathematical morphology described before, the possibility theory plays its main role in analyzing the single images and in classifying them but the interpretation phase requires a-priori knowledge mainly based on statistics and coded in a probabilistic language. Without entering in a philosophical context, in general, in our life, we take into account the statistics of previous events of a given type to foresee the probability of their next appearance but we use our subjective perception and knowledge to assign a degree of possibility to that appearance.

In Italy, at the end of Sixties, a movie comedy was entitled "*certainly, most certainly, ... indeed probable*!"; some years later, a theater actor based its cabaret piece on the statement "*is possible*, *is not possible*, *it is impossible*!". The co-existence of these sentences is part of the "fuzzy" component of the human being, ... and my life is surely a bit "fuzzy".

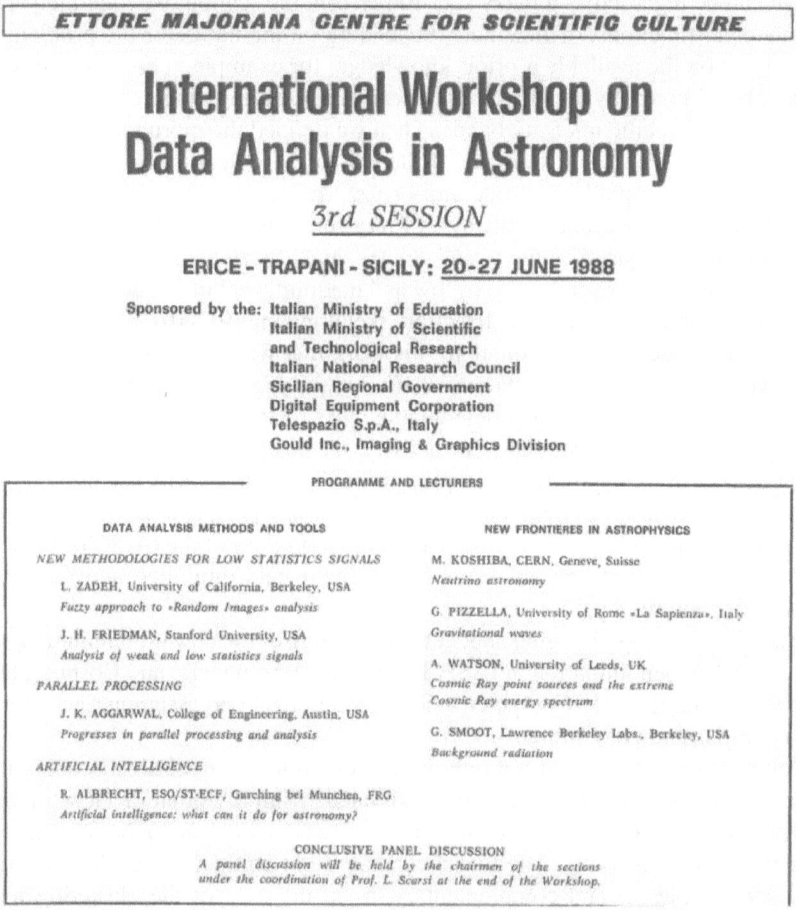

ETTORE MAJORANA CENTRE FOR SCIENTIFIC CULTURE

International Workshop on Data Analysis in Astronomy

3rd SESSION

ERICE - TRAPANI - SICILY: 20-27 JUNE 1988

Sponsored by the: Italian Ministry of Education
Italian Ministry of Scientific and Technological Research
Italian National Research Council
Sicilian Regional Government
Digital Equipment Corporation
Telespazio S.p.A., Italy
Gould Inc., Imaging & Graphics Division

PROGRAMME AND LECTURERS

DATA ANALYSIS METHODS AND TOOLS

NEW METHODOLOGIES FOR LOW STATISTICS SIGNALS

L. ZADEH, University of California, Berkeley, USA
Fuzzy approach to «Random Images» analysis

J. H. FRIEDMAN, Stanford University, USA
Analysis of weak and low statistics signals

PARALLEL PROCESSING

J. K. AGGARWAL, College of Engineering, Austin, USA
Progresses in parallel processing and analysis

ARTIFICIAL INTELLIGENCE

R. ALBRECHT, ESO/ST-ECF, Garching bei Munchen, FRG
Artificial intelligence: what can it do for astronomy?

NEW FRONTIERES IN ASTROPHYSICS

M. KOSHIBA, CERN, Geneve, Suisse
Neutrino astronomy

G. PIZZELLA, University of Rome «La Sapienza», Italy
Gravitational waves

A. WATSON, University of Leeds, UK
Cosmic Ray point sources and the extreme Cosmic Ray energy spectrum

G. SMOOT, Lawrence Berkeley Labs., Berkeley, USA
Background radiation

CONCLUSIVE PANEL DISCUSSION
A panel discussion will be held by the chairmen of the sections under the coordination of Prof. L. Scarsi at the end of the Workshop.

Fig. 58.1. Program: 3rd International Workshop on Data Analysis in Astronomy

References

1. Di Gesú, V., Maccarone, M.C.: A Method to Classify Celestial Objects Shapes Based on the Possibility Theory. In: Sedmak, G. (ed.) ASTRONET 1983, pp. 355–363 (1983)
2. Di Gesú, V., Maccarone, M.C.: Features Selection and Possibility Theory. Pattern Recognition Letters 1, 63–72 (1986)
3. Di Gesú, V., Maccarone, M.C.: An Approach to Random Images Analysis. In: Cantoni, V., Di Gesú, V., Levialdi, S. (eds.) Image Analysis and Processing II, pp. 111–118. Plenum Press, New York (1988)
4. De Luca, A., Termini, S.: A Definition of Non-probabilistic Entropy in the Setting of Fuzzy Set Theory. Information and Control 20, 301–312 (1972)
5. Di Gesú, V., Maccarone, M.C.: Vision Problems in Sparse Images. In: Pieroni, G.G. (ed.) Issues on Machine Vision, pp. 135–147. Springer, Berlin (1989)
6. Maccarone, M.C., Di Gesú, V., Tripiciano, M.: PDB: A Pictorial Data Base Oriented to Data Analysis. Software Practice and Engineering 23(1), 55–73 (1993)
7. Maccarone, M.C., Tripiciano, M., Di Gesú, V.: Fuzzy Mathematical Morphology to Analyse Astronomical Images. In: Proceedings of the 11th ICPR, vol. 3, pp. 248–251 (1992)
8. Di Gesú, V., Maccarone, M.C., Tripiciano, M.: Mathematical Morphology Based on Fuzzy Operators. In: Lowen, R., Roubens, M. (eds.) Fuzzy Logic: State of the Art, pp. 477–486. Kluwer Academic Publ. (1993)
9. Maccarone, M.C.: Fuzzy Mathematical Morphology: Concepts and Applications. Vistas in Astronomy 40(4), 469–477 (1996)
10. Di Gesù, V., Maccarone, M.C.: A Fuzzy Approach to Compute the Medial Axis of Gray Level Images. In: Proceedings of the 7th IFSA World Congress, Prague, Czech Republic, pp. 221–226 (1997)
11. Maccarone, M.C.: The CCMA Scientific Network: Three Years Activity Report. In: Maccarone, M.C., Murtagh, F., Kurtz, M., Bijaoui, A. (eds.) Advanced Techniques and Methods for Astronomical Information Handling, pp. 9–20. Observatoire de la Côte d'Azur, France (1997)
12. Zadeh, L.A.: Fuzzy Sets. Information and Control 8, 338–352 (1965)
13. Zadeh, L.A.: Outline of a New Approach to the Analysis of Complex Systems and Decision Processes. IEEE Transactions on Systems, Man and Cybernetics SMC-3(1), 28–44 (1973)

59

On Fuzziness and the Interpretability-Accuracy Trade-Off: A Personal View

Luis Magdalena

59.1 Introduction

Nowadays, one of the significant topics in fuzzy systems design is that of the interpretability-accuracy trade-off. We can consider that the presence of this question as a key aspect of fuzziness is quite new, as a matter of fact, only at the beginning of this century, the question has been formalized and analyzed. But, on the other hand, most of us, who have been working in fuzzy systems design for many years, have crossed the border line between interpretable and accurate fuzzy systems, several times in our career.

Those of us who work as engineers, trying to solve real world problems, are somehow driven by user requirements. This is neither good nor bad, it is simply a fact. And both, interpretability and accuracy, usually appear as important items in those requirements. As a result, we have moved across the line, producing in some cases systems that were *only* accurate models, disregarding interpretability, while in other cases we were quite respectful with interpretability even jeopardizing accuracy, always trying to fulfill the needs of the end user of the system.

My career on fuzziness has been mostly related to fuzzy systems design, and with the balance between those two concepts, the search for the equilibrium position among them, and my different perceptions of fuzzy systems during those years, depending on where the balance was in every new project or idea.

59.2 Accuracy

I arrived to fuzzy systems in late eighties, and I did clearly on the side of accuracy. As a PhD student, being part of a big project where several Soft Computing techniques (neural, evolutionary and fuzzy) were applied in the operation of a fossil power plant in Spain, I was involved in the design of a model using a Takagi-Sugeno-Kant fuzzy system. The model was mainly built on the basis of data obtained from the power plant plus the use of a pre-existing model being too complex (too time consuming) to be applied on-line in the power plant. I can say that in such a situation, accuracy was almost the only requirement. So, in this initial stage, my maxim was: *First accuracy, and then ... more accuracy*.

R. Seising et al. (Eds.): *On Fuzziness: Volume 1*, STUDFUZZ 298, pp. 401–406.
DOI: 10.1007/978-3-642-35641-4_59

On top of that model, a recommendation system for the operators was built. But not even there, interpretability was considered. It is quite strange, because now, with my present background, I think that such a project was the prototype of those situations where interpretability is quite important. Recommendation systems for expert operators are usually required to be as much interpretable as possible, but in that moment, nobody even mentioned the word "interpret" during the project.

59.3 Interpretability

Anyway, it did not take long to me discovering the other side of the frontier. In parallel with this industrial project, I was involved in my own research to produce my PhD dissertation. The main topic was walking robots, and one of the important tasks was fuzzy controlling the walk of a humanoid biped robot [1]. I was not responsible for the hardware (the robot itself), but for the control algorithms to produce a *humanoid walk* of the robot. The idea was to produce a set of fuzzy control rules acting on the motors of the robot to produce a sequence of movements emulating a stable and regular human walk. In this case knowledge extraction was focused on through a completely different approach. First, there was no physical system available for training, and second, there were no available experts on biped robot humanoid walking. To cope with that situation, first, as part of my work I programmed a quite complex kinematic and dynamical model of the robotic structure using a *conventional* approach by means of differential equations, and second, as the main source of knowledge I used the literature on biomechanical studies of human walk, complemented with my own performance as *expert* in human walking (as most of us could be considered due to our *daily experience*). As a result, I had different sources to produce some expert knowledge, plus a model of the system to control, to be used for testing purpose.

This second project was by far much more complex than that of the power plant (at least my part of the whole project), mainly because there was not a clear definition of the steps to follow in solving the problem. Knowledge extraction being not a real expert, with trial and error on the mathematical model, was hard, but I finally succeeded to create a set of rules producing a sequence of movements that could be recognized as somehow human walk of the robot model. Human walk is at the end a sequence of simple movements: compass gait of the legs, flexion and extension of the knees, a certain balance of the trunk to compensate other movements and maintain equilibrium, and some movements of the feet to produce additional impulse (being not taken into account in this case since the considered robot had no feet). So, at the end, most of the rules were grouped producing sequences of the form *accelerate a certain joint (to flex or extend it)*, then *once achieved a desired speed maintain it*, and finally *once reached the objective flexion/extension stop the movement*, being the speed and the value of the maximum flexion or extension, the main informations to be defined.

Once completed a first stage in the construction of the fuzzy controller, that of generating a set of rules able to produce a regular walking sequence, the second part was even more complex: generalizing the obtained knowledge to produce different gait patterns with longer or shorter steps and faster or slower walking speed. Designing new knowledge bases by hand was not feasible, so I opted for using genetic algorithms to produce a population of different descriptions, on the basis of the few variations obtained in the first stage [2]. Finally, the different descriptions should be *merged* to produce a single overall description able to control the robot while walking with gaits of different characteristics. And for that final merging, the ability to understand the different descriptions was crucial, so, a certain level of interpretability was required.

59.4 The Trade-Off

As a result, my first two experiences with fuzzy systems offered me the two views of the interpretability-accuracy trade-off. A first problem where the only constraint was accuracy, and a second one where accuracy made no sense since there were neither a model to replicate nor a (clear) pattern to reproduce, being in this case interpretability a key issue to ease the further required analysis.

During those initial years I realized both sides of the interpretability vs. accuracy question, and at the same time, I became involved in Hybrid Fuzzy Systems, one of the elements that in the nineties produced the great explosion of the *accurate fuzzy modeling* wave. And I was quite involved in that wave, particularly in the Genetic Fuzzy approach. After obtaining my PhD in 1994, I continued my Thesis work on Genetic Fuzzy Systems and started my collaboration with Francisco (Paco) Herrera. A couple of special sessions plus a tutorial in Prague (IFSA'97) [3], a special issue in 1998 [4], and finally, the big year of 2001, with a special issue of Information Sciences [5], a mini-track at IFSA-NAFIPS in Vancouver and a book [6], this time having also Oscar Cordón and Frank Hoffmann as part of the team.

It must be said that, in those initial days, Hybrid systems were perceived as a method for the automatic design of fuzzy systems from data. And due to the influence of machine learning, the knowledge induced from data was usually evaluated in terms of accuracy.

At the same time, the concept of Soft Computing was growing, involving the idea of synergy and the view that Neuro Fuzzy or Genetic Fuzzy approaches were designed to integrate in a single system the best qualities of each component: knowledge representation (fuzzy) and learning capabilities (genetic and neural). And I was really in agreement with this view. I really wanted to merge knowledge representation and learning. But after some years of work in the field, some of us realized that many genetic fuzzy systems had lost the *respect* for their fuzzy origins and the knowledge representation capabilities were forgotten. Thus, I wonder where were fuzzy roots of those Hybrid Systems were?

In the meantime, the year 2000 was my "graduation year" in organizing scientific conferences, being General Co-chair of IPMU 2000 conference, that Julio Gutiérrez, Enric Trillas and myself organized in Madrid [7].

Fig. 59.1. IPMU2000 VIP dinner, Madrid, July 2000. From right to left, back to front: Augustine Esogbue, Luis Magdalena and his wife, Ron Yager and his wife, Julio Gutiérrez, Settimo Termini and his wife, Bernadette Bouchon-Meunier, Janusz Kacprzyk, George Klir and his wife, with Lotfi between them.

We were entering into the new century, and just after concluding our work in the book Genetic Fuzzy Systems, three of its four authors started the discussion about that problem: Were hybrid systems maintaining their fuzzy soul? Interpretability and accuracy were in the core of that analysis, and the effect was the compilation, jointly with a fourth colleague, of two edited volumes considering different interpretability and accuracy issues [8, 9]. In these two volumes published in 2003, many researchers concerned by these questions expressed their ideas, and the role of accuracy and interpretability in fuzzy systems design was widely considered.

I have to say that after my previous experience, at that moment I positioned myself on the side of interpretability, without disregarding accuracy, but having always in mind the central idea that one of the key aspects of fuzzy systems design, and one of the main advantages of fuzzy systems, was interpretability. Additionally, I clearly kept the idea that interpretability was not a given property of fuzzy systems, but something you have to take care of during the whole design process. So, my approach from that time is that of designing interpretable fuzzy systems trying to achieve the higher level of accuracy without accepting a significant loss of interpretability. That was my position regarding interpretability-accuracy trade-off.

At that time, two objectives were on the table, and different policies to pursue them, as well as different levels of achievement were considered. Finally, with the growing of multi-objective optimization, a new option appeared, that of a simultaneous improvement of both aspects (interpretability and accuracy). The use of multi-objective optimization techniques considering interpretability and accuracy as the two objectives to be optimized, was the opportunity to generate in a single process a wide range of solutions with different interpretability-accuracy balances. The ideas of non-dominated solutions (solutions that are not simultaneously outperformed in the two, or more, optimization criteria by any other solution), and Pareto front (the set of all non-dominated solutions), were the key to produce in a single run of the optimization process, many solutions with different trade-offs, where the end-users were able to find the best option according to their needs [10].

59.5 New Challenges

Apparently everything was solved, but recently, additional challenging questions have arisen. Assuming we search for interpretability either during a conventional design process or as part of a multi-objective optimization process, we need to describe and quantify the property under study. So, how do we measure interpretability? And to create and index, we need to answer another question: What is interpretability? Then, once the general problem of interpretability-accuracy trade-off seemed to be close to be solved, the most difficult question appeared: the meaning and the measurement of the concept of interpretability.

After solving many different issues, we still have to tackle with a challenging open question. Defining and measuring interpretability is not an easy task. Interpretability is mostly a subjective concept, involving at the same time syntactical and semantical aspects, being strongly affected by the background of the interpreter, by the application field, by the application purpose, etc. New concepts as comprehensibility or understandability, and readability, expressing different aspects of interpretability, are now on the table. Approaches considering global and local interpretability are used. Concepts as cointension are included in the analysis. In summary, interpretability is still an open issue [11], and its role in properly understanding fuzziness is quite important.

References

1. Magdalena, L., Monasterio-Huelin, F.: Fuzzy Controlled Gait Synthesis for a Biped Walking Machine. In: 1993 IEEE International Conference on Fuzzy Systems, pp. 1334–1339 (1993)
2. Magdalena, L., Monasterio-Huelin, F.: A Fuzzy Logic Controller with Learning through the Evolution of Its Knowledge Base. International Journal of Approximate Reasoning 16(3-4), 335–358 (1997)
3. Herrera, F., Magdalena, L.: Genetic Fuzzy Systems. In: Genetic Fuzzy Systems. Seventh IFSA World Congress (IFSA 1997). Lecture Notes of the Tutorial: Genetic Fuzzy Systems, vol. 13, pp. 93–121. Tatra Mountains Mathematical Publications, Prague (1997)
4. Herrera, F., Magdalena, L.: Introduction: Genetic Fuzzy Systems. International Journal of Intelligent Systems 13(10-11), 887–890 (1998)

5. Cordón, O., Herrera, F., Hoffmann, F., Magdalena, L.: Recent Advances in Genetic Fuzzy Systems. Information Sciences 136(1-4), 1–5 (2001)
6. Cordón, O., Herrera, F., Hoffmann, F., Magdalena, L.: Genetic Fuzzy Systems: Evolutionary Tuning and Learning of Fuzzy Knowledge Bases. Advances in Fuzzy Systems – Applications and Theory, vol. 19. World Scientific (2001)
7. Bouchon-Meunier, B., Gutierrez-Rios, J., Magdalena, L., Trillas, E., Yager, R.R. (eds.): Proceedings of the Eight International Conference on Information Processing and Management of Uncertainty in Knowledge-based Systems (Volumes I, II and III). Universidad Politécnica de Madrid (2000)
8. Casillas, J., Cordón, O., Herrera, F., Magdalena, L. (eds.): Interpretability Issues in Fuzzy Modeling. STUDFUZZ, vol. 128. Springer, Heidelberg (2003)
9. Casillas, J., Cordón, O., Herrera, F., Magdalena, L. (eds.): Accuracy Improvements in Linguistic Fuzzy Modeling. STUDFUZZ, vol. 129. Springer, Heidelberg (2003)
10. Casillas, J., Herrera, F., Pérez, R., del Jesus, M.J., Villar, P. (eds.): Special Issue on Genetic Fuzzy Systems and the Interpretability-Accuracy Trade-off. International Journal of Approximate Reasoning 44(1) (2007)
11. Alonso, J.M., Magdalena, L.: Editorial: Special issue on interpretable fuzzy systems. Information Sciences 181(20), 4331–4339 (2011)

60
On Fuzziness

Trevor Martin

Q: *Is the world fuzzy?*
A: *Mu.*

In Hofstadter's *Gödel, Escher and Bach – an Eternal Golden Braid*, this and many similar questions can be answered by "Mu", indicating that the question is essentially meaningless (and hence cannot be answered) because it depends on incorrect assumptions. Let us refine the question:

Q: *Do people perceive and describe the world in fuzzy terms?*
A: *Most of the time.*

I have no formal evidence for this answer – just my own experience, and my understanding of other people's experience. The fundamental idea of fuzzy made sense to me the first time I came into contact with it and I still find that compelling note of truth in its basic notion. Logic, with its well-defined terms and categories is perfect for studying idealised abstractions of the world. But if we want to simulate, or model, or interact with the way people perceive and describe the world, we need to confront the fact that everyday human language is not (and cannot be) as precise as formal mathematics. There are good reasons for this, mainly that the lack of exact definitions allows us to communicate efficiently using natural language, without needing armies of experts to hammer out the exact meaning of every single term in a sentence. Each person can use common sense in their interpretation of words, and almost all of the time there is sufficient shared understanding for the communication to "just work". A weather forecast that tells you it will be "a showery day with a chilly breeze" rests on well-understood but imprecisely defined terms - and allows you to make a decision on which coat to wear, whether to take an umbrella, etc without knowing exactly how much rain will fall, or for how long, or what the temperature will be outside.

Fuzzy also provides a good model for the way we (or, at least, I) make sense of observations on the world - by grouping together things that are approximately the same, treating them as though they were identical and then making allowances for the fact that they are not all the same. We reason about classes of objects or events - trees, busy roads, games, TV comedy, interesting papers - without being able to say precisely what the members of the class all have in common. They are "the same"

R. Seising et al. (Eds.): *On Fuzziness: Volume 1*, STUDFUZZ 298, pp. 407–414.
DOI: 10.1007/978-3-642-35641-4_60

because we think of them in the same class. We don't need the exact definition of what is meant by "the same" – we interpret using common sense. In recent years, this has been presented as "computing with perceptions" – but, to me, it was always fuzzy's raison d'etre.

60.1 A Bit of History

I first came into contact with fuzzy in 1983, as a post-doc. At that time, the PhD I had just completed in quantum mechanics was not a big asset in the job market; instead, the world was excited about AI. Major national and international projects such as the Japanese 5th generation project, the UK's Alvey project, and the European Strategic Program on Research in Information Technology (ESPRIT) were starting or already running. I shared in some of that excitement, having read Hofstadter's *Gödel, Escher and Bach*, with its mix of top-down axiomatic reasoning and bottom-up emergent behaviour – although I couldn't quite figure out how to get the "right" set of axioms.

This interest led me into a post-doc position with Jim Baldwin and Bruce Pilsworth, developing Fril. At that time, Fril was entering its third generation – Baldwin's early work with Zhou led to a relational language, able to handle fuzzy uncertainty in relations (for example, the degree to which Jack liked Jill) and in attribute values, such as allowing someone's age to be represented as "young" or "old" rather than a precise numerical value. The first version of Fril was implemented as an interpreted language in Lisp, and ran on the mainframe computers that were common at the time. A second version of Fril – again, essentially a relational language – was written in Forth and ran on an IBM PC. This was used to develop a monitoring system for electricity generating plant – one of the first fuzzy knowledge-based systems to be implemented. During the latter project, it had become apparent that whilst a pure relational language could model the knowledge required for a real-world system, it was less effective in the messier aspects of programming including interacting with a user, monitoring use of resources and progress of a computation, and debugging. These problems could be solved by escaping into the underlying implementation language (lisp, forth) but such a mechanism was messy and error-prone. As an alternative, Baldwin and Pilsworth came up with the idea of incorporating a procedural component, modelled on the Prolog (Programming in Logic) language. Prolog had successfully combined a declarative programming paradigm with an efficient procedural reading, to give a language that could be presented as "writing in logic" but also functioned as a general purpose programming language. Part of the beauty of Prolog (and Lisp) was the uniformity of program and data - both were expressed in the same form, meaning that program could be manipulated as data and data could be executed as program. This made "meta-programming" a simple task, and meant that once the language core was implemented, extensions could be written efficiently in the language itself.

Fig. 60.1. From left to right: Janusz Kacprzyk, Lotfi A. Zadeh, George Klir, Qiang Shen, Enrique Ruspini, Philippe de Wilde at FUZZ-IEEE 2007 banquet in London, on the ship MV Symphony

60.2 Conferences

My first exposure to the wider fuzzy world was at the Symposium on Fuzzy Sets held in Cambridge, UK in 1984, where I presented a paper entitled "FPROLOG – a fuzzy prolong interpreter" – later expanded and published in Fuzzy Sets and Systems. Many of the fuzzy pioneers were there – Zimmermann, Yager, Mamdani, Kacprzyk, Bouchon, Dubois and Prade to name but a few – but Zadeh was not there. It must have been a productive meeting, as it sowed the seeds for the IPMU conference series and also led to the formation of the International Fuzzy Systems Association with the bi-annual IFSA congress series. A year later, at the First IFSA Congress in Palma de Mallorca, I met Lotfi Zadeh for the first time. I was impressed by his vision and knowledge – but also by the fact that he was happy to talk to young newcomers to the field, and was genuinely interested to hear about our work. Over the years, I have seen this many times – Lotfi always makes time to talk to young researchers, to listen, to give advice and encouragement.

Fig. 60.2. Lotfi A. Zadeh and Ronald R. Yager at FUZZ-IEEE 2008 in Hong Kong, during the harbour cruise

60.3 General Purpose Fuzzy Programming Languages

Back in the early 80s, there was little available in the way of fuzzy programming languages. Mamdani's pioneering (1970s) work on fuzzy control reportedly used a PDP-8 with 8K of 12 bit memory and paper tape as backup storage. The 1990's fuzzy boom, with its dedicated packages for development of fuzzy control software and hardware, was some way in the future.

In the 1980s, the well-funded researcher might have a VAX computer, with a fortran compiler and facilities for assembler code (although use of the C language was spreading). There were specialised – and expensive – expert systems such as Mycin, Prospector plus the generalised shells derived from these expert systems, intended to be "refilled" with knowledge from other domains. Additionally, there were fuzzy additions to production systems (rule-based languages) – notably, Siler's FLOPS which extended the OPS-5 production system and fuzzy-CLIPS, developed by Togai from NASA's CLIPS system. As with the early versions of Fril, these systems were good for processing facts and rules and were able to model uncertainty but were less good at handling the procedural code needed to implement a fully-functional knowledge-based system.

Version 3 of Fril was conceived as a solution to this problem – it would have a declarative reading and allow "natural" expression of knowledge as rules and facts, with uncertainty included – but it would also allow efficient coding of procedural features. A prototype was written in Lisp as an interpreter, allowing easy integration with the earlier relational code - but when it became clear that the interpreted nature of the language got in the way of execution speed, Fril was redesigned as a compiled language. The whole system was based on an abstract machine, with a bytecode emulator written in C – making the language (relatively) easy to port between different platforms.

In 1986, a Bristol-based IT company (Equipu) expressed interest in Fril and a new venture (Equipu-AI Research) was formed to develop Fril commercially. Following a takeover of Equipu in 1988, the venture became Fril Systems Ltd a wholly independent company with 3 directors (Baldwin, Martin, Pilsworth). At its peak it had 5 employees and made a small but profitable existence selling the Fril software to clients such as BT, the Japanese LIFE project, the UK Defence Research Agency, Reuters, British Aerospace amongst many others. Further business came from consultancy and add-ons to Fril such as an expert system shell and a fuzzy conceptual graphs package. The company continued until 2001, when we were faced with (i) a need to invest a large amount of time (and money) in order to update the implementation, and (ii) the absence of said time and money. We took the logical but difficult step and closed Fril Systems, although the language is still running on current operating systems and can be found out there on the web.

I think Zadeh appreciated what we were doing, and often mentioned Fril in his talks and papers - including his response to Elkan's views on fuzzy. In general, though, my impression is that Zadeh was more focussed on illustrating new ideas than on the detail of how to code robust implementations, and so he probably didn't fully appreciate some of the finer aspects of Fril. His early papers on PRUF showed how fuzzy could capture some of the intrinsic flexibility of human language - but writing the software to implement those ideas (as we did for the demo software in the Fril book) brought into sharp focus the computational resources needed to model even relatively simple statements about tall Swedes.

60.4 The Need for Fuzzy Thinking Today

From the first time I met Jim Baldwin and came into contact with fuzzy, I have been convinced that it reflects the way we view and talk about the world. I am not referring to the narrow field of fuzzy control – as Zadeh himself has frequently stated, fuzzy logic itself is anything but fuzzy, it is a precise mathematical framework for dealing with uncertainty. Often, in my view, it is far too precise, and many aspects of fuzzy - particularly fuzzy control – exhibit the very over-precision that first inspired Zadeh. I have lost count of the number of papers I have seen which quote "defuzzified" values to 6 or 7 significant figures. Fuzzy is not just about mathematics – it is a way for machines to simulate aspects of human reasoning. The broader interpretation of fuzzy is a way of reproducing human-style "back of the envelope" approximations – not just applicable to numerical problems, but also to symbolic and quantitative reasoning. This idea owes a lot to time spent in discussion with Jim Baldwin, as well as Zadeh's talks and papers. Even before Zadeh spoke of the four facets of fuzzy logic, it was clear that the prime purpose of fuzzy was to reflect aspects of human reasoning. Fuzzy control is a natural outlet for the idea, but by no means the only application.

Fig. 60.3. Lotfi A. Zadeh and Yasuhiko Dote at the 4th IEEE Conference on Soft Computing as Transdisciplinary Science and Technology, 2005 in Muroran, Japan

To me, the most pressing current problem in computer science arises from the explosion of digital data and computer-based interactions, and the consequent need to do something useful with the resultant data. Quite simply, the human-scale and the internet-scale do not match. Although statistics has given us well-established methods for aggregation, summarisation and (to an extent) visualisation of numerical data, the majority of new data is not numbers – it is words, photos, videos, diagrams, relations and structures.

Condensing this to manageable levels is key to maximising the potential of the human – computer combination. Humans make sense of the flood of sensory data by reducing it to simple concepts, with imprecise definitions. The need to bridge the gap between "hard" data and "soft" human concepts is the foundation for us to make use of the rapidly increasing data available to us.

The semantic web was an attempt to impose structure on the explosion of data, by adding meta-data – tags whose meaning is defined by fixed vocabularies and ontologies. The fact that the semantic web vision remains largely unrealised suggests that it is flawed in some way. Contrast the take-up of semantic web ideas with the less rigorous – almost anarchic – ethos of web 2.0, where users add tags which have an informal meaning derived from the shared view and use of the tags. The flaw in the semantic web vision lies in its need for precisely defined terms and for exact matches between those terms and the content of a document or web page. Coupled with the fact that most queries are expressed using the ambiguous words of natural language, the area should have been perfect for fuzzy – but despite islands of good work, fuzzy (and uncertainty in general) remains a sideline to the main stream of the semantic web. It is implicit in the structure of Web 2.0, because of the underlying natural language basis.

This is not the only area in which fuzzy has not been accepted as widely as it should. The real world does not map naturally to the crisp and artificial categories required by the relational model, but there are few, if any, large scale commercial applications of fuzzy databases. The ideas were there, but the software demonstrators and full scale applications have not emerged.

60.5 The Future Is Still Fuzzy

We are in the age of "big data" where statistical methods and conventional machine learning techniques reign supreme. It is a world where huge bodies of data can be amassed and analysed with relative ease – and the "wisdom of the crowd" can be tapped freely.

And yet – almost paradoxically – there is still a role for fuzzy. Statistical methods work well in cases where there is plenty of data and where the future continues in a similar vein to the past. If we have 3 possible treatments for a disease and several good studies of their efficacy in different circumstances (i.e. sufficient measurements of related attributes), then statistical machine learning can predict both the best treatment for a specific patient and its likelihood of success. Given a new brand of washing powder, we can predict sales because it is a known product in a known market. But given a brand new class of product - disruptive innovation such as an iPod or a Sony Walkman in an earlier generation - there is simply no data to interpolate. Without data and an understanding of its structure, machine learning is helpless - but humans can function with little or no data and in cases where they do not completely understand a system. Human judgment, instinct, and reasoning (with all its inherent fuzziness) is often sufficient. In Donald Rumsfeld's terms, machine learning needs to be firmly positioned in the "known knowns", where the structure of a problem is clear and there is full knowledge of attributes and their values (even if expressed as distributions). Fuzzy is needed when we move outside that zone of well-defined attributes and values.

Statistical machine learning loses its usefulness when we move towards the "known unknowns" (we have no idea what values an attribute can take, just that it exists), the "unknown knowns" (we don't know whether an attribute is applicable) and the "unknown unknowns" (attributes we are not even aware of). For example, if we are monitoring behaviour in a computer network, how do we know what constitutes an attack? Specific patterns that have been seen before are easy to categorise, but novel behaviour requires judgment and insight. Human ingenuity and inventiveness ensures a stream of new and more devious ways to attack. Not only is the definition of "attack" fuzzy, it must also evolve continually. Fuzziness allows us the flexibility of language; machine learning is more like Star Trek's Borg, able to adapt once it has a rigorous definition and has seen sufficient examples but prone to discarding "noise" before that point is reached.

As always, the key to the success of fuzzy lies in finding and implementing good applications which demonstrate its capabilities. Lotfi Zadeh has a range of examples showing where traditional methods are inadequate. The challenge today is to show that fuzzy can handle those examples in an understandable and efficient way. The power of fuzzy is that it reflects the way we view the world and communicate our views in everyday terms. Fuzzy gives us the power to make computers that work with those same terms, rather than the artificial and arbitrary categories imposed by binary logic. We should use that power to its full extent.

Fig. 60.4. Lotfi A. Zadeh at the 4th IEEE Conference on Soft Computing as Transdisciplinary Science and Technology, 2005 in Muroran, Japan

61

Zadeh Fuzzy Probability, De Finetti Subjective Probability and Prevision

Antonio Maturo and Aldo G.S. Ventre

61.1 Introduction and Motivation

The concepts of *fuzzy event* and *fuzzy probability* are introduced by Zadeh in [13] and [16]. In particular *linguistic probabilities* are considered. In the same period de Finetti in [3] assumes a point of view close to that of Zadeh with the consideration of qualitative probability.

The theories of Zadeh and de Finetti differ substantially in expressing linguistic (resp. qualitative) probabilities with numbers. While de Finetti, treating the subjective probability, assumes that probabilities are real numbers belonging to $[0,1]$, Zadeh, introducing the fuzzy probability, sees probabilities as fuzzy numbers contained in $[0,1]$.

Both take into account an uncertainty in the probability. In de Finetti this uncertainty is related to the fact that the probability assessments vary with the subject (the expert) who decides their values, and this variability is not measured. In Zadeh, instead, the left and right spreads of fuzzy probabilities indicate the possible oscillations of the probability with respect to a central value to which is attached the utmost confidence.

Both authors are based on the principle of consistency : if $\{E_1, E_2, ..., E_n\}$ is a finite partition of the certain event and $\{p_1, p_2, ..., p_n\}$ is an assignment of (crisp) probability, then the sum of p_i is 1. In the fuzzy modellization of Zadeh every p_i is a value of a fuzzy probability P_i and there exists a function $\mu_i : x \in P_i \to [0,1]$ that denotes the degree to which x belongs to P_i. The number $\mu_1(p_1) \wedge \mu_2(p_2) \wedge ... \wedge \mu_n(p_n)$ is the degree to which $(p_1, p_2, ..., p_n)$ belongs to the Cartesian product of the supports of the P_i.

Fuzzy modeling provides an algebraic structure that parallels the linguistic and semantic structure of the human reasoning [12], [14], [15]. Fuzzy models play an essential role in building "sufficiently exact" descriptions of the physical situations, complex systems and events. In this sense, fuzzy modeling operates a synthesis of the information to be conveyed [4] and the knowledge to be understood. The simplification of a description in an economic or humanistic system, reducing the necessary imprecision to a level of relative unimportance [2], leads to a lucid and intuitive decision making act. We believe that guessing the outcome of a complex process through the fog was felt by de Finetti when introducing the idea of "betting" in order to decide, i. e. cut off, by an intuitive, immediate vision.

R. Seising et al. (Eds.): *On Fuzziness: Volume 1*, STUDFUZZ 298, pp. 415–420.
DOI: 10.1007/978-3-642-35641-4_61

61.2 Decomposable Measure and Fuzzy Prevision

A generalization of finitely additive probability (de Finetti) and fuzzy probability (Zadeh) is given by *fuzzy measure* [1], [10], [11], [17]. As for the finitely additive probability, the codomain of a fuzzy measure is the real interval $[0,1]$, but the *additivity* is replaced by the weaker condition of *monotonicity*.

Decomposable fuzzy measures with respect to t-conorms are considered in several book and papers, (see, e.g., [10], [11], [1], [5]). Especially in assigning scores in decision making, they are a palatable compromise between the too much general concept of fuzzy measure and the very particular one of finitely additive probability. The additivity is replaced by the weaker property of additivity w. r. to a t-conorm.

Definition 1. *Let U be a set and $\mathscr{F}$ a field of subsets of U. A fuzzy measure m on $\mathscr{F}$ is said to be* a measure decomposable *w. r. to a t-conorm $\oplus$ if:*

$$A \cap B = \emptyset \Rightarrow m(A \cup B) = m(A) \oplus m(B).$$

The concept of *finitely additive probability* has been extended in [3] to the *coherent prevision*. While a finitely additive probability is a function defined in a set of events, a coherent prevision is defined in a set of (de Finetti) random numbers. The events are particular random numbers with codomain contained in $\{0,1\}$. The idea of fuzzy prevision is virtually contained in [16] where Zadeh formalizes the properties of linear combinations of fuzzy probabilities.

An important advantage of such an extension consists in the possibility of replacing the union of events with the sum of random numbers (resp. fuzzy random numbers). Then finitely additive probability is framed in the more general environment of vector spaces of random numbers [3]. In this way also a very useful geometrical interpretation is obtained, based on hyperplanes, convex sets and join spaces (see, e.g., [9], [6]).

A concept of *fuzzy prevision* can also be introduced as an extension of the concept of fuzzy measure, in an analogous way that de Finetti coherent prevision is an extension of the finitely additive probability [3]. Moreover we present the concept of *decomposable fuzzy prevision* that may be the happy medium between a coherent prevision and a fuzzy prevision.

Definition 2. *A* (de Finetti) random number *is a function $X : \Pi \to R$, where Π is a partition of the certain event. The set Π is the* domain *of X, the set R of the real numbers is the* codomain *and the set $X(\Pi)$ is the* range. *X is said to be* bounded *(resp.* finite*) if its range is bounded (resp. finite).*

Definition 3. *Let S be a non empty set of bounded random numbers. A* (de Finetti) prevision *on S is a function $P : S \to R$ such that:*

P1 $\forall a,b \in R, \forall X \in S,\ a \le X \le b \Rightarrow a \le P(X) \le b$ (mean property)*;*
P2 $\forall X,Y \in S, X+Y \in S \Rightarrow P(X+Y) = P(X)+P(Y)$ (additivity)*.*

The prevision P is said to be coherent *if there exists a prevision P^*, extension of P, defined on the vector space $V(S)$ generated by S.*

A prevision P on S reduces to a *finitely additive probability* on S if every element X of S assumes only values belonging to the set $\{0,1\}$.

The concept of prevision on a set of random numbers can be extended from many different points of view. Let us consider three cases:

RFN *Random Fuzzy Numbers Extension.* The de Finetti random numbers are replaced by more general *random fuzzy numbers*, defined as functions $X : \Pi \to H$, where Π is a partition of the certain event and H is a set of fuzzy numbers containing R and closed with respect to the (Zadeh) addition and the multiplication of an element of H by a real number [14]. The idea is that a fuzzy prevision is an extension to random fuzzy numbers of the prevision with analogous properties as in definition 3 (see, e.g., [7]).

JSE *Join Space Extension.* The atoms of a set of de Finetti random numbers are interpreted as point of a join space (see, e.g., [9], [6], [8]). The Euclidean coherence conditions given in [3] are replaced by join-coherence conditions, defined as follows: a prevision is coherent if belongs to the convex hull, of a join space, generated by the atoms. In particular, if the set of atoms is finite, the convex hull is a a polytope [9], [6]. The choice of the join space is equivalent to assume a particular point of view to evaluate the effects of decisions.

FME *Fuzzy Measure Extension.* In a similar manner in which the coherent prevision is defined as an extension to random numbers of a finitely additive probability, we introduce a fuzzy prevision as an extension to the random numbers of a fuzzy measure, or in particular, of a measure decomposable with respect to a t-conorm. Then a fuzzy prevision is defined as a real function having as domain a set S of random numbers, and such that its restriction to a set of events (i.e., random numbers with range contained in $\{0,1\}$) is a fuzzy measure. In this line of thinking we introduce the following definition.

Definition 4. *Let S be a family of de Finetti random numbers. We define* fuzzy prevision, *of type FME, on S, any function $P : S \to R$ such that:*

FP1 $\forall a,b \in R, \forall X \in S,\ a \le X \le b \Rightarrow a \le P(X) \le b$ (mean property)*;*
FP2 $\forall X,Y \in S, X \le Y \Rightarrow P(X) \le P(Y)$ (monotonicity)*.*

Let us introduce the following definitions of *additive generator* on R and *g-operation*, as generalizations of the concepts given in [11].

Definition 5. *We define* additive generator *on R every function g defined in a closed interval $[0,b_g]$ of $[0,+\infty]$, with codomain $[0,+\infty]$, and such that $g(0) = 0$ and g is strictly increasing and continuous.*

Definition 6. *Let g be an additive generator on R with base interval $[0,b_g]$. We define* pseudoinverse *of g the function $g^{(-1)}$, defined in $[0,+\infty]$, with codomain the base interval $[0,b_g]$ of g, and such that $g^{(-1)}(y) = g^{-1}(y)$ if $y \in [g(0),g(b_g)]$; $g^{(-1)}(y) = b_g$ if $y \ge g(b_g)$.*

Definition 7. *Let g be an additive generator on R with base interval* $[0, b_g]$. *We define* operation generated *by g, we call it the g-*operation, *the operation* $\oplus$ *defined as follows:*

$$\forall x, y \in [0, b_g] : x \oplus y = g^{(-1)}(g(x) + g(y)). \quad (61.1)$$

The g-operation $\oplus$ *is* strict, *if* $g(b_g) = +\infty$ *and* nonstrict, *if* $g(b_g)$ *is finite.*

From (61.1) the following theorem follows:

Theorem 2. *The g-operation* $\oplus$ *given by (61.1) is increasing in each argument, associative, commutative, and with 0 as neutral element. In particular, if* $b_g = 1$, *then* $\oplus$ *reduces to an Archimedean t-conorm.*

We say that the de Finetti random numbers X and Y with domain Π are orthogonal, we write $X \perp Y$ if, $\forall z \in \Pi, X(z)Y(z) = 0$. The orthogonality between two random numbers plays the role of on extension of the incompatibility between two events. Let us introduce a definition for *decomposable fuzzy prevision* as a generalization of decomposable fuzzy measure.

Definition 8. *Let g be an additive generator on R with base interval* $[0, b_g]$, *and* $\oplus$ *the correspondent g-operation. Let S be a set of de Finetti random numbers with range contained in* $[0, b_g]$ *and P a fuzzy prevision on S. We say that P is a* $\oplus$-decomposable fuzzy prevision *on S if*

$$\forall X, Y \in S : X \perp Y, P(X + Y) = P(X) \oplus P(Y). \quad (61.2)$$

61.3 An Application of Fuzzy Prevision to Social Sciences

We consider the problem to build a Social and Cultural Center. We assume there is a set $\mathscr{A} = \{A_1, A_2, ..., A_m\}$ of *alternative projects* and a set $\mathscr{O} = \{O_1, O_2, ..., O_n\}$ of *objectives* to be satisfied.

For example, the objectives may be the following: O_1 = promote social inclusion, O_2 = improve local development, O_3 = achieve functional regeneration of the urban context. The alternatives for the cultural center may be: A_1 = in the historical center, A_2 = in the residential area of urban completion, A_3 = in the area of commercial and industrial expansion.

It seems reasonable to represent the objectives O_j as subsets (or event, in the probabilistic notation) of a universal set U, called *the certain event*. A decision maker D associates to every pair (A_i, O_j) a real number P_{ij} that represents the *score* which, in the opinion of D, measures the degree of fulfillment of the objective O_j if the alternative A_i is realized.

Our proposal is to associate to every pair (A_i, O_j) a nonnegative random number $X_{ij} : \Pi_{ij} \to R$, where Π_{ij} is a partition of the certain event, O_j^c, the contrary of O_j, belongs to Π_{ij} and $X_{ij}(O_j^c) = 0$. Every $E \in \Pi_{ij} - \{O_j^c\}$ is a particular *aspect* of the objective O_j to be satisfied and $X_{ij}(E)$ is the *gain* or *utility*, with respect to the aspect E of the objective O_j, if the alternative A_i is chosen.

Moreover we propose to interpret the score P_{ij} as the prevision of X_{ij} or at least a fuzzy prevision. Then the function

$$P : S = \{X_{ij}, i \in \{1,2,...,m\}, j \in \{1,2,...,n\}\} \to P_{ij} \in R$$

is a de Finetti prevision (or fuzzy prevision) on S and coherence conditions dependent on definitions and theorems of previous Sections must be satisfied.

Finally, if P is decomposable w. r. to g-operation $\oplus$, it seems reasonable to assume that the global scores $S(A_i)$ of the alternatives A_i are obtained by the formula:

$$S(A_i) = P_{i1} \oplus P_{i2} \oplus ... \oplus P_{in}. \tag{61.3}$$

A practical application of the problem posed in this section has affected the city of Pescara in 2005. The administration of Pescara has chosen to establish a social and cultural center (called *T. Dezi Center*) in an urban area of type A_3. The choice of that location is important for the objective O_3, in an attempt to overcome the conditions of marginalized urban area.

References

1. Banon, G.J.F.: Distinction Between Several Subsets of Fuzzy Measures. Fuzzy Sets and Systems 5, 291–305 (1981)
2. Bellman, R.E., Giertz, M.: On the Analytic Formalism of the Theory of Fuzzy Sets. Information Sciences 5, 149–157 (1973)
3. de Finetti, B.: Teoria delle probabilità. Einaudi, Torino (1970); published in English as Theory of Probability. J. Wiley, New York (1974)
4. Goguen, J.A.: The Logic of Inexact Concepts. Synthese 19(3/4), 325–373 (1969)
5. Klir, G.J., Yuan, B.: Fuzzy Sets and Fuzzy Logic: Theory and Applications. Prentice Hall, New Jersey (1985)
6. Maturo, A.: A Geometrical Approach to the Coherent Conditional Probability and its Fuzzy Extensions. Scientific Annals of University of A. S. V. M., "Ion Ionescu de la Brad", Iasi, vol. XLIX(2), pp. 243–256 (2006)
7. Maturo, A., Ventre, A.G.S.: On Some Extensions of the De Finetti Coherent Prevision in a Fuzzy Ambit. Journal of Basic Science 4(1), 95–103 (2008)
8. Maturo, A.: Join Coherent Previsions. Set-valued Mathematics and Applications 1(2), 135–144 (2008)
9. Prenowitz, W., Jantosciak, J.: Join geometries. Springer, New York (1979)
10. Sugeno, M.: Theory of Fuzzy Integral and its Applications. Ph. D. Thesis, Tokyo (1974)
11. Weber, S.: Decomposable Measures and Integrals for Archimedean t-conorms. Journal of Mathematical Analysis and Applications 101(1), 114–138 (1984)
12. Zadeh, L.A.: Fuzzy Sets. Information and Control 8, 338–353 (1965)
13. Zadeh, L.A.: Probability Measures of Fuzzy Events. Journal of Mathematical Analysis and Applications 23, 421–427 (1968)

14. Zadeh, L.A.: The Concept of a Linguistic Variable and its Applications to Approximate reasoning I. Information Sciences 8, 99–249 (1975)
15. Zadeh, L.A.: The Concept of a Linguistic Variable and its Applications to Approximate reasoning II. Information Sciences 8, 301–357 (1975)
16. Zadeh, L.A.: The Concept of a Linguistic Variable and its Applications to Approximate reasoning III. Information Sciences 9, 43–80 (1975)
17. Zadeh, L.A.: Fuzzy Sets as a Basis for a Theory of Possibility. Fuzzy Sets and Systems 1, 3–28 (1978)

62

From Ordinary Triangular Norms to Discrete Ones

Gaspar Mayor

In the early eighties I met professor Nadal Batle after many years. At that time he became Rector of the University of the Balearic Islands and I was teaching mathematics to future primary school teachers. Because there were no doctors of Mathematics in our University, Batle encouraged me to do a thesis on some topic that seemed interesting to us. In our first working meeting, I heard talk about fuzzy sets for the first time. Advised by professor Batle, I contacted shortly thereafter with professor Enric Trillas, pioneer and master of the theory of fuzzy sets in Spain, who opened to me the doors of this exciting theory and made it possible I meet professor Alsina and professor Valverde. I recall with some nostalgia my visits to the School of Architecture in Barcelona where I learned many things from them and spent unforgettable moments from all points of view. My early research interest was that of triangular norms. At that time I thought it was an interesting subject which had much to say and today I still seem exciting research topic ...

> *"It is important to know as much as possible about t-norms and, in particular, to have a large repertoire of them at hand."*
>
> Berthold Schweizer & Abe Sklar

62.1 Triangular Norms

The notion of triangular norm (t-norm for short) first appeared in 1942 in a brief paper due to Karl Menger and entitled *Statistical Metrics* [12]. In this paper, Menger introduced the concept of a space in which distances between points are given by probability distributions functions. Menger's triangle inequality has the form

$$F_{pr}(x+y) \geq T(F_{pq}(x), F_{qr}(y))$$

where p, q, r are points in the underlying space; x, y are real numbers; F_{pr}, F_{pq}, F_{qr} are the probability distribution functions associated with the pairs (p,r), (p,q), (q,r) respectively, and T is a function from the closed unit square $[0,1]^2$ to the closed unit interval $[0,1]$, which Menger called the triangular norm of the statistical metric. He assumed that for all a,b,c,d in $[0,1]$:

R. Seising et al. (Eds.): *On Fuzziness: Volume 1*, STUDFUZZ 298, pp. 421–427.
DOI: 10.1007/978-3-642-35641-4_62

(i) $T(a,b) = T(b,a)$
(ii) $T(a,b) \leq T(c,d)$ whenever $a \leq c$ and $b \leq d$
(iii) $T(1,1) = 1$ and $T(a,1) > 0$ whenever $a > 0$

Statistical metrics spaces may be viewed as generalizations of classical metric spaces in which the distances between points are described by probability distribution functions rather than by real numbers. In a statistical metric space, with each pair of points (p,q) there is associated a probability distribution function F_{pq} whose value $F_{pq}(x)$ is interpreted as the probability that the "distance" between p and q is less than x.

In the course of the work on statistical metric spaces carried out by Berthold Schweizer and Abe Sklar [13–15], they returned to Menger's triangle inequality and realized that condition (iii) should be replaced by the stronger one: $T(a,1) = a$ for all a in $[0,1]$, and, in order to extend the triangle inequality to a polygonal inequality, they established that T be associative. Thus their requirements on T became:

(i) $T(a,b) = T(b,a)$
(ii) $T(a,b) \leq T(c,d)$ whenever $a \leq c$ and $b \leq d$
(iii) $T(a,1) = a$
(iv) $T(T(a,b),c) = T(a,(T(b,c))$

There exist uncountably many t-norms. The following are the four basic t-norms T_M, T_P, T_L and T_D (denoted by M, Π, W and Z, respectively, in [16]):

$$T_M(a,b) = min\{a,b\} \quad \text{(minimum)}$$
$$T_P(a,b) = ab \quad \text{(product)}$$
$$T_L(a,b) = max\{a+b-1,0\} \quad \text{(Łukasiewicz t-norm)}$$
$$T_D(a,b) = \begin{cases} min\{a,b\}, & \text{if } max\{a,b\} = 1 \\ 0, & \text{otherwise} \end{cases} \quad \text{(drastic product)}$$

Since then, a triangular norm is known as a function T satisfying the above set of axioms. In [15] triangular conorms were introduced as dual functions of t-norms: $S(a,b) = 1 - T(1-a, 1-b)$; however, an independent axiomatic definition can be given: A triangular conorm (t-conorm for short) is a function S from $[0,1]^2$ to $[0,1]$ which is commutative, associative, monotone and has 0 as neutral element, i.e., it satisfies conditions (i), (ii), (iv) and $S(a,0) = a$. The four t-conorms corresponding to the above basic t-norms are:

$$S_M(a,b) = max\{a,b\} \quad \text{(maximum),}$$
$$S_P(a,b) = a+b-ab \quad \text{(probabilistic sum),}$$
$$S_L(a,b) = min\{x+y,1\} \quad \text{(bounded sum)}$$
$$S_D(a,b) = \begin{cases} max\{a,b\}, & \text{if } min\{a,b\} = 0 \\ 1, & \text{otherwise} \end{cases} \quad \text{(drastic sum).}$$

Despite the fact that t-norms were first introduced in the context of statistical metric spaces, they are an important tool for the interpretation of the intersection of fuzzy sets. Triangular norms also play an important role in decision making, in statistics, in the theories of non-additive measures, etc., without forgetting that they are interesting mathematical objects for themselves.

In 1965, Lotfi A. Zadeh published the paper "Fuzzy Sets" [17]. In this foundational work, Zadeh says *"A fuzzy set is a class of objects with a continuum of grades of membership"* and more precisely *"A fuzzy set A in a space X is characterized by a membership (characteristic) function $f_A(x)$ which associates with each point in X a real number in the interval $[0,1]$, with the value $f_A(x)$ at x representing the "grade of membership" of x in A."*

In Fuzzy Sets, Zadeh extended the basic notions about ordinary sets to fuzzy sets; in particular, the *intersection* of two fuzzy sets A and B with respective membership functions $f_A(x)$ and $f_B(x)$ is a fuzzy set C, $C = A \cap B$, whose membership function is related to those of A and B by

$$f_C(x) = min\{f_A(x), f_B(x)\}, x \in X.$$

The notion of union of fuzzy sets was defined in a similar manner: $C = A \cup B$ with $f_C(x) = max\{f_A(x), f_B(x)\}$. If each fuzzy set and the associated membership function are denoted by the same capital letter, then equations defining intersection and union can be written:

$$(A \cap B)(x) = min\{A(x), B(x)\}, \qquad (A \cup B)(x) = max(A(x), B(x))$$

In 1973, Bellman and Giertz [4] proved that, under reasonable assumptions (distributivity, monotonicity and boundary conditions), the minimum and the maximum are the only functions F and G generating intersection and union by means of

$$(A \cap B)(x) = F(A(x), B(x)), \qquad (A \cup B)(x) = G(A(x), B(x))$$

for all $x \in X$ and for all fuzzy sets $A, B : X \to [0,1]$.

Zadeh is the first to introduce in 1976 nondistributive dual pairs (F, G) [18]: $F = T_L$, $G = S_L$ and $F = T_P$, $G = S_P$; but papers that explore in depth the use of t-norms and t-conorms to model intersections and unions of fuzzy sets are not published until the early eighties. It is worth quoting in this regard the role played by the Linz Seminars on Fuzzy Set Theory [5] and also the pioneering work entitled *On Some Logical Connectives for Fuzzy Sets Theory* [2] by C. Alsina, E. Trillas and L. Valverde.

In [15], Schweizer and Sklar showed, from Aczél's fundamental representation theorem in [1], that every strict (strictly increasing on $(0,1]^2$ and continuous) t-norm admits the representation

$$T(a,b) = f^{-1}(f(a) + f(b)) \tag{62.1}$$

for all a, b in $[0,1]$, where the function $f : [0,1] \to [0,\infty]$ is continuous and strictly decreasing with $f(0) = \infty$ and $f(1) = 0$. They called f an additive generator of T and showed that any two such generators differ by a positive multiplicative constant. In [7], C. H. Ling extended representation 62.1 to a wider family of t-norms: Any continuous Archimedean ($T(x,x) < x$ for all x in $(0,1)$) t-norm T admits the representation $T(a,b) = f^{(-1)}(f(a)+f(b))$ for all a,b in $[0,1]$, where the function $f : [0,1] \to [0,\infty]$ is continuous and strictly decreasing with $f(1) = 0$ and $f^{(-1)}$ is the pseudo-inverse of f, i.e. $f^{(-1)}(t) = sup\{x \in [0,1]; f(x \geq t)\}, 0 \leq t \leq \infty$.

The construction of t-norms from other t-norms is crucial to obtain a description of the structure of continuous, non-Archimedean t-norms: If $(Ti)_{i\in I}$ is a family of t-norms and $((\alpha_i,\beta_i))_{i\in I}$ is a family of non-empty, pairwise disjoint open subintervals of $[0,1]$, then the following function $T : [0,1]^2 \to [0,1]$ is a t-norm:

$$T(a,b) = \begin{cases} \alpha_i + (\beta_i - \alpha_i)T_i\left(\frac{a-\alpha_i}{\beta_i-\alpha_i}, \frac{b-\alpha_i}{\beta_i-\alpha_i}\right), & \text{if } (a,b) \in [\alpha_i,\beta_i] \\ min\{x,y\}, & \text{otherwise} \end{cases}$$

This t-norm is called the ordinal sum of the "t-norms" $(T_i,(\alpha_i,\beta_i))_{i\in I}$. Of course, each t-norm T can be viewed as a trivial ordinal sum with one summand $(T,(0,1))$. Also, the minimum T_M can be considered an ordinal sum of t-norms with index set $I = \emptyset$.

Then, using a result of K. H. Mostert and A. L. Shields [8], Ling showed that every continuous t-norm is the ordinal sum of continuous Archimedean t-norms. No doubt the representation given in 62.1 and subsequent extensions have allowed a deep knowledge of continuous t-norms. For the general class of all t-norms no representation theorems exist so far. Such a characterization of arbitrary t-norms would be closely related to the solution of the famous, still unsolved general associativity functional equation [3].

62.2 Discrete Triangular Norms

According to the fact that in most practical situations it is necessary to discretize the unit interval, we need to deal with logics where the set of truth values is modelled by a finite set $L = \{0,1,\dots,n\}$. The intersection of two $L-$valued sets $A,B : X \to L$, can also be interpreted from the so called discrete triangular norms [6, 9, 10]

$$(A \cap B)(x) = T(A(x),B(x)), x \in X.$$

As it is expected, we use in the definition of discrete triangular norms (discrete t-norms, for short) the set of axioms provided by Schweizer and Sklar, once adapted to this finite setting. Thus our requirements on $T : L^2 \to L$ are:

(i) $T(a,b) = T(b,a)$
(ii) $T(a,b) \leq T(c,d)$ whenever $a \leq c$ and $b \leq d$
(iii) $T(a,n) = a$
(iv) $T(T(a,b),c) = T(a,T(b,c))$

The following are the three basic discrete t-norms T_M, T_L and T_D:

$$T_M(a,b) = min\{a,b\} \qquad \text{(minimum)}$$
$$T_L(a,b) = max\{a+b-n,0\} \qquad \text{(Łukasiewicz t-norm)}$$
$$T_D(a,b) = \begin{cases} \min\{a,b\}, & \text{if } max\{a,b\} = n \\ 0, & \text{otherwise} \end{cases} \qquad \text{(drastic t-norm)}$$

Note that discrete triangular conorms can be also introduced as dual functions of discrete t-norms: $S(a,b) = n - T(n-a,n-b)$.

Two fundamental classes of discrete t-norms have been considered: the class of smooth discrete t-norms ($T(a+1,b) - T(a,b) \leq 1$) and the class of Archimedean discrete t-norms ($T(a,a) < a$ for all $a \neq 0,n$). In [9] Mayor and Torrens show that the only smooth and Archimedean discrete t-norm is the Łukasiewicz t-norm $T_L(a,b) = max\{a+b-n,0\}$, and then they were able to characterize the class of smooth discrete t-norms: A t-norm T on $L = \{0,1,\ldots,n\}$ is smooth if and only if there exists a subset I of L, $I = \{0 = \alpha_0 < \alpha_1 < \ldots < \alpha_r < \alpha_{r+1} = n\}$ such that T is given by

$$T(a,b) = \begin{cases} max\{\alpha_i, a+b-\alpha_{i+1}\}, & \text{if } (a,b) \in [\alpha_i,\alpha_{i+1}]^2 \; i: 0,1,\ldots,r \\ min\{a,b\}, & \text{otherwise} \end{cases}.$$

That is, a discrete t-norm is smooth if and only if it is an ordinal sum of smooth Archimedean discrete t-norms. Taking into account that smoothness is the proper equivalent of the continuity of ordinary t-norms, the given representation of smooth discrete t-norms is in full analogy to the Ling's representation of ordinary continuous t-norms. As in the case of ordinary t-norms, no representation theorems exist so far for the class of all discrete t-norms.

An additive generator $f : L \to [0,\infty)$ of a discrete t-norm T is a strictly decreasing function with $f(n) = 0$ such that $T(a,b) = f^{(-1)}(f(a)+f(b))$ for all a,b in L, where $f^{(-1)}$ is the pseudoinverse of f, defined by $f^{(-1)}(t) = min\{x \in L; f(x) \leq t\}, 0 \leq t < \infty$. Observe that we define the pesudoinverse of f in a different way with respect to the continuous case; this allow us to generate t-norms with nontrivial idempotent elements ($T(a,a) = a$, $0 < a < n$). We show in [11] that every smooth discrete t-norm has an additive generator; in particular, the minimum T_M can be additively generated by the function $f = (2^n - 1, 2^{n-1} - 1, \ldots, 3, 1, 0)$ where $f(0) = 2^n - 1$, $f(1) = 2^{n-1} - 1, \ldots, f(n-2) = 3, f(n-1) = 1, f(n) = 0$.

In fact, the problem of the existence of additive generators for discrete t-norms is related to the following one: Characterize finite sets of nonnegative integers A, with $0 \in A$, such that the binary operation on A defined by $x * y = max\{z \in A; z \leq x+y\}$ be associative. Thus, this matter becomes an additive number theory problem ...

Acknowledgement. I would like to thank everyone who helped me for so many years in my research. Also those who have helped me in everything else.

Fig. 62.1. Rudolf Felix, Anne Baumeister, Lotfi Zadeh, Gaspar Mayor, Enric Trillas, Janusz Kacprzyk Lazlo Koczy, Pedro Albertos at the First EUSFLAT Conference 1999, Bahia Mediterráneo Restaurant in Palma de Mallorca

References

1. Aczel, J.: Sur les opérations définies pour nombres réels. Bulletin de la Societe Mathematique de France 76, 59–64 (1949)
2. Alsina, C., Trillas, E., Valverde, L.: On Some Logical Connectives for Fuzzy Sets Theory. Journal of Mathamatical Analysis and Applications 93, 15–26 (1983)
3. Alsina, C., Frank, M.J., Schweizer, B.: Associative Functions. Triangular Norms and Copulas. World Scientific, Hackensack (2006)
4. Bellman, R.E., Giertz, M.: On the analytic formalism of the theory of fuzzy sets. Information Sciences 5, 149–156 (1973)
5. Klement, E.P.: Some remarks on t-norms, fuzzy σ-algebras and fuzzy measures. In: Klement, E.P. (ed.) Proceedings of the Second Internatinal Seminar on Fuzzy Set Theory, Linz, pp. 125–142 (1980)
6. Klement, E.P., Mesiar, R., Pap, E.: Triangular Norms. Kluwer Academic Publ., Dordrecht (2000)
7. Ling, C.H.: Representation of associative functions, vol. 12, pp. 189–212. Publicationes Mathematicae Debrecen (1965)
8. Mostert, P.S., Shields, A.L.: On the Structure of Semigroups on a Compact Manifold With Boundary. Annals of Mathematics 65, 117–143 (1957)
9. Mayor, G., Torrens, J.: On a Class of Operators for Expert Systems. International Journal of Intelligent Systems 8, 771–778 (1993)
10. Mayor, G., Torrens, J.: Triangular Norms on Discrete Settings. In: Klement, E.P., Mesiar, R. (eds.) Logical, Algebraic, Analytic, and Probabilistic Aspects of Triangular Norms. Elsevier, Amsterdam (2005)

11. Mayor, G., Monreal, J.: Additive Generators of Discrete Conjunctive Aggregation Operations. IEEE Transactions on Fuzzy Systems 15, 1046–1052 (2007)
12. Menger, K.: Statistical Metrics. Proceedings of the National Academy of Sciences of the United States of America 28, 535–537 (1942)
13. Schweizer, B., Sklar, A.: Espaces Métriques Aléatoires. Comptes Rendus de l'Académie des Sciences de Paris. Sér. A 247, 2092–2094 (1958)
14. Schweizer, B., Sklar, A.: Statistical Metric Spaces. Pacific Journal of Mathematics 10, 313–334 (1960)
15. Schweizer, B., Sklar, A.: Associative Functions and Statistical Triangle Inequalities, vol. 8, pp. 169–186. Publicationes Mathematicae Debrecen (1961)
16. Schweizer, B., Sklar, A.: Probabilistic Metric Spaces. North Holland, New York (1983)
17. Zadeh, L.A.: Fuzzy Sets. Information and Control 8, 338–353 (1965)
18. Zadeh, L.A.: A Fuzzy-algorithmic Approach to the Definition of Complex or Imprecise Concepts. International Journal of Man-Machine Studies 8, 249–291 (1976)

Author Index

MIX
Papier aus verantwortungsvollen Quellen
Paper from responsible sources
FSC® C105338

If you have any concerns about our products,
you can contact us on
ProductSafety@springernature.com

In case Publisher is established outside the EU,
the EU authorized representative is:
Springer Nature Customer Service Center GmbH
Europaplatz 3, 69115 Heidelberg, Germany

Printed by Libri Plureos GmbH
in Hamburg, Germany